# MACHINES IN OUR HEARTS

# MACHINES

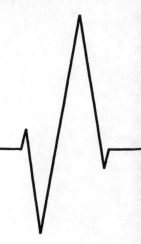

KIRK JEFFREY

# IN OUR HEARTS

## THE CARDIAC PACEMAKER,
## THE IMPLANTABLE DEFIBRILLATOR,
## AND AMERICAN HEALTH CARE

The Johns Hopkins University Press
*Baltimore and London*

The Johns Hopkins University Press
2715 North Charles Street
Baltimore, Maryland 21218-4363
www.press.jhu.edu

LIBRARY OF CONGRESS CATALOGING-IN-PUBLICATION DATA

Jeffrey, Kirk.
Machines in our hearts : the cardiac pacemaker, the
implantable defibrillator, and American health care /
Kirk Jeffrey.
p. cm.
Includes bibliographical references (p.     ) and index.
ISBN 0-8018-6579-4 (acid-free paper)
1. Cardiac pacing — History. I. Title.
RC684.P3 J444 2001
617.4′12059′09 — dc21
00-009627


A catalog record for this book is available from
the British Library.

FOR FRANCES

# CONTENTS

List of Illustrations and Tables   **ix**

Acknowledgments   **xi**

Introduction   **1**

CHAPTER 1     Heart Block and the Heart Tickler   **14**

CHAPTER 2     The War on Heart Disease and the
Invention of Cardiac Pacing   **36**

CHAPTER 3     Heart Surgeons Redefine Cardiac Pacing   **58**

CHAPTER 4     The Multiple Invention of
Implantable Pacemakers   **83**

CHAPTER 5     Making the Pacemaker Safe and Reliable   **107**

CHAPTER 6     The Industrialization of the Pacemaker   **136**

CHAPTER 7     The Pacemaker Becomes a Flexible Machine   **161**

CHAPTER 8     Slowing the Pace:
The Industry's Time of Troubles   **186**

CHAPTER 9     Competition through Innovation:
Accelerating the Pace of Change   **209**

CHAPTER 10     Preventing Sudden Cardiac Death:
The Implantable Defibrillator   **235**

CHAPTER 11    The 1990s and Beyond:
"When Life Depends on Medical
Technology"  **263**

APPENDIX A    Device Reliability, Qualification Tests,
and Improvements  **291**

APPENDIX B    Number of Implantations  **294**

APPENDIX C    ICHD Pacemaker Identification Code  **296**

Abbreviations  **297**

Notes  **299**

Bibliographical Note  **353**

Index  **361**

## ILLUSTRATIONS AND TABLES

The First Published Electrocardiographic
Tracing of the Normal Heartbeat   **16**

ECG of Complete Heart Block   **17**

Structure of the Heart   **20**

Albert S. Hyman's Pacemaker, 1932   **21**

Paul Zoll Demonstrates His External Pacemaker,
Mid-1950s   **38**

C. Walton Lillehei with a Young Patient
Recuperating from Open-heart Surgery, 1961   **62**

Warren Mauston with his External Pulse
Generator, 1959   **63**

Frank Henefelt and the First Chardack-
Greatbatch Pacemaker, 1960   **88**

A Chardack-Greatbatch Implantable Pacemaker
of the Early 1960s   **112**

P.S. Walking the Hallway of Montefiore
Hospital, Fall 1958   **113**

CPI's Microlith-P Pacemaker and
Model 2000 Programmer, 1978   **162**

Dog Being Defibrillated, 1975   **238**

ECG of Ventricular Fibrillation   **239**

Guidant Model 2901 Programmer, 1997   **267**

TABLE 1            Technologies of Heart-rhythm Management   9

TABLE 2            The Search for a Long-term Pacing System,
                   1956–1960   84

TABLE 3            Implantations Reported, January 1961–
                   August 1963   108

TABLE 4            Sales Income and Net Income at Medtronic,
                   1960–1972   139

TABLE 5            Some Indicators of Growth in American
                   Cardiovascular Medicine   155

TABLE 6            Selling and R&D Expenses at Medtronic,
                   Cordis, and Intermedics, 1980   195

TABLE 7            Estimated Market Shares of Pacemaker
                   Manufacturers, 1975 and 1980   217

TABLE 8            Market Introduction of DDD Pacemakers   223

TABLE 9            The Pacemaker/ICD Manufacturers in 1998   275

## ACKNOWLEDGMENTS

I am grateful to many people and organizations for encouraging and facilitating my work on the technologies and practices of heart-rhythm management. John G. Truxal and Marian Visich Jr. started me on a first-rate intellectual adventure when they commissioned me in 1989 to write a brief monograph about the pacemaker for the series they edited in the Alfred P. Sloan Foundation's New Liberal Arts Program. Faculty development grants from Carleton College freed me to conduct research and interviews and to attend medical conventions. I thank Elizabeth McKinsey, the dean of the college, for her support and interest. Arthur Norberg extended the hospitality of the Charles Babbage Institute for the History of Computing at the University of Minnesota during a formative stage of my research in 1991.

Help, suggestions, and information were provided by numerous physicians — Drs. Agustin Castellanos, William Chardack, Howard W. Frank, Seymour Furman, W. Bruce Fye, Jerry C. Griffin, Samuel W. Hunter, Richard E. Kerber, C. Walton Lillehei, James D. Maloney, Victor Parsonnet, Ronald E. Vlietstra, and Paul M. Zoll — and nurse-specialist Susan L. Song. Drs. Parsonnet, Laurence M. Epstein, and David W. Hayes permitted me to observe pacemaker and ICD implantations at close range.

Several engineers, inventors, and business leaders discussed innovation and competition in the heart-rhythm management industry with me. I particularly want to thank Earl E. Bakken, Alan D. Bernstein, Wilson Greatbatch, Ron Hagenson, Jerry Hartlaub, Thomas E. Holloran, J. Walter Keller, and Peter Tarjan. Chicago patent attorney Edward W. Remus gave me a crash course on patenting and patent litigation in the medical-device industry.

As work on the book progressed, it often seemed that heart-rhythm management was changing so rapidly that it would elude my attempt to make sense of it. Patrik Hidefjäll, Dr. Joel D. Howell, Kai N. Lee, Robert C. Post, David J. Rhees, and anonymous readers for *Technology and*

*Culture* and the Johns Hopkins University Press helped me clarify my interpretive ideas.

At the office of the North American Society of Pacing and Electrophysiology (NASPE), Dorothy Kelleher and Janet Giroux made the NASPE Oral History Collection available for my use. Several pacemaker and ICD patients talked to me about their experiences with heart-rhythm disorders and implanted devices. Elmer A. Braun, Wesley Johnson, and Helen Johnson graciously consented to my using parts of their accounts in this book.

I am greatly indebted to the staff of the Biomedical Library at the University of Minnesota–Twin Cities, and to Steve Rasmussen, director of the Medtronic Library in Fridley, Minnesota, and his staff. I would also like to thank the staffs of the Bakken Library and Museum of Electricity in Life, Minneapolis; the Guidant/CPI library in Arden Hills, Minnesota; the Baker Library at Harvard Business School; the library at Heart House in Bethesda, Maryland; the James J. Hill Reference Library in St. Paul; and the Laurence McKinley Gould Library at Carleton College. Kristine Altenhafen, Ruth Freiman, J. Walter Keller, Nikki Lamberty, Karen Larson, Carol Lindahl, Dr. Berndt Lüderitz, Patti Peltier, and George Szarka helped me track down hard-to-find people, publications, and photographs. Bruce Thomas of the Carleton College Physics Department helped me breadboard Wilson Greatbatch's blocking oscillator circuit from the Chardack-Greatbatch implantable pacemaker of 1960. Careful readings of the manuscript by Linda Picone and Frances Long caught numerous awkward or unclear sentences.

This book draws on papers that I have previously published in the journals *Pacing and Clinical Electrophysiology* (NASPE), *Cardiology Clinics* (W. B. Saunders), *Technology and Culture* (Society for the History of Technology), *Invention and Technology* (Forbes), and *Circulation* (American Heart Association), and the book *Exposing Electronics* (Harwood).

The opportunity to form friendships with men and women involved with cardiac pacing and defibrillation has been for me the most satisfying part of working on this book. Drs. Arthur Linenthal and Stafford I. Cohen, both associates of Paul Zoll, faithfully kept track of my progress and offered many helpful insights. It was a privilege to collaborate on a historical paper with Dr. Victor Parsonnet and to work with Dr. Seymour Furman on the NASPE Oral History Committee. Patrik Hidefjäll, a fellow explorer in the world of implantable medical devices, contacted me in 1995 while working on his doctoral dissertation in technology studies at the Linköping University in Sweden. When he visited the United States later that year, we jointly interviewed several key industry leaders. We have shared ideas ever since; I could not have written this book without his incisive criticism.

Most physicians and industry people have resisted the temptation to advise me on how to organize and interpret the information I was gathering. I know that the result will not please them in some respects, but I hope that they will recognize this book as an observer's serious attempt to understand and do justice to the medical-industrial field they have done so much to create. They will note, too, that my last word on heart-rhythm management is an affirmative one.

My large and loving family have followed the progress of this book and contributed to it in countless ways, not least by permitting me to take advantage of their goodwill for years on end. Above all I thank my wife, Frances Long, for her patience and support during the book's long gestation.

In the spring of 1976, Elmer A. Braun, of Charleston, West Virginia, underwent a physical examination when his former employer changed insurance carriers for its retired workers. An electrocardiogram (ECG) revealed that something wasn't right with Braun's heartbeat. A complex network of specialized cells initiates the heartbeat high in the right atrium and passes it on through the two atria and then to the major pumping chambers, the ventricles. In his heart, the electrical signals were starting out in the normal fashion but were delayed en route to the left ventricle. His "disorder" showed up only in the ECG — he felt fine. "I knew that I had something," he later wrote, "but I had no symptoms of the problem."[1]

Braun's doctor in Charleston recommended that he have a more thorough diagnostic workup, so a few weeks after the insurance exam, he traveled to the Cleveland Clinic for more tests. These lasted four days and included stress testing through exercise, in which the heart is taxed by having the patient walk and jog on a treadmill. He also underwent a coronary angiography: a cardiologist advanced a catheter through an artery from his groin into the aorta and coronary arteries, then injected a contrast medium so that the arteries would show up clearly in X-ray images. (The disorder causing Braun's heart-rhythm problem, known as left bundle branch block, was highly associated with coronary artery disease.) At the end of the four days, the doctors "told me that someday I would need a pacemaker and that I would know that day as I would 'get out of breath.' "[2]

Five years passed, and then, while walking his dog one fall day, Braun noticed that he was short of breath. That afternoon he went to a local clinic to see a cardiologist, who put him into a wheelchair and pushed him across the street to the hospital: his heart rate had fallen from the normal range of 60 to 100 beats per minute to 39. Two days later, a pacemaker team consisting of a surgeon and a cardiologist made a small incision in Braun's upper chest and opened a vein, threaded a wire lead down into his right ventricle, and connected it to a pulse generator that "looked like the original Zippo lighter" and weighed 95 grams. Together, the generator and the lead constituted a cardiac pacemaker. Powered by a solid-state lithium battery, the device fired tiny electrical stimuli into his heart muscle to trigger regular contractions. It was set to deliver 72 impulses per minute, but would withhold its stimuli unless Braun's normal heartbeat fell below that rate. The generator, product of three decades of invention on two continents, did not stay outside Braun's body but went inside. The surgeon created a pocket beneath the layers of skin and fat but above the pectoral muscles, attached the end of the lead to the generator, positioned the generator in the pocket, and, layer by layer, closed the wound. Braun requested that the generator be implanted in his left chest because "I didn't want it to interfere with my golf swing."[3]

A day or two after the implantation, a cardiologist checked Braun's ECG to make sure that the pacemaker was indeed stimulating his heart correctly. Braun left the hospital and resumed his normal life, except that he took his pulse every day and transmitted a simplified ECG tracing by telephone to his doctor every three months. The clinic had provided him with the trans-telephone device, which picked up his heartbeat through two leads clipped to his index fingers, amplified the signal, and transmitted it over the telephone to a recorder at the doctor's office. By checking this rhythm strip, the doctor could make sure that the pacemaker was functioning normally and could determine its stimulus rate. When the rate began to fall off slightly, that indicated that the pacemaker battery was approaching the end of its worklife.

Braun's doctor had said that the pacemaker should last about eight years. Eight years and four months after the original implantation, its battery began to show signs of impending exhaustion. Braun checked into the hospital again for a replacement pulse generator. During that eight-year interval, pacemaker manufacturers had learned how to downsize their pulse generators, so Braun's second generator was much smaller and thinner than the first and weighed just 26 grams. The implantation team also gave him a lead for his right atrium to go with the existing ventricular lead. Braun agreed to a base rate of 60 because "they convinced me that I would sleep better" at 60 than at 72. Through the atrial lead, the pacemaker could sense the natural rate at which Braun's heart wanted to pump at any given moment and then send stimuli to the ventricle: "If my activity requires that my heart should pulse at 75 or 80, etc., the pacemaker will respond." The dual-chamber setup also enabled the pacemaker to coordinate the action of the upper and lower chambers by giving the ventricles time to fill with blood from the atria before firing its ventricular stimulus. This improved the efficiency of the ventricles' pumping. The new unit had a memory that stored records of episodes when Braun's heart raced or otherwise misbehaved so that the doctor could download and study the information later. Elmer Braun reported all this in a letter to his grandson written in February 1990, a few weeks after he had received the second pacemaker. He added a postscript: "I played golf last Wednesday and everything felt fine."[4]

Braun's experience as a pacemaker patient in the 1980s was representative on a number of counts. First, he had a slow heart rate but was not in extremis and convulsing in a hospital emergency room. The pacemaker saved him that experience; had it not been available, he might later have encountered far more terrible symptoms than shortness of breath. For this patient, the entire experience of receiving a pacemaker was a fairly mild one: three days in the hospital; an implantation procedure lasting about an hour and requiring only local anesthesia; a quick return to everyday life. Most of the earliest pacemaker patients, those on whom the technology of pacing was first tried back at the end of the 1950s, had been pacemaker-dependent. In contrast, the typical candidate for an implanted pacemaker today is not device-dependent: he or she would not die suddenly if the device were to fail for some reason.

The rituals of implantation and recovery in Elmer Braun's case superficially resembled those of acute-care medicine. In effect, implantation of a pacemaker enabled his doctor to treat a chronic degenerative disorder of the heartbeat as if it were an acute illness. It worked — Braun left the hospital "cured." Appearances can deceive, however. Braun's underlying problem had not been eliminated and would always require pacing. In Lewis Thomas's phrase, cardiac pacing is a halfway technology, a treatment that manages but does not cure the disease.[5]

Braun's account mentions numerous encounters with technologies of modern medicine. As used today, this term *medical technology* extends well beyond tools and machines. One standard definition speaks of "the drugs, devices, and medical and surgical procedures used in medical care, and the organizational and supportive systems within which such care is provided." I invite the reader to go through Braun's account and note down the tools and machines, drugs, procedures, and organizational systems mentioned or implied (for example, implantation of a pacemaker implies the use of a local anesthetic such as lidocaine). The narrower term *medical devices* still covers "an almost unencompassable range of technologies," as William W. Lowrance has observed, "from rather simple classical aids, such as crutches and eyeglasses, to novel high-technology instruments and implantable organs, and from inexpensive devices used intimately by individuals, to capital hardware used in large institutions for the benefit of many thousands. This makes the topic exceedingly difficult to analyze as a category."[6]

Some of Braun's contact with the technology of medicine took place in the setting of a doctor's office, some in a hospital, some in his own home. He does not tell us explicitly about his feelings on encountering the machines and procedures — whether he felt reassured, indifferent, apprehensive. But overall, technology is ubiquitous in his account. Here at the end of the twentieth century, when we discover that we may have a serious medical problem we expect to enter a technology-dominated world. We expect the doctor to do something active, to deploy tools and machines for our benefit; we think of technology as the center of modern medicine and share a faith in the power of human inventions to identify and control our physical maladies. Arthur Caplan has suggested that we hardly believe a sickness to be "real" unless we can visualize it and reach it with our machines and medications.[7]

## A MACHINE IN THE HEART/THE HEART AS A MACHINE

This book gives an account of the invention of the cardiac pacemaker and the subsequent development and transformation of this machine. By placing a piece of electrical equipment inside Elmer Braun's body and leaving it there, Braun's cardiologist deployed technology in a way that no one had ever attempted until the late 1950s. The pacemaker was born in 1952 as an appliance the size of a breadbox that stood on a hospital cart and plugged into a wall socket. As it grew up, it shrank. Within a few years, medical researchers and engineers had transformed it into a little device that was completely implanted within the patient's body with one component actually threaded down a vein into the heart's interior. Today we have a number of implanted machines, such as defibrillators and nerve stimulators, that manage some physiological function, but the pacemaker was the very first of these. Surgeons carried out the earliest implants in human beings between 1958 and 1960.

Pacemakers (or pacers) in the 1990s are no larger than wristwatches with one or two leads instead of a wristband. In the early days of implantable pacers, the devices were thicker and heavier than an old pocket watch; people in fact sometimes called them "heart tickers." But a pacemaker today can do far more than send little ticks of electricity to the heart. Most pacers implanted in the 1990s coordinate the pumping action of the upper and lower chambers (the atria and ventricles) and change their rate depending on the patient's activity level. Some can intervene to slow down a dangerously fast heartbeat. We live in "the age of the smart machine"; this phrase certainly applies to the newer pacers, for they include microprocessors and have become, in effect, computers.[8] (Braun's second pacer had a microprocessor.) Once implanted, a pacemaker can be reprogrammed, its behavior completely reconfigured. In the near future, these tiny machines may be smart enough to diagnose the patient's heart-rhythm problems and choose how to respond by themselves, without the doctor's needing to intervene at all.

The idea of entrusting one's heartbeat to a small computer, a machine, implanted within the body, is deeply disturbing to some people, though others are thrilled to see machines contribute in such a direct and intimate way to human well-being. I simply want to emphasize, at this point, that pacemakers have always been machines. By *machine*, I mean an apparatus designed and built by human beings that uses power to carry out some task or function. Every machine will sooner or later wear out, perhaps to be replaced by a new and "improved" machine. Pacemakers have grown more sophisticated, changed from heart tickers to heart managers—but they have always been machines.

The pacemaker and its younger cousin the implantable cardioverter-defibrillator (ICD) fit the economists' definition of merit goods—products that have greater significance to society than ordinary consumer goods.[9] Any life-sustaining technology probably qualifies as a merit good, but pacemakers support the human *heart*, which gives them added cultural significance. Pacemakers, artificial valves, transplants, and emergency procedures like defibrillation would probably be less salient in our popular culture if they ministered to some other organ such as the bladder or the thyroid gland. Although all the organs of the body are necessary to sustain life, European and American cultures have imbued the heart with special cultural meanings. For centuries, we imagined the heart as the seat of the human emotions, particularly humane feelings toward others, and of the soul itself. This mystique long antedates William Harvey's discovery (1628) that the heart pumps blood throughout the body. The English language retains scores of metaphorical expressions of the ancient belief that the heart is the mysterious center of our emotional natures, our very identities.[10]

If prescientific and romantic notions about the heart persist even today, how did it happen that the U.S. public after World War II so readily accepted open-heart surgery, heart transplants, and machine substitutes for the heart? For years, our most famous doctors were the heart surgeons: Dwight Harken, Charles Bailey, C. Walton Lillehei, Michael DeBakey, Norman Shumway. Yet open-heart surgery, the pacemaker, and all the other advances in heart care since 1945 were founded on a disenchanted understanding of the heart. Behind them all lies the belief that the heart is nothing but a pump, a complex "machine" within the body. Thinking of the heart as a pump leads on to the possibility that we can open it up and repair it when it breaks down, perhaps by replacing worn-out parts of the machinery.[11]

Prescientific attitudes about the heart linger at the periphery of modern culture. For example, men and women scheduled for heart transplants often ask if their new hearts will cause changes in their personalities, particularly if the donor was of the other sex or a different ethnic origin. I agree with medical journalist Lynn Payer, however, that a mechanistic, disenchanted understanding of the heart (and of the body generally) has largely supplanted the earlier notions, both among physicians and in the culture at large. This shift had taken place by 1900, if not well before; it made possible the laboratory studies of the heartbeat as an electrical phenomenon and Willem Einthoven's invention of the electrocardiograph at the end of the nineteenth century. Heart surgery and the other new treatments of the postwar period did not inspire but rather built upon this understanding of the heart as a piece of machinery. And yet the older feeling lingers that people and machines have a special status if they minister to disorders of the human heart.

## THE CARDIAC PACING/ELECTROPHYSIOLOGY COMMUNITY
The cardiac pacemaker and the ICD are artifacts of human devising. We can define *cardiac pacing* as the larger set of doctrines and practices centered on the pacemaker. The pacemaker and pacing are inseparable, two halves of the same walnut, and this book moves back and forth from one to the other. In the early years of the field, surgeons implanted most pacemakers; later, cardiologists gained a dominant position. In the 1990s, the management of ICDs and pacemakers has fallen to *clinical cardiac electrophysiology* (EP), a subspecialty of cardiology concerned with the heartbeat and its disorders. The community of pacemaker and ICD specialists has its own professional organizations, journals, textbooks, annual conventions, training fellowships, and competency examinations. More broadly still, pacing and clinical EP have grown into a subculture complete with creation myths; revered elders; complex networks of friendship and rivalry encompassing physicians, engineers, and employees of the medical-device manufacturing companies; and a distinctive terminology bewildering to the outsider.

The growth and transformation of pacing — its progress, if you will — has not been owing to some kind of inevitable unfolding of the possibilities latent in the technology and cannot be ascribed solely to the bright ideas of a few medical or engineering geniuses. I believe that the ongoing development of cardiac pacing has come about through competition and conflict within the technological community, and at times between that community and outside groups such as the Food and Drug Administration (FDA). The core group for innovation has always been surprisingly small, consisting at any one time of no more than a few hundred people. Beyond this core stands a much larger aggregation of men and women who are involved with pacing and electrophysiology but who have not contributed significantly to innovation.[12]

Differences of outlook and opinion run through the history of the field. Medical researchers of the 1950s debated whether it was even desirable to try to manage the human heartbeat over weeks and months with electrical stimuli. They also disagreed over how best to get the stimulus to the heart and whether it would be necessary to synchronize the pumping of the upper chambers of the heart with the action of the larger, more powerful lower chambers. They tried many things, some of which worked better than others. Out of these debates and experiments came the standard pacemaker design of the late 1960s, the ancestor of today's implantable devices.

Pacemaker implantation had attained the status of an established medical procedure by about 1975; then the FDA, Congress, the Health Care Financing Administration (HCFA, which administers Medicare), hospitals, and managed-care plans one by one began to interest themselves in its future development. Their demands caused leaders of the pacing community time and again to rethink the technology and the medical practices of pacing. Much the same pattern is evident today with the ICD, a newer technology.[13]

The pacing/EP community has always included many physicians and engineers based outside the United States, for research and inventive activity in western Europe and elsewhere have contributed in important ways to the growth and redefinition of cardiac pacing. Non-American physicians and engineers publish in U.S. journals and give papers at American professional meetings; Americans travel halfway around the world to attend conventions and conferences on the treatment of heart-rhythm disorders.[14] Some European companies, notably the German firm Biotronik, have marketed their products for decades in the United States. But because the U.S. system of health care differs in important ways from health care in other prosperous democratic nations, I have generally elected to limit the scope of this book to the American side of what is truly an international story.[15]

## THE RHYTHM-MANAGEMENT INDUSTRY

This book shifts its focus midway from the physicians and engineers who invented cardiac pacing and created a technological community to the manufacturing firms that have the greatest degree of control over the technology today. The manufacturers supplanted research physicians as the prime directors of technological change during the 1970s. The upshot is that when it comes to cardiac pacing and defibrillation, doctors are in effect working in alliance with large corporations in determining how best to treat patients. This has been true in pacing since the late 1960s, and in my view is the inevitable result of the fact that designing and manufacturing implantable pacemakers and defibrillators requires technical knowledge and skills that reside only with business firms; that is, not in doctors' offices.

Today observers speak not of the pacemaker industry but of the heart-rhythm management industry. *Rhythm management* encompasses cardiac pacing ("brady pacing" for *bradycardia*, an unduly slow heart rate, and, more recently, anti-tachycardia pacing to interrupt a *tachycardia*, a dangerously rapid rate). The term also refers to two other invasive technologies: the implantable cardioverter-defibrillator, a device that delivers a shock to the heart to halt tachycardia or to terminate chaotic electrical activity ( *fibrillation*); and catheter ablation, a means of destroying the cells that cause the heartbeat to go awry (see table 1).

The ICD somewhat resembles a pacemaker physically, but it functions quite differently. The ICDs of the 1980s sensed the onset of ventricular fibrillation (VF) and delivered, not an ongoing series of tiny electrical stimuli, but a powerful shock to halt all electrical activity and enable a normal rhythm to reestablish itself. Newer ICDs can also cardiovert rapid but organized heartbeats (tachycardia) and perform brady pacing. Because ICD technology in some respects is an offshoot of pacemaker technology, and because the ICD market will soon surpass the market for pacemakers, I have included a chapter on the invention and industrialization of the ICD (chapter 10). Catheter ablation, invented in the 1980s, is also now a well-established technology of heart-rhythm management. Drugs that manage some rhythm problems by themselves or in conjunction with electrical devices are also available, but the drug manufacturers are not usually considered to be part of the heart-rhythm management industry.

The invention of the pacemaker and later transformations of the device brought the manufacturers into new relationships with physicians, creating new activities and relationships that focused on cardiac pacing. Manufacturing firms kept in touch with academic researchers, customers, and even competitors. Some business leaders in the pacemaker industry realized at

**TABLE 1.** TECHNOLOGIES OF HEART-RHYTHM MANAGEMENT

| Problem | Technology | Description |
| --- | --- | --- |
| Slow ventricular rate; inability of heart to speed up appropriately (bradycardia) | Bradycardia pacing | Implanted pacemaker delivers tiny electrical stimuli to trigger the heart to beat at a more rapid rate and (in some cases) to coordinate atria and ventricles. |
| Rapid atrial or ventricular rate (tachycardia) | Anti-tachycardia pacing | Pacer or ICD delivers several energy pulses that disrupt the abnormal rhythm. |
| Fixed rapid atrial rate; some forms of ventricular tachycardia | Cardioversion | Rhythm device delivers shocks synchronized to disrupt the established rhythm. |
| Random electrical activity (no organized beats) in atria or ventricles | Defibrillation | Defibrillator delivers a shock that terminates all electrical activity in the fibrillating chambers and enables an organized beat to reestablish itself. |
| Certain forms of fixed, recurrent tachycardia | Ablation | Cells in the heart that are triggering or propagating the rhythm disturbance are identified and destroyed (ablated) with radiofrequency energy delivered through a catheter electrode. |

an early point that establishing and nurturing relationships with physicians was vital because no company could possess, in-house, all the relevant knowledge it needed to innovate. A company's connections to friendly physicians could help it identify promising new ideas and, equally important, discard those that were not so promising. These relationships also proved vital when companies tried to market successive pacemaker models to their consumers, the doctors who performed implantations. Beginning with chapter 3, I show how one firm, Medtronic, nurtured these all-important ties to physicians.

## CYCLES OF INNOVATION IN PACING AND DEFIBRILLATION
Heart specialists, working usually in close collaboration with electrical engineers, began to invent cardiac pacemakers at the end of the 1940s; inventive work on the implantable defibrillator got under way in the late 1960s.

In both pacing and defibrillation, innovation continues today — indeed, manufacturers have radically reengineered and redefined the pacemaker in the last few years. Because invention and reinvention have characterized cardiac pacing for nearly half a century, the meaning of the words *cardiac pacemaker* has never entirely settled down. It is impossible to identify any one person or research group as the primary inventor of the pacemaker, and there seems to be no point at which we can say, "There — they've finished inventing the pacemaker." It is more fruitful to think of the pacemaker as "an embodiment of human activity," to borrow sociologist Susan Bell's felicitous phrase, and to recognize that the activity is ongoing.[16]

To help organize the story of rhythm-management technology it is useful to recall the concept of *cycles of innovation*, which Louis Galambos introduced in a recent study of vaccine development in the pharmaceutical industry.[17] Four cycles of innovation are apparent in the history of cardiac pacing since 1952, each arising from advances in electronic circuitry: first the vacuum tube, then transistorized circuitry with discrete components, later hybrid integrated circuits, most recently the microprocessor. A cycle may be said to have ended when the possibilities of its technology "platform" had been fully exploited or when the industry had decisively shifted to a more advanced one. Newcomers or firms with small shares of the market have generally led these shifts from an established platform to a new one because the inauguration of a new cycle of innovation clearly presented great opportunities for firms to enter the industry or to gain significantly in market share. It is unclear whether some new cycle of innovation will eventually take over once the manufacturers have exhausted the possibilities of the microprocessor as a tool in the competition for market share. Eventually the pacemaker, once a glamour product, will become a commodity: all brands will offer essentially identical features, all secrets of design and production will stand revealed. Prices will plummet. Of course, the manufacturers strive to postpone that day, and thus far they have succeeded handily.

During each cycle of innovation, the leading doctors and engineers in cardiac pacing loosely shared a common vision of "the ideal pacemaker" and tried to see it embodied in hardware.[18] In the earliest cycle, Paul Zoll and other innovators thought of the pacemaker as a piece of hospital equipment for emergency use. Once transistors had come onto the scene and pacemakers became implanted devices, innovators overwhelmingly concerned themselves with designing and building pacers that would reliably pace the heart for a year or two, or longer, if possible. By the mid-1970s, with reasonable reliability achieved, integrated circuits enabled pacemaker designers to add adjustable features. Doctors began to communicate with implanted pacemakers by radiofrequency signals using handheld devices called programmers. More recently still, microprocessor technology has arrived in the pacemaker industry and there is talk of "automatic" pacemakers that carry out some of the work of diagnosing heart-rhythm problems and then reconfigure themselves appropriately.

## THE PROTEAN PACEMAKER

Why should any reader who is not a cardiologist wish to know about the pacemaker and the ICD? Certainly, mechanical technologies are more accessible to the nonexpert, but there is something breathtaking about the audacious idea of taking over from nature the management of the human heartbeat and the ensuing quest to design electronic devices that would emulate nature ever more perfectly. Then, too, people love medical miracle stories, hackneyed though they be: the dying patient, the desperate search for a new treatment, the heroic doctor, the miraculous recovery. The history of cardiac pacing and defibrillation discloses many such stories. One of the best (chapter 3) concerns Arne Larsson, of Stockholm, Sweden, the first human being to be given an implantable pacemaker to keep his heart going. That was in 1958. Larsson is still living near Stockholm. He is 85 years of age and on his twenty-sixth pacemaker. He has his own web site.[19]

I also believe that by understanding the rapid dissemination of cardiac pacing and defibrillation into U.S. medicine, the repeated transformation of these medical devices, and the struggles over control of the future of heart-rhythm management, we learn something important about the dynamics of growth and change in American health care. Thomas Hughes's term *technological momentum* is helpful here, for the industry and the medical field of heart-rhythm management, taken together, has acquired its own momentum not unlike the momentum of the electric-power systems that Hughes studied.[20] Heart-rhythm management has a kind of mass consisting of numerous people and institutions and billions of dollars of invested capital, a trajectory in the sense of broadly-agreed-upon goals, and a built-in tendency to grow. In the United States, policymakers often try to affect the direction and velocity of these large systems through regulation. But the FDA and HCFA, the two agencies most directly concerned with cardiac pacing and electrophysiology, have not greatly slowed or redirected the momentum of heart-rhythm management.[21]

The pacemaker has proved itself an extraordinarily protean artifact, for it successfully manages chronic disorders of heart rhythm that were not even defined when Paul M. Zoll announced the first clinically effective external pacemaker in 1952. Doctors' understanding of "the pacemaker" and "cardiac pacing" changed repeatedly as medical researchers defined new rhythm disorders for which pacing appeared the appropriate therapy. Knowledge gained in the laboratory passed to clinicians who, in turn, informed biomedical engineers of new needs and opportunities for pacing that required new pacing hardware. The very success of clinical cardiac pacing stimulated further basic research into conduction disorders of the heart, bringing the process of knowledge transmission full circle. The future development of the ICD may follow the same track.[22]

All of this has taken place in an environing society that eagerly awaits new medical technology. By the 1950s, Americans had developed a firm conviction that technological solutions lay near at hand for many of the diseases that plagued humanity. This confidence arose from earlier successes at controlling infectious disease through public-health measures and the development of the "miracle drugs" of the 1940s and 1950s. The public consequently valued and supported medical research and innovation and, as we noted earlier, particularly admired the open-heart surgeons and other specialists in cardiovascular medicine. In the 1950s, the surgeons' inventiveness and boldness with new invasive procedures captivated the mass public and their political leaders.

The field and the technology of pacing were born during that era of American medical history. The era lasted from World War II until the 1970s and was marked by strong public and elite enthusiasm for research, by an interest in heroic and high-tech solutions to problems of disease, and, as Daniel M. Fox points out, by an emphasis on supply—on subsidizing more hospitals, medical schools, physicians, and treatments, while giving only limited attention to questions of access and cost.[23] Like all other aspects of health care, heart-rhythm management is today recasting itself to survive in quite a different era marked by powerful pressures toward cost containment, though not toward distributional equity, and toward some limitations on the historical autonomy that the medical profession enjoyed. Heart-rhythm management, in fact, has enacted on a smaller stage many of the larger tensions and struggles under way within postwar medicine.

As this book was nearing completion, I spoke by telephone with Elmer Braun at his home in Charleston, West Virginia. He was then in his early eighties and on his fourth pacemaker. When one of his leads developed a crack in the insulation in 1995, his cardiologist replaced the damaged lead and the pulse generator. Extracting a nonfunctioning lead from the heart can be a dangerous procedure, so the old lead was disconnected but left in place, one end of it firmly lodged in Braun's heart. In August 1997, tests showed another faulty lead; again the doctors gave him a new lead and a new generator. When we spoke, he had two good leads and two nonfunctioning ones. The lead failures seem not to have troubled him: "Everything is fine," he reported. "I feel great." He had recently had lens implants for cataracts and was about to be fitted with new glasses. He told me, "I'm like an old car — you replace the muffler, replace this, replace that, and it keeps on running."[24]

## CHAPTER 1

## HEART BLOCK AND THE
## HEART TICKLER

Modern cardiac pacemakers and other implantable devices that manage disorders of heart rhythm are a result of decades of scientific research about the workings of the normal heartbeat. But there was never a simple and certain route "from bench to bedside." Even in a science-based field like cardiology, important laboratory research may not, for decades, lead to improved treatments. Sometimes, too, the process works the other way: new treatments can come first and inspire new scientific research.

The inventors of the first heart-pacing devices in the 1920s probably knew of the latest experimental findings about the heartbeat and its disorders, but this new knowledge apparently did not contribute to the design of their pacemakers or inspire new ideas about the clinical uses for the inventions. A generation later, when Paul Zoll invented a practical cardiac pacemaker in the early 1950s, he undoubtedly drew on an accumulation of laboratory research with which he was thoroughly familiar and to which he had made contributions of his own — but his attempt to treat a serious rhythm disorder by firing electrical pulses into the heart got underway because of his experiences as a clinical cardiologist (a cardiologist who treats patients) during and after World War II.[1]

### SLOW PULSE WITH FAINTING FITS
In February 1846, William Stokes (1804–78), of Dublin, a leading physician of his day and a recognized expert on diseases of the chest and use of the stethoscope, visited the Meath Hospital to see a patient named Edmund Butler. Butler was 68 years old and reported that he had been healthy until about three years earlier, "at which time he was suddenly seized with a fainting fit. . . . This occurred several times during the day, and always left him without any unpleasant effects. Since that time he has never been free from these attacks for any considerable length of time." Based on Butler's own account, Stokes gives a vivid description of the attacks:

There is little warning given of the approaching attack. He feels, he says, a lump first in the stomach, which passes up through the right side of the neck into the head, where it seems to explode and pass away with a loud noise resembling thunder, by which he is stupefied. This is often accompanied by a fluttering sensation about the heart. He never was convulsed or frothed at the mouth during the fit, but has occasionally injured his tongue. The duration of the attack is seldom more than four or five minutes, but sometimes less; but during that time [he] is perfectly insensible.[2]

To Stokes, Butler "seemed the wreck of what was once a fine, robust man. He lay generally in a half drowsy state, but when spoken to was perfectly lively and intelligent." Stokes timed his patient's pulse rate at 28 to 30 per minute. Butler's heartbeat had "a dull, prolonged, heaving character"; the arteries "pulsate visibly all over the body." Stokes also noticed a lack of connection between Butler's pulse rate and the rate at which he could hear the heart beating: over the course of one minute, he detected 36 pulses at the wrist but could hear only 28 definite heartbeats. Stokes inferred that occasional inaudible "semi-beats" must be occurring between the regular contractions of the ventricles. "On listening attentively," he could hear "occasional abortive attempts at a contraction." When he reexamined Butler three months later, Stokes found that the patient's right external jugular vein was visibly throbbing at more than twice the rate of the ventricular contractions.

In the century or so before Stokes, a few medical men had noted similar collections of symptoms centering on "slow pulse with fainting fits." Stokes's observations of the mysterious condition were more detailed than any earlier physician's, but he conceded that he could not explain everything he had observed. On one point, though, Stokes was firm: the episodes of unconsciousness were unrelated to epilepsy, injury to the spinal cord, or some other neurological source. He ascribed the attacks to "circumstance tending to impede or oppress the heart's action." Stokes followed his countryman Robert Adams (1791–1875) in attributing the condition to "fatty degeneration of the heart." This implied that weakness of the heart muscle caused the slow and "heaving" heartbeat in the same way that weak arm and shoulder muscles might force a person doing push-ups to slow down and stop.

Butler's condition eventually improved slightly: the attacks of dizziness continued, but he reported that he fainted less often. His heart rate, however, remained abnormally slow. After a few months, Stokes lost track of him. He knew from other case histories, though, that many elderly people

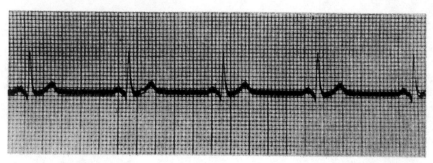

**The First Published Electrocardiographic Tracing of the Normal Heartbeat.** Willem Einthoven took this tracing using a string galvanometer; it was published in 1902. The ECG is read from left to right. Elapsed time from one heavy vertical line to the next is 0.2 second. The R waves indicating ventricular depolarization (sharp upward spikes) occur at intervals of just over one second, indicating a ventricular rate of 57 beats per minute. The P waves (small rounded deflections just before the R waves) indicate atrial depolarization. They appear normal in shape and each occurs about .12 second before an R. This P-R interval and the 1:1 ratio of P waves to R waves indicate normal conduction of the heartbeat from the atria to the ventricles. (Courtesy of Dr. W. Bruce Fye)

with similar symptoms experienced a gradual slowing down of their hearts and fainting attacks that lasted longer and longer. Sooner or later, their hearts would slow to the point that they died. Then doctors could, in some cases, examine the victims' hearts and other organs at autopsy. Thus Adams in 1819 had seen a patient who had a "remarkable slowness of the pulse" and went about "oppressed by stupor, having a constant disposition to sleep." This man had suffered numerous "apoplectic attacks," the most recent just three days earlier. Six months later, the patient died and Adams attended the autopsy; he noted that each chamber of the heart was "covered with a layer of fat" and that fat appeared to have replaced the muscular walls in places.[3]

Stokes's understanding of the malady depended on the diagnostic techniques available to him, and these techniques centrally involved his own, trained, senses. He examined each patient carefully with his eyes and listened to the patient's account of his illness, but he tried to go beyond the subjective impressions to which these sources gave rise. As doctors had done for centuries, he felt the patient's pulse and described it qualitatively; for example, it had a "prolonged, sluggish character." He must have used the stethoscope, the instrument invented by Laennec in 1819 that had greatly enhanced the physician's ability to detect and localize sounds in the heart and lungs, particularly the sounds of the heart valves closing.[4] When he could, Stokes examined organs after a patient's death—though not,

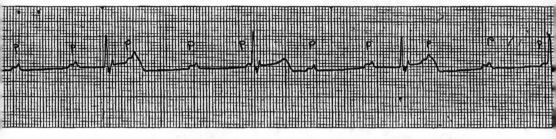

ECG of Complete Heart Block. Here the atria (P waves) and the ventricles (larger QRS complexes) are beating entirely independently—the atria at 75 per minute, the ventricles at 30. Note how the P waves "march" through the QRS complexes. Atrial signals to the ventricles are blocked, and cells in the ventricles just below the AV node are acting as a backup pacemaker but at a slow rate. (Kevin R. Brown and Sheldon Jacobson, *Mastering Dysrhythmias: A Problem-Solving Guide* [1st ed., © 1987], 100, 109; reprinted courtesy of Delmar, a division of Thomson Learning [fax: 800/730-2215])

apparently, with the aid of a microscope. None of these technologies provided a graphical record such as a waveform that the physician could analyze and save to compare with similar records from other patients, and none replaced the physician's traditional reliance upon his own, cultivated senses of touch, hearing, and sight.[5]

In common with many medical writers of the nineteenth century, Adams and Stokes both organized their analysis of the mysterious disorder by emphasizing structural anomalies in the heart—the layer of fat—and suggesting plausible correlations between these and the symptoms they had observed. Stokes noted subtle symptoms that escaped most doctors of that era, but viewed in another way his discussion of the Butler case left many loose ends, symptoms for which it might have been difficult to find obvious structural correlates. At the end of the nineteenth century, the intermittent episodes of syncope (loss of consciousness) came to be known as Stokes-Adams attacks.

As physiologists around the turn of the century studied the origins of the heartbeat and brought new scientific instruments into play, they began to redefine diseases of the heart by focusing less on abnormalities of heart anatomy and more on the behavior and functional capacity of the heart as a living, moving organ. The modern understanding of a case like Butler's would be based on this physiological understanding of the heart and on interpretations of various measurements and graphical tracings available to physicians in the twentieth century—just as Stokes's concept, "fatty degeneration of the heart," was an interpretation of symptoms and sounds available to a trained and careful observer of the nineteenth.[6]

## PHYSIOLOGISTS AND THE FRAMING OF HEART BLOCK

The disorder that Stokes had described as "fatty degeneration of the heart with remarkable slowness of the pulse" was uncommon but dramatic and intriguing. It caught the attention of a new generation of experimental physiologists, particularly in Britain. By the first decade of the twentieth century, these laboratory scientists had developed a sophisticated understanding of how the heart beats and how disorders of the heartbeat could lead to the symptoms that had afflicted Edmund Butler; physiologists had even learned to create the condition in laboratory animals. Appropriately enough, the physiologists completed the process of framing the new disease by giving it a distinctive name: complete heart block. The fainting spells, or Stokes-Adams attacks, were now understood as acute episodes signaling the deterioration of the heartbeat and the slowing of the circulation of blood.[7]

Several streams of research and discovery contributed to the emerging understanding of complete heart block. One of these concerned the origin of the heartbeat. By the turn of the century, scientists had resolved an old debate over whether the heartbeat arose from inherent properties of the heart muscle itself (the myogenic theory, the prefix *myo* referring to muscle tissue) or was controlled by the network of nerve cells running through the heart muscle (the neurogenic theory). Here the key figure was Walter Gaskell (1847–1914), a student of the renowned physiologist Michael Foster, of Cambridge University. In 1883, Gaskell reported on a series of experiments involving the tortoise heart that demonstrated that "the ventricle contracts in due sequence with the auricle [the atrium] because a wave of contraction passes along the auricular muscle and induces a ventricular contraction when it reaches the auriculo-ventricular groove," the border between the upper and lower chambers. The contractions were "clearly both myogonic [myogenic] and automatic." Later investigators reproduced Gaskell's findings in mammals. Then in 1907, the British anatomist Arthur Keith (1866-1955) described a small mass of fibrous tissue embedded in the wall of the upper part of the right atrium. He suggested that the heartbeat originated in this node, which is today generally known as the sinoatrial node, or simply the sinus node.[8]

Doctors and scientists had long known that, in a heartbeat, the atria contracted slightly ahead of the ventricles. In his experiments with tortoise hearts during the early 1880s, Gaskell showed that by gradually enlarging a slit between the atria and the ventricles, he could make it more and more difficult for the "wave of contraction" to pass to the ventricle. Eventually the ventricles received no signal from the atria. They might continue to beat, but at a slow rate dissociated from the atrial beats. Gaskell spoke of having imposed a partial or complete "block" between the upper chambers and the lower.

Investigators had also understood since midcentury that electrical activity was somehow involved in the heart's contractions. The British physiologist Augustus D. Waller (1856-1922) managed to record the electrical variation accompanying the heartbeat by positioning electrodes on the chest or arms of a human subject and connecting them to a laboratory instrument called a capillary electrometer. As the column of mercury in the electrometer rose and fell with changes in the electrical potential, Waller captured these changes on a moving photographic plate. But as historian Robert G. Frank Jr. points out, the lines in these first human electrocardiograms did not explain themselves: "they had to be assigned meaning." Frank has reconstructed Waller's probable thought process. Waller realized in 1888 that he could think of the heart as a dipole, a system with a positive electric charge at one end and a negative charge at the other. It followed that the electrical activity must consist of a diphasic variation as each section of the heart wall became electrically negative, then positive. Waller went on to establish that this electrical change was propagated, wavelike, through the atria and on to the ventricles and that it slightly preceded the physical contraction of the chambers.[9]

Waller's insights proved fundamental, but the capillary electrometer provided images of poor quality and was too complex to use outside the laboratory. Moreover, as Frank has observed, the idea that diseases with striking physical symptoms might originate in a defect of the heart's electrical system was an entirely new and unfamiliar notion. It would take many years before clinicians recognized and absorbed the new *electrophysiological* approach to the heartbeat.[10]

Willem Einthoven (1860–1927), a young physiologist at the University of Leiden, extended Waller's investigations and improved on his laboratory apparatus. Einthoven's contributions to an understanding of the heartbeat are so fundamental that we can regard them as a distinct stream of scientific investigation, though it arose directly out of the work of Gaskell and Waller. Einthoven saw Waller demonstrate his equipment and technique at Basel in 1889. Within weeks, he had turned his research lab entirely to the study of the waveforms that the electrometer revealed. Over the next few years, Einthoven refined Waller's technology and devised a mathematical procedure for deriving the "true" waveform from the highly distorted tracing yielded by the electrometer. Einthoven noted three major deflections in

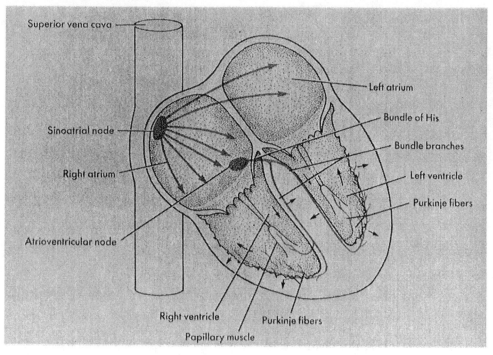

**Structure of the Heart.** This simplified cutaway shows the four chambers of the heart and the main structures of the conduction system by which each heartbeat is conducted through the atria and on to the ventricles. (Courtesy of Medtronic, Inc.)

the typical waveform. He labeled these with the letters *P, QRS,* and *T,* and established their normal amplitude and duration. Physicians throughout the world continue to use these identifiers today.[11]

By 1900, Einthoven had also discovered the relationship between the deflections of the waveform and the physical events in the heartbeat. He recognized first that the P wave represented the contraction of the atria — more precisely, the P wave corresponded to the wave of depolarization that triggered the atria to contract. The brief QRS portion of the waveform indicated that the wave had reached the ventricles, while the T portion signaled repolarization and relaxation of the ventricles. Einthoven noted that the normal interval from the onset of P to the beginning of the QRS (the P-R interval) in human beings was 0.12 to 0.20 seconds, but during the latter part of this interval the waveform was flat. Evidently the electrical wave, or "signal," paused before reaching the ventricles. The QRS in a normal heartbeat lasted just half as long as the P-R interval, indicating that the wave spread quickly throughout the ventricles. All this Einthoven

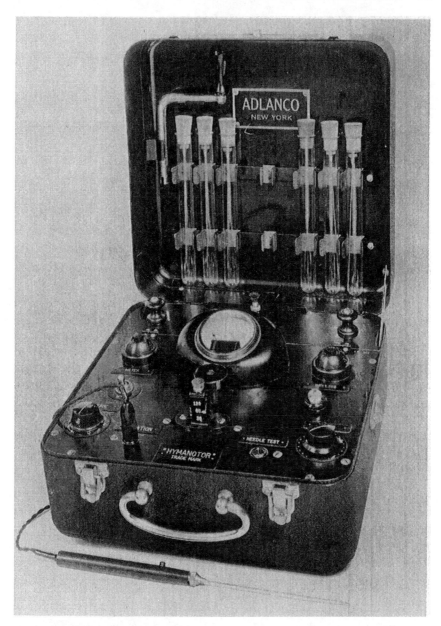

Albert S. Hyman's Pacemaker, 1932. Adlanco, a subsidiary of the German firm
Siemens, built this pacemaker (the "Hymanotor") for Dr. Albert S. Hyman of New York City.
Tested with animals both in the United States and in Germany, it probably was never used
with human subjects. Note the needle electrode in front of the pacemaker case. (Reprinted
from Dennis Stillings, *The Early History of Attempts at Electrical Control of the Heart: Harvey
to Hyman* [Minneapolis: Medtronic, Inc., 1979], n.p.)

gleaned from painstaking studies of electrocardiograms (ECGs) using his capillary electrometer.[12]

If the ECG were ever to become a tool that might assist in the effort to understand diseased conditions of the heart, and if it were ever to be useful in clinical diagnosis, some new kind of instrument would be needed — something that would produce a graphical record that scientists and clinicians could read directly without the labor of reconstructing the distorted waveform to yield a "true" electrocardiogram. Between 1898 and 1901, Einthoven invented a new instrument, the string galvanometer, the direct ancestor of today's electrocardiograph. The "string" was a silver-coated filament connected by wires at each end to large jars filled with liquid. Into these jars, the patient, seated between the jars, would place his hands and forearms. The silvered string was suspended between the poles of an electromagnet. Electrical currents in the heart caused the string to vibrate between the poles, and this motion, magnified four hundred times, was projected onto a moving photographic plate. Since the deflection of the string was proportional to the amplitude of the current, this new device yielded a more comprehensible record of the electrical activity of the heartbeat.[13]

During the years of Einthoven's studies of the ECG, a third stream of physiological research was leading to a more complete understanding of the structures within the heart that passed on the electrical wave. At Carl Ludwig's famous physiological laboratory in Leipzig, anatomist Wilhelm His Jr. (1863–1934) discovered a "muscle bundle" connecting the septal walls in the atria and ventricles in mice, dogs, and human beings. The bundle "arises from the posterior wall of the right auricle" and "forks itself into a right and left limb," each extending into a ventricle. His and others later showed that damage to the bundle could cause dissociation of the atrial and ventricular contractions. By 1900, physiologists had agreed that this structure, which is known today as the bundle of His or the AV bundle, conducted the cardiac impulse from the atria to the ventricles and that degeneration of the bundle or damage to it could block the impulse from reaching the two main pumping chambers of the heart, the ventricles. At its lower end, the bundle of His forms branches that pass the impulse on to the right and left ventricles. Either of the branches, or both, could block the impulse, as in Elmer Braun's case of left bundle branch block.

To test this understanding of the function of the bundle of His, the U.S. physiologist Joseph Erlanger (1874–1965) put a tiny clamp around the bundle in a dog; by tightening it, he could produce first-degree, second-degree, and finally complete heart block. (In first-degree block, the electrical impulse is delayed but still reaches the ventricles. This delay shows up in an ECG as a prolonged P-R interval. In second-degree block, some impulses fail to reach the ventricles. In complete or third-degree block, no impulse gets through.)[14]

Just at the time of Erlanger's report (1906), Sunao Tawara (1873–1952), a young researcher working in the pathology lab at the University of Freiburg, announced that he had found another structure in the wall of the right ventricle above and connected to the bundle of His. Subsequent work established that this atrioventricular or AV node delayed the electrical impulse before passing it on to the bundle of His and the ventricles, thereby giving the ventricles a little more time to fill with blood from the atria before the onset of ventricular contraction. With Keith's discovery of the sinus node in 1907, physiologists had assembled the basic understanding of the heartbeat that all later researchers, clinicians, and inventors have taken as their point of departure. In the process, they had also arrived at an understanding of the disorder known as complete heart block. As the British heart specialist Thomas Lewis pointed out, "The conduction of impulses depends upon the integrity of a very limited tract of tissue, and upon the integrity of this tract alone." Scientists understood that when the electrical signal was blocked at the AV node or the bundle of His, the ventricles were able to beat slowly on their own; but this *idioventricular rate* tended to slow down as time passed. The fainting episodes (Stokes-Adams attacks) that William Stokes had observed indicated that the ventricles were not delivering an adequate supply of oxygenated blood to the brain. It was not so clear in the first decade of the century just what was going wrong with people's conduction systems. At autopsy, some victims of heart block were shown to have "lesions" of various descriptions in the node or the bundle; but the causes of these injuries remained a mystery.[15]

## EDMUND BUTLER REVISITED

By World War I, physicians interested in diseases of the heart—assuming that they kept up with the newest research findings—were equipped to interpret cases like that of William Stokes's 1846 patient in a new way. They knew that the heart pumps blood to all the organs and tissues of the body and that it has its own electrical circuitry to initiate and coordinate this pumping action. Blood returning to the heart from the lungs and the body enters the two upper chambers, the auricles, or atria. A heartbeat begins when the atria contract and force blood through one-way valves into the two lower chambers, the ventricles. The ventricles contract a fraction of a second after the atria, the right ventricle pumping blood to the lungs to pick up oxygen and the left ventricle pumping oxygenated blood out to the body.

Each heartbeat is initiated by the sinus node, a small bundle of cells that is spontaneously rhythmic — it has the capacity to initiate an electrical impulse on its own, the rate depending on the body's activity level, temperature, and other factors. The sinus node is thus the heart's natural pacemaker. The electrical signals from the sinus node pass through the atria, causing those chambers to contract. The signals are delayed briefly at the AV node, then proceed through the bundle of His and the bundle branches and on throughout the ventricles.

Heart block, then, arose from "the failure on the part of the auricles to evoke a response in the ventricles"; one physician wrote that this "is now absolutely proved." Postmortem examination of some hearts showed "lesions in the auriculo-ventricular bundle [i.e., the AV node and the bundle of His]," which "must be regarded as analogous to an important junction of a busy railway system." But "as to the causes of the changes in the bundle . . . we are entirely in the dark." This collection of statements from a paper published in 1906 expressed an up-to-date clinician's understanding of heart block 60 years after Stokes. The writer, Edinburgh's G. A. Gibson, published tracings of the radial pulse, but no ECG, in a patient with heart block; he said little about electrical conduction in the heart, in fact. But his discussion differed in important ways from Stokes's. Gibson referred confidently to structures within the heart that Stokes had not known about and conveyed at least a rough understanding that these structures passed the heartbeat on and that the heartbeat was something like an electrical wave.[16]

Stokes had commented on Edmund Butler's lethargy and drowsiness; later generations of clinicians would understand these symptoms as owing to the extremely slow beating of the ventricles. Stokes noted that "the impulse of the heart is extremely slow, and of a dull, prolonged, heaving character." To physicians of the twentieth century, this vivid description would immediately indicate that instead of crisp, snappy ventricular beats, Stokes had heard idioventricular beats in which the two great chambers were contracting oddly because the electrical wave was not rushing through the heart walls in the normal way. Stokes did not say so, but it is also likely that Butler's ventricles were unable to pick up their rate when he tried to stand up and move around. Occasionally Butler's idioventricular rate probably dropped below 28; when it fell too far, Butler's brain would be starved of oxygen and he would experience one of his "fainting fits."[17]

The revised understanding of the heartbeat and of heart block did not exactly build on earlier knowledge from William Stokes's day; it would be more accurate to say that the experimental physiologists of the late nineteenth century had asked a set of questions that had not occurred to Stokes.

The new emphasis on the heart's function and behavior and the new technology of the laboratory, above all the electrocardiograph, developed in parallel. Over the course of three decades, roughly 1900 to 1930, clinicians adopted the new physiological way of thinking about the heart and learned to ascribe meanings to ECG tracings. But the new knowledge, embedded in a concept of the heart as a living, functioning entity, did not by itself lead to drastically improved treatments for disorders of heart rhythm.[18]

## CLINICIANS AND HEART BLOCK: THE PROBLEM OF TREATMENT

Physiologist Julius H. Comroe Jr., in a lengthy study of the scientific background to various "advances" in medical treatment, lists several eminently practical principles that physiologists had presented in the scientific literature by the 1920s. Laboratory research had established that a stopped heart could sometimes be revived, that you could touch or electrically stimulate a beating heart without necessarily courting disaster, that it was possible to induce the ventricles to contract by delivering electrical stimuli directly to them rather than by stimulating the sinus node, and that the ventricles could deliver an adequate supply of blood to the body even without the assistance of the atria. In the nineteenth century, occasional investigators had commented on one or another of these points. The brilliant Scottish scientist J. A. MacWilliam (1857–1937) had firmly asserted all four in a single scientific paper in 1889, and Joseph Erlanger had confirmed them in 1912. Comroe is correct in saying that the scientific basis for cardiac pacing of a primitive sort had been laid by World War I.[19]

The new understanding of the heartbeat and its disorders stands as a major achievement of laboratory physiology in the late nineteenth century, but it did not lead directly to improved treatment for complete heart block or other problems of heart rhythm. Indeed, the key ideas that Comroe abstracted from the scientific literature of the early twentieth century stand out only because the cardiac pacemaker had been invented by the 1970s, when he prepared the list. No one in the 1910s or 1920s drew up Comroe's list of key findings about the heart. No one at that time reasoned from such a list that it was now possible to invent the pacemaker. Background knowledge, by itself, never guarantees that something will be invented or, once invented, that it will come into regular use. The knowledge that a stopped heart could perhaps be revived remained purely experimental and had no medically useful result until well after 1920. David C. Schechter, a historian of cardiac electrostimulation, observes that "the treatment of complete heart block with electricity was an arbitrary undertaking" until after World War II. Investigators sometimes applied electricity to the chest or heart, but "the [electrical] current was meant to be a sort of 'tonic.' There was assuredly no thought of mechanically usurping the function of the natural pacemaker [by delivering electrical pulses similar to those of the sinus node]."[20]

MacWilliam was probably the first scientist to suggest that repeated electrical stimuli might speed up the heartbeat. A physiologist at the University of Aberdeen who had concentrated his research on a different heartbeat disorder, MacWilliam suggested in his 1889 paper that it might be possible to restore a normal heartbeat in "inhibited" hearts by applying "a periodic series of single induction shocks sent through the heart at approximately the normal rate of cardiac action." MacWilliam reported that he had actually tried this technique in his laboratory using anesthetized cats whose hearts had been exposed. It was a simple matter to slow or stop the heartbeat by rubbing the vagus nerve in the neck. He then inserted a needle electrode into the apex of the right ventricle and applied "a periodic series of induction shocks (regulated by a metronome)." He found that "each systole [contraction] causes the ejection of a considerable amount of blood into the aorta and pulmonary artery, and a marked rise of the blood-pressure at each beat." Further, "when a regular series of stimuli are employed, the contraction power [of the ventricles] becomes rapidly improved; the beats increase in force, and often approach the normal strength." Though he did not try the technique in a human subject, MacWilliam suggested that it ought to be possible to drive the heartbeat by placing a large sponge electrode "well moistened with salt solution" on the chest over the heart and a second electrode on the back. "The shocks employed should be strong, sufficient to excite powerful contraction in the voluntary muscles."[21]

MacWilliam was engaged in studies of ventricular fibrillation (VF), the lethal breakdown of the organized heartbeat that was sometimes brought on when a person was struck by lightning or suffered a serious injury. He went on to make important contributions to the understanding of fibrillation, but his 1889 paper suggesting a way to manage a slow heartbeat by means of repeated electrical pulses was forgotten — an isolated bit of work that had no influence. For nearly 40 years, no scientist or clinician attempted pulsed electrostimulation to speed up the heartbeat, even though a parallel stream of research showed that powerful shocks could terminate as well as bring on fibrillation. The inventors of the earliest pacemakers seem not to have known of MacWilliam's paper, which resurfaced as a historical curiosity only in 1972.[22]

By the 1920s, researchers had used the ECG to investigate and define several heart-rhythm disorders that were known to occur in human beings — notably, atrial fibrillation and heart block. They could not explain why these problems developed, but they could identify the symptoms in living patients and could give an account of the probable course of each disorder over time. But the new understanding of the heartbeat and its disorders made little difference to the clinical treatment of rhythm disturbances (*arrhythmias*) in the 1920s. The size and cost of Einthoven's string galvanometer restricted it at first to a small number of hospitals and research laboratories. Moreover, as Stanley Reiser and others have pointed out, until investigators had correlated the ECG tracings with more familiar clinical symptoms and with autopsy results, it was difficult to assign clinical "meanings" to changes in the waveform. For another 30 years, to the extent that doctors tried to treat heart block at all, they relied solely on drugs. When two doctors, one of them a recognized expert on the heartbeat, independently invented artificial pacemakers, the inventors did not make the connection between physiologists' findings about heart-rhythm disorders and their little machines that delivered electrical stimuli. They used the inventions not to treat heart block but for other purposes.[23]

### THE FIRST INVENTION OF CARDIAC PACING

In September 1929, at a meeting of the Australasian Medical Congress in Sydney, Australia, a physician-anesthetist from Sydney, Mark C. Lidwill (1878–1968), described a piece of electrical equipment that he had invented to drive the human heart. Apparently Lidwill intended his machine not for cases of heart block but for emergencies when surgical patients went into cardiac arrest while under general anesthesia. No photograph or diagram of Lidwill's device is known to exist. The inventor said that it was portable and plugged into a standard wall socket. "One pole is applied to a pad on the skin, say the left arm, and is saturated with strong salt solution," he explained. "The other pole[,] which consists of a needle insulated except at its point, is plunged into the ventricle and the machine is started." The operator, presumably an anesthetist, could vary the pacing rate and the voltage.

Lidwill told his audience that using a slightly earlier model of the apparatus, probably in 1925 or 1926, he had tried to revive several stillborn infants. One of the infants had "recovered completely and is now living and quite healthy." This child did not respond to other treatments such as an injection of epinephrine (adrenaline) into the heart, so Lidwill "plunged" the needle electrode first into the right atrium and then, when atrial pacing failed, into the right ventricle. After ten minutes of ventricular pacing, he turned the pacemaker off and the child's heart continued on its own. Thus a baby in Australia was the first human being to undergo successful pacing of the heart, and Lidwill's machine was the first artificial pacemaker. Lidwill suggested that the pacemaker might be appropriate for cardiac arrest from anesthesia, drowning, certain types of gas poisoning, in acute illness such as diphtheria when the heart stopped, and perhaps when an underlying disease of the heart disrupted the normal rhythm and brought it to standstill. He recognized that pacing might succeed only occasionally, "but one life in fifty or even a hundred, is a big advancement where there is no hope at all."[24]

Lidwill's brief report appeared in the transactions of the meeting, but nothing more has been discovered about the first pacemaker. Just why the inventor and his associate, physicist Edgar H. Booth, did not continue their research in electrostimulation is not known. Lidwill's work seems another example of an isolated investigation that aroused no interest and led nowhere — except that one investigator halfway around the world read the published account and took note.

Albert S. Hyman (1893–1972), a practitioner cardiologist in New York City who also pursued clinical research, was thinking along the same lines as Lidwill. Though he did not have an academic appointment, Hyman had rubbed shoulders with distinguished medical scientists during the formative years of his career; this had apparently launched him on studies of sudden standstill of the heart. During his freshman year at Harvard College in 1911–12, Hyman's faculty adviser had been the famous physiologist Walter B. Cannon. Nearly 60 years later, Hyman recalled his first visit to Cannon's laboratory at Harvard Medical School: "I saw the pioneer experimental work then being carried out on the isolated turtle heart. I saw the complicated apparatus employed to register the various activities of the [heart], and I became deeply impressed with the physiologic processes involved in the simple beating of the animal heart." Hyman also recounted an incident in which an experimental heart stopped beating. "As we turned to leave the room, I timidly asked Dr. Cannon, 'That heart looked exactly like all the other hearts which were working. What happened, and why can it not be started again?' Dr. Cannon looked at me with a wistful smile. . . . 'Young man,' he said, 'that is indeed a very important question. All that I can say at this moment is that I am glad that you asked it; and if the seeds of inquiry have been properly implanted by this incident, perhaps some day you may be able to produce the answers.'" This story, while sounding too pat to be literally true, suggests that Hyman wished to have his investigations into electrostimulation of the heart remembered as part of the mainstream of laboratory research into cardiac physiology.[25]

Hyman continued to study the heartbeat during his four years at Harvard Medical School (1914–18). An anecdote from these years sounds more believable. On February 26, 1918, he was working in the emergency room at Boston City Hospital when a 44-year-old man was brought in with a compound fracture of the right femur — the thighbone. The victim was in shock as a result of his terrible injury and, while Hyman examined him, the man's heart stopped beating. An injection of epinephrine got it started again; but four minutes later, "a gross irregularity developed, followed by complete asystole [failure to contract]. This time epinephrine had no effect, and the man died." Hyman began a notebook and eventually collected 81 cases in which physicians had injected stimulants directly into healthy but asystolic hearts; there had been only 13 successful outcomes. Cardiac arrest tested the limits of doctors' professional knowledge and skill. In most cases, they were unable to stave off "complete surrender to death." Yet many patients who died of sudden cardiac arrest, like the man at Boston City Hospital, had undiseased hearts. These hearts had stopped beating on the operating table or from shock after the body had suffered a grievous injury.

After his residency and a period of study in Europe, during which he came under the influence of British cardiologist James Mackenzie, Hyman set up a practice in New York City and lined up financial support from a small foundation that enabled him to continue clinical research on the problem of "sudden death in persons who had a more or less normal cardiovascular system." At first, he focused on the problem of just where to inject stimulants: Into the wall of the heart (the *myocardium*)? Or through it directly into a chamber? And which chamber? Experiments with laboratory animals led Hyman to conclude, by early 1929, that what kicked the heart back into action was not the stimulant but the needle's puncturing the heart wall. The needle set up an "action current of injury," a familiar phenomenon in muscle tissue. Study of ECGs revealed that the first beats after the needle stick had a distinctive pattern. "In successful cases these ectopic [abnormal] beats were sufficient to restore a sufficient volume output of the left ventricle to flood the coronary system, which in turn led to the restoration of normal sinus nodal activity."[26]

As Hyman understood the underlying mechanism of resuscitation, during the first seconds of standstill the threshold of electrical conductivity in the heart was lowered and heart muscle temporarily became more irritable, more likely to respond to a stimulus. If the physician could react swiftly enough, he stood a good chance of reviving the heart with a single thrust. But this period of heightened irritability was temporary. "As the electrodynamic balance of the entire heart becomes more and more disturbed, a single prick of the needle may not be sufficiently powerful. . . . Two or even three or more needle thrusts may be required." Multiple punctures were a desperate procedure threatening hemorrhage into the pericardium, the sac enclosing the heart. Could some other repeated irritant or stimulus work just as well? "If the electric difference of potential could be rhythmically developed during this period of cardiac standstill," he wrote, "it would appear . . . that regular contractions of the heart would follow, and under such artificial stimulus production an automatic activity of the entire heart might be maintained."[27] Early in 1929 — before Lidwill's published report but after the first clinical use of Lidwill's device — Hyman made a seminar presentation in which he proposed the concept of pacing the heart by pulsed electrical stimuli. By 1930, he knew of Lidwill's pacemaker, but the Australian physician's paper gave few details of its design. Hyman's own pacemaker took shape in 1929–30; his patent application is dated March 12, 1930. It seems likely that Hyman found encouragement in Lidwill's report but designed his pacemaker on his own, assisted only by his brother Charles.

When he announced his invention of an artificial pacemaker in 1932, Hyman emphasized that it was not enough simply to direct bursts of electrical energy into the heart muscle. He listed three specific requirements (italicized in the list below) for an artificial pacemaker. From an old photograph, the patent application, and Hyman's own description, we can determine how he tried to satisfy each requirement.[28]

— Because its intended use was to resuscitate the arrested heart, "*[the pacemaker] must be prepared at all times for instantaneous action*" (italics added). The device should be a stand-alone piece of hardware that the physician could count on even if it had languished in the corner for months. A photograph depicts a small machine with a spring motor and a conspicuous hand-crank. Sturdily built, the pacemaker measured about 12 inches long by 12 wide and was about 15 inches in height. It weighed 16 pounds and fit into a carrying case. Overall, it bore little resemblance to the pacemakers of the 1950s and later. For a power source Hyman experimented with flashlight batteries, but he chose instead a small, spring-driven mechanism that would spin the generator to produce a uniform electrical current for six minutes. The crank would rewind the spring, Hyman assuming that six minutes of artificial pacing would be more than enough time to restore a normal heart rhythm.

— *The stimuli must emulate the heart's own pulses in regularity, duration, and amplitude.* Hyman used a small direct-current generator that seemed to meet this requirement. To create pulses in the current, the brothers employed a rotary interrupter-disk geared at ratios that would permit pacing stimuli of 2 to 3 millivolts at rates of 30, 60, or 120 per minute. To enable the physician to observe the pacing speed, the inventors added neon lamps powered by the same current but activated in alternation with the pacing pulses. When the pulse was on the lamps were off, and vice-versa.

— *The stimuli should emanate from a site close to the sinus node in the wall of the right atrium.* Lidwill had attempted atrial and ventricular stimulation; Hyman, on the other hand, paid attention to the upper chambers only. Stimulating the right atrium was the logical thing to do since he was trying to revive hearts that had intact conduction systems (no heart block) but had been stopped by some accidental cause. Hyman designed a bipolar needle electrode that the physician was supposed to insert between the first and second ribs into the right atrium close to the sinus node. A generation later, inventors of pacemakers did not try to pace the atrium as Hyman had done, but chose the right ventricle instead, for reasons discussed in chapter 2.

The Hyman pacemaker had technical problems that would have rendered it unusable without a great deal of development work. Doctors recognized that Hyman would have them locate precisely a small region of the heart and plunge a needle into it, and this in conditions of the greatest possible stress, with the patient convulsing and dying. Hyman seemed confident that he could insert a needle into the wall of the right atrium without causing serious injury to the heart, but many doctors must have regarded this procedure as dangerous. The inventor apparently was not well versed in electrical matters. None of Hyman's writings discusses the rate of the pacemaker stimuli (number per minute) or the duration and amplitude of each stimulus — crucial information for determining the effect the machine would have on the heart. In common with other researchers of that era, Hyman also made no distinction between standstill and fibrillation of the ventricles. In both, the circulation of blood ceases and the victim collapses, but pacing a fibrillating heart — firing small bursts of electricity into it — will have no effect. The two conditions, standstill and fibrillation, are difficult to distinguish without an ECG tracing.

Wolfgang Schivelbusch comments that "a technological innovation becomes a historically significant one only when there is an actual economic demand for it." No demand developed in the 1930s or the 1940s for an artificial pacemaker that would perform emergency resuscitation. Why was this? David C. Schechter, a physician who has studied the history of cardiac electrostimulation, points out that electrotherapeutics had long been associated with medical quackery. In an era when government regulation of medical devices did not exist, physicians were bombarded with claims about health gadgets and had developed extreme skepticism. Some doctors might also have recalled reading about the macabre experiments of the early nineteenth century in which scientists had applied electrical currents to the bodies of recently hanged or beheaded criminals to study the action of the nerves and muscles.[29]

More broadly, the medical profession in the 1930s regarded cardiac arrest as an unusual and generally irreversible condition. Most doctors apparently considered it a hopeless situation when it occurred or believed that injections of stimulants stood as good a chance as anything to revive the stopped heart. Whatever methods doctors tried, they reported few successes. Paul Dudley White's authoritative textbook *Heart Disease* devoted less than a page to cardiac arrest, an event that now is believed to kill hundreds of thousands of Americans every year. Hyman was not able to persuade his colleagues that arrest was a treatable condition. In this sense, the pacemaker seems to have been an answer to a question that very few medical people were asking at that time. When artificial cardiac pacing later did gain acceptance, it was not as a stand-alone procedure but as one technique in a larger set of treatments for diseases of the heart. The development of open-heart surgery in the years after World War II was the event that gave doctors, for the first time, an incentive to find artificial means to pace the heart.[30]

Rapid dissemination of new medical thinking and new treatments into widespread use is a phenomenon of our own era. The need for X rays and the ECG strikes us as obvious today; but when they were first introduced at the turn of the century, clinicians often failed to incorporate them effectively into medical practice for years because the new equipment did not readily harmonize with existing ideas about disease and treatment. It would have been surprising indeed had physicians eagerly accepted the Hyman pacemaker, even if its inventor had reported successful use in human subjects.[31]

A radically new idea or artifact stands a better chance of a respectful hearing if it enjoys the support of influential leaders and opinion shapers. Albert S. Hyman was poorly situated to exercise influence in the community of heart specialists. Hyman held a medical degree from Harvard, had worked briefly with James Mackenzie, and had coauthored a textbook on clinical electrocardiography. However, as a Jewish physician, he was informally but effectively "excluded from New York's medical elite." With no chance to secure an appointment at a major teaching hospital in the city, Hyman maintained a private practice and held a part-time staff position at the 154-bed Beth David Hospital in Manhattan. In 1926 he had helped to found the Sir James Mackenzie Cardiological Society, later renamed the New York Cardiological Society (NYCS), an organization that focused on the needs of practitioners rather than academic cardiologists. Twenty years later, the NYCS became the springboard from which Franz Groedel launched the American College of Cardiology. But throughout the 1920s and 1930s, it was a distinctly marginal professional society. Academic and physiological cardiologists remained firmly in control of cardiology — through their appointments at leading hospitals and medical schools, through national reputations as scientific and clinical researchers, and through control of the American Heart Association. A secondary figure in U.S. cardiology, Hyman stood little chance of winning much support for his ideas about pacing the heart.[32]

A coincidence of timing may have contributed to public uneasiness about Hyman's invention and reinforced its association with the troubling history and mythology of electroresuscitation: the horror film *Frankenstein* (1931), in which a mad scientist uses electricity to bring to life a monster constructed from human parts, had appeared just a few months before Hyman announced his invention. If nondoctors heard of it at all, did they associate Hyman's work with Victor Frankenstein's transgression? In her novel *Frankenstein* (1818), Mary Shelley never directly says whether Frankenstein (a chemistry student, not a physician) employs electricity to "infuse a spark of being into the lifeless thing" that he has created — but the movie leaves nothing to the viewer's imagination.[33]

The story of Hyman's pacemaker has two intriguing sequels. Late in 1931, while his animal experiments were under way, the inventor contacted a New York subsidiary of the German company Siemens AG, which he probably knew as a maker of ECG machines and other medical equipment. Siemens agreed to build six of the Hyman pacemakers and sent the first of these back to Hyman in December 1932. In a letter many years later, a Siemens spokesman wrote: "In spite of several changes, made by our New York agency on the sample apparatus in accordance with Dr. Hyman's wishes, no useful results were achieved." Siemens also sent a unit to Dr. Siegfried Koeppen, a young laboratory scientist at the Medizinisch Policlinic of Leipzig University who had been searching for ways to revive people whose breathing and circulation had ceased from accidental electrocution. In ten months of experiments, Hyman's pacemaker proved unsuccessful: Koeppen could not even revive a rabbit. When the inventor did not reply to letters asking for advice on test conditions, Koeppen canceled further work at the end of 1934; privately, he wrote that Siemens had been taken in by a *"plumper Schwindel"* — a clumsy fraud.[34]

By World War II, Hyman had begun to realize that the pacemaker, originally intended to restart healthy hearts that had stopped beating, might also be used to revive Stokes-Adams victims. In 1942, at a small conference at the Brooklyn Naval Hospital, Hyman described a clinical case of heart block: a retired naval officer, age 76, had a ventricular rate of 32 and complete dissociation between the atria and ventricles. When his ventricular rate dropped below 24 beats per minute, the officer began to experience Stokes-Adams attacks. Hyman said that on February 12, 1942, he and a Dr. Richman had restored this patient to consciousness by using the needle electrode to pace the right *atrium*. This is a puzzling report: in complete heart block, atrial stimuli would not be propagated to the ventricles. Perhaps the doctors had mistakenly inserted the electrode into the right *ventricle* and unintentionally achieved ventricular pacing, or the patient's heart had independently resumed a faster rate. Since an ECG record is not known to exist, the report cannot be verified. Four years later, Hyman and several associates read a "preliminary report" at a local medical society meeting in which they announced that they had used a Hyman pacemaker successfully to pace three elderly patients for short periods at New York City Hospital. These cases may be authentic. But, as before, the doctors paced the right atrium so the clinical significance of this work is difficult to pin down. They never published their report and Hyman never again championed artificial cardiac pacing.[35]

This account of the emergence of a new understanding of the heartbeat and the invention of crude pacemakers between the two world wars illustrates a point that shows up repeatedly in the history of cardiac pacing and defibrillation. The properties of a new medical device, in themselves, do not determine whether the device will come into general use or be rejected. The reaction of the broader community of investigators and clinicians is crucial in determining the fate of a new invention. Given existing medical knowledge about cardiac arrest and existing medical routines of the 1920s and 1930s, the pacemaker had no place in clinical medicine. The complete nonacceptance of the Hyman pacemaker — heart doctors' refusal even to test Hyman's claims or to try to improve on his invention — is striking. By the late 1960s, matters had changed fundamentally. The new pacemakers of the 1960s also had serious flaws, but many physicians took up pacing anyway and devised ways to improve the devices.[36]

Physicians who specialize in disorders of the heartbeat today remember Drs. Lidwill and Hyman. Papers recalling their work occasionally appear in cardiology journals, and the North American Society of Pacing and Electrophysiology (NASPE) awards two Mark C. Lidwill Traveling Fellowships every year and the Albert S. Hyman Fellowship in Cardiac Pacing and Electrophysiology in alternate years. A small, out-of-focus photograph of Hyman's pacemaker in its earliest embodiment regularly turns up in the introductory "historical overviews" in books on cardiac pacing. Then in 1999, Dr. Seymour Furman, editor of the medical journal *PACE (Pacing and Clinical Electrophysiology)*, acquired a copy of the original Hyman patent from 1930 and asked George Szarka, the pacemaker engineer at his hospital, to build a full-sized working model from the drawings and description. The machine, painted shiny black, attracted attention as a display piece at NASPE's annual scientific meeting in May 1999: the motor made a satisfying whirring sound; the little lamps flashed on and off.[37]

In these activities we can glimpse a subtext to the story of Lidwill and Hyman that is important to the community of pacemaker physicians: as remembered, the story affirms that it was not engineers in corporate research labs who had invented the earliest cardiac pacemakers but doctors — two clinicians whose interest was above all in treating patients.

**CHAPTER 2**

## THE WAR ON HEART DISEASE
## AND THE INVENTION OF
## CARDIAC PACING

As a doctor, Paul M. Zoll had a two-track career: he did laboratory research at the Beth Israel Hospital in Boston and he also maintained a large general practice, even making house calls. The medical world took notice when Zoll announced in 1952 that he had successfully kept a patient alive through numerous episodes of ventricular standstill using a bedside device that delivered electrical pulses to the heart.[1] Zoll's external pacemaker had imperfections, but these lacks inspired heart researchers to improve on the original design—and put cardiac pacing on a path of continuous technological change. By the end of the 1950s, pioneer open-heart surgeons and a handful of tiny manufacturing firms were also paying attention to cardiac pacing. These new participants began to reshape the pacemaker and redefine its uses; in the process, they drove pacing a considerable distance from what Paul Zoll had originally had in mind. Even at this early point in its history, "cardiac pacing" had a quicksilver quality: the meaning of terms like *pacing* and *pacemaker* began to undergo a process of redefinition that has never yet stopped.

### A NEW ERA IN HEART MEDICINE
Medical thinking about diseases of the heart changed rapidly during the 1930s and 1940s. Electrostimulation of the heart, largely ignored in the early 1930s, reemerged in an era of intensified research and high expectations for new treatments. The idea of attacking acute maladies of the heart by getting inside the heart and putting instruments and devices into it came into favor after World War II.[2] This shift in outlook occurred in a number of fields of heart medicine: in techniques of diagnosis, in surgery of the heart, and in the treatment of rhythm disturbances like fibrillation and heart block. The announcements of laboratory discoveries or successful new procedures in one field of heart medicine aided and challenged researchers in every related field by answering some questions, raising still more, and conjuring new possibilities. The momentum of research and discovery depended, in

turn, on funding from the federal government and private sources. The policy of encouraging medical research expressed ordinary Americans' high hopes and enlarged expectations; these were nourished throughout the 1940s and 1950s by the incontestable achievements of wartime medicine and by a stream of popular accounts about medical breakthroughs.

The new postwar treatments required new facilities and complex equipment that could only be maintained and put to use in the hospital. As mortality from childhood infectious diseases declined rapidly in the 1940s, hospitals began to redefine themselves as care centers for heart disease, stroke, and cancer — diseases of middle age that, by 1945, had already become the leading causes of death in the United States. In this context of hospital-based coronary care, the modern invention of cardiac pacing took place. Paul Zoll's external pacemaker of 1952 did not invade the human body, but later pacemakers certainly did. By an interesting irony, the pioneer of clinical cardiac pacing specialized in internal medicine rather than surgery, always viewed implantation with unease, and eventually revealed himself as a man of deeply conservative instincts.[3]

Albert Hyman had not invented his pacemaker to treat heart block with Stokes-Adams attacks but to restart healthy hearts that had stopped because of injury from a fall or an electric shock. The background scientific knowledge needed to build and use Hyman's invention had been minimally adequate by the 1920s. But beginning in the late 1930s, specialists in cardiac electrophysiology made important discoveries about the behavior of the heart during a Stokes-Adams attack.[4] Around the same time, improved knowledge of ventricular fibrillation led to new attempts to terminate this lethal arrhythmia with electric shocks. From a somewhat bizarre practice associated with eighteenth-century Leyden jars and nineteenth-century quackery, electricity came to hold a place, admittedly still tenuous, as a possible treatment for heart-rhythm disorders. These developments were crucial for Zoll's invention of the external pacemaker in 1952 and the diffusion of pacing into U.S. hospitals.

## NEW THINKING ABOUT STOKES-ADAMS ATTACKS

In the decade after Hyman's pacemaker, cardiologists refined their ideas of what happened in a Stokes-Adams attack, although today's understanding of the underlying causes did not emerge until the 1960s. A group in London led by John Parkinson (1885-1976) proposed in 1941 that the term *Stokes-Adams disease* be reserved for repeated episodes of *syncope* (failure of the circulation and loss of consciousness) in the presence of some degree of heart block. Obtaining an ECG tracing of an actual Stokes-Adams attack

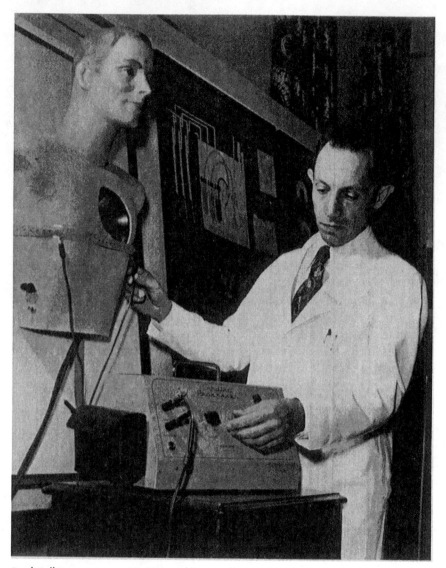

**Paul Zoll Demonstrates His External Pacemaker, Mid-1950s.** A leather strap held the two chest electrodes in place on either side of the heart. (Courtesy of Ruth Freiman, archivist, Beth Israel Deaconess Medical Center, Boston)

was difficult because the event would occur without warning. However, Parkinson and his colleagues were able to bring together 64 ECGs recorded during Stokes-Adams attacks — 56 culled from published articles and 8 from their own hospital. In about half the cases, syncope was the consequence of ventricular standstill or an extremely slow ventricular rate, but in the other half the victim collapsed because the ventricles had speeded up sharply or gone into fibrillation. Parkinson suggested that patients whose Stokes-Adams attacks included tachycardia or fibrillation had a poor chance of long-term survival.[5]

The accepted treatment for a Stokes-Adams attack in the 1920s and 1930s had been the drug epinephrine, a stimulant that caused the heart to accelerate and beat more strongly. This made sense as long as the cause of the attack was an unduly slow heart rate, but epinephrine would intensify tachycardia or fibrillation. The Parkinson report thus spurred a search for new drug treatments for the different varieties of Stokes-Adams. Parkinson did not mention electrical stimuli as a possible treatment, but this paper helped provide a firm footing for Paul Zoll's work a few years later. Zoll could conclude that a stimulator delivering pulses to the heart could manage about half of all severe Stokes-Adams attacks, the half in which standstill of the ventricles or a very slow rate was the underlying cause.[6]

Parkinson made Zoll's work possible in a more fundamental sense as well, for when Zoll introduced the pacemaker in 1952 as a device that would treat Stokes-Adams attacks, he was invoking a clearly defined disease entity about which there was impressive recent information. Parkinson had legitimated Stokes-Adams *disease* as a condition in its own right. This choice of terms called attention to Stokes-Adams and made it seem more terrible and more real than had its former designations as a "syndrome" or a "complication" of complete heart block. The new ontological status for Stokes-Adams disease probably helped Zoll win attention and acceptance for the external pacemaker several years later. But the fact that physicians could not determine the mechanism of a Stokes-Adams attack without obtaining an ECG tracing while the event was in progress also led Zoll to move on from inventing a pacemaker to introducing a cardiac monitor and a closed-chest defibrillator.[7]

## VENTRICULAR FIBRILLATION AND DEFIBRILLATION IN THE HOSPITAL

The history of cardiac pacing and defibrillation have been intertwined. New knowledge about fibrillation gained during the 1930s and 1940s led to the first successful defibrillation in 1947, when a surgeon in Cleveland halted an episode of ventricular fibrillation (VF) in a hospital operating room by delivering a powerful electric shock to a patient's heart. Five years later, Zoll's pacemaker enabled a patient to live through a succession of Stokes-Adams attacks.

John A. MacWilliam had vividly described VF in 1887. When he applied strong electrical currents to the ventricles of a dog's heart, "the ventricular muscle is thrown into a state of irregular [arrhythmic] contraction, whilst there is a great fall in the arterial blood pressure. The ventricles become dilated with blood as the rapid quivering movement of their walls is insufficient to expel their contents; the muscular action partakes of the nature of a rapid incoordinated twitching of the muscular tissue." Its complex and uncoordinated character made fibrillation "fundamentally different from a rapid series of normal contractions." VF did not stop spontaneously: it would kill an animal or a human being within a few minutes because the fibrillating heart could not pump blood. MacWilliam mentioned several "influences" that could induce VF — above all, electric shock — and suggested in 1889 that VF was the underlying mechanism in many sudden and otherwise unexplained deaths, including death during chloroform anesthesia. His papers formed the basis for twentieth-century research into this lethal arrhythmia, although for many years the work he had done received little attention.[8]

At the end of the nineteenth century, physiologists at the University of Geneva demonstrated with lab animals that it was possible to terminate VF by passing a strong electric current through the heart. But electric counter-shock, as it would later be termed, faced "practical difficulties" in translation to clinical use. One physician said: "In the first place, there is no experimental evidence . . . to show what voltage and what strength of current should best be employed in this method of resuscitation. In the second place, there would usually be great difficulty in providing the current at the required voltage for use on the spot and within a few minutes." He might have added that from visible symptoms or the feel of the pulse, an observer cannot distinguish between fibrillation and standstill of the heart.[9]

Reports that electric shocks had revived the apparently dead can be found as far back as the 1750s, a few years after the invention of the Leyden jar, the earliest electric condenser;[10] but systematic investigation into ways to halt VF got underway only in the 1920s. Two circumstances appear to have led to this research. During the decade, a widespread consensus took shape among heart specialists about the existence of a clinical event that they called coronary thrombosis, formation of a blood clot in the coronary arteries. By depriving a portion of the heart muscle of its supply of oxygenated blood, the clot caused the muscle to die. Coronary thrombosis is one cause of *acute myocardial infarction*, or heart attack.[11] MacWilliam argued in 1923 that when a person died of a heart attack, the mechanism of death was ventricular fibrillation. This observation had far-reaching implications, for it meant that fibrillation did not have to be instigated by some powerful and sudden outside force such as electrocution: "The abrupt onset of fibrillation may come without preceding signs of cardiac failure, as tested by exercise, or without immediately premonitory evidences of cardiac disturbance."[12]

At the same time, electric-power companies began to sponsor research to learn more about the effect of electrical currents on the human body and to discover how to resuscitate human beings after accidental shocks. One group of researchers succeeded in defibrillating dogs using a 60-cycle alternating current of about one ampere applied for durations of 0.1 to 5 seconds.[13] Others demonstrated that a shock was most likely to induce fibrillation if it fell during the "relative refractory period" corresponding with the downslope of the T wave of the ECG. Physiologist Carl J. Wiggers later designated this brief interval the *vulnerable period* of the cardiac cycle.[14] This new information about the importance of the timing and duration of an electrical stimulus armed investigators to pace the heart. It had the greatest possible influence on Paul Zoll, who repeatedly cited Wiggers and others working on defibrillation. After 1940, investigators understood clearly not only that electricity delivered to the heart need not induce VF, but that they could set the amplitude, duration, and timing of electric stimuli so as to reduce greatly the chances of bringing on such a disaster.

The work of the late 1920s and 1930s represented a major shift in the practical understanding of fibrillation and defibrillation. Efforts at open-chest defibrillation in the operating room began in 1938 when surgeon Claude Beck, of Western Reserve University in Cleveland, twice defibrillated human hearts. Both patients, however, had suffered brain damage and they died shortly afterward. Beck finally succeeded in 1947 when he performed an emergency thoracotomy (surgical opening of the chest) on a young man whose heart had gone into VF during a routine operation. Beck massaged the heart with his hand to keep the circulation going while others set up the defibrillating apparatus, then used alternating-current (AC) countershock to defibrillate the patient.[15]

By demonstrating with a human being that electrostimulation could be both effective and safe, open-chest defibrillation contributed to the climate of possibility in which Zoll and others invented pacemakers during the 1950s. Beck himself went from demonstrating defibrillation in the hospital to advocating a far more disturbing practice. In the late 1940s, he opened a campaign to train medical and dental personnel and even laypersons to cut open the chest of a person suffering from cardiac arrest and massage the heart by hand — on the street, on the golf course, or in the office. Paul Zoll later commented that open-chest cardiac massage outside the hospital was "abhorrent"; he was disturbed by the media's fascination with the idea. Throughout the 1950s, Zoll took care to reassure the medical community that cardiac pacing was a relatively gentle, closed-chest treatment used only at bedside within the hospital.[16]

## CARDIAC CATHETERIZATION—
## A NEW DIAGNOSTIC TECHNOLOGY

Beginning with Claude Bernard's experiments on horses from 1844 to the 1870s, a number of European investigators had introduced catheters into the blood vessels of animals or dying human beings and advanced them toward the heart. But cardiologists generally "credit" Werner Forssmann (1904–79), a surgical trainee in Eberswalde, Germany, with performing the first successful right-heart catheterization. In 1929, Forssmann opened a vein in his own arm and guided a urological catheter into the right atrium of his heart. With the catheter in place, and without discomfort, he then walked to the X-ray department at his hospital to have films made of his chest. Forssmann also developed the technique of injecting radiographic contrast material through a catheter into the right atrium to permit X-ray images of the right heart.[17]

During and after World War II, cardiologists and thoracic surgeons, led by André Cournand (1895–1988) and others at Columbia University, transformed catheterization of the heart into a standard tool for evaluating the function of the heart in normal and sick patients. Opening the vein at the crook of the patient's arm, they would advance a flexible catheter up the venous system, through the large vein known as the superior vena cava, and into the right atrium of the heart.[18] In 1942, Cournand moved the catheter past the tricuspid valve into the right ventricle, and in 1944, he moved it on into the pulmonary artery.[19] Contrary to some expectations, clots did not form along the catheter, respiration and the heartbeat did not materially change, and cardiac output (the amount of blood pumped by the left ventricle) was not affected as he performed these maneuvers.

Cournand's group introduced various technical improvements in catheter design to create a tool that was flexible yet stiff enough to be maneuverable. They also developed new recording devices that could graph the blood pressure at various sites within the heart. Through studies of traumatic shock from battlefield wounds during World War II, they demonstrated that cardiac catheterization was a safe and enormously revealing diagnostic technology. By providing information about blood pressure and flows within the heart, the catheter technique enabled the cardiologist to judge the pumping efficiency of the heart's chambers and the adequacy of the circulation. Catheterization, a procedure that invaded the sanctum of the human heart, began to move from the research laboratory into large hospitals by the late 1940s. Within a few more years, it had become an ordinary in-hospital procedure, part of the standard workup for most patients scheduled to undergo heart surgery. More than thirty thousand heart catheterizations were performed in U.S. hospitals in 1961.[20]

## HEART SURGERY AND HYPOTHERMIA

Cardiac pacing was closely associated with open-heart surgery in the 1950s, with the surgeon making way for the pacemaker both literally and metaphorically. Formerly, thoracic surgeons "had been taught never, never, never to go through the pericardium [the membrane enclosing the heart].... It was a sacred area — you didn't invade that." On the eve of World War II, the young surgeon Dwight E. Harken (1911–93) noticed again and again his seniors' "reluctance to touch, even retract, the heart." New treatments of the postwar era, like pacing, depended on a new understanding of the heart as a hardy organ that a surgeon could safely touch, manipulate, and even cut open. This "acceptance of the *idea* of heart surgery" diffused rapidly after the war — actually, somewhat in advance of specific surgical procedures.[21]

Heart surgery had lagged behind other forms of surgery before the 1940s. Surgeons had developed a few operations to correct specific problems and had tried these in human beings. Some patients recovered and were able to lead normal lives, but mortality rates in heart surgery remained unacceptably high, often 50 percent or higher. On the eve of World War II, improvements in anesthesia, blood transfusion, and techniques of suturing blood vessels finally created conditions for the expansion of heart surgery. Surgeons first performed operations around the heart (extracardiac surgery). They learned that they could temporarily clamp the aorta, the major artery leading out of the heart, while surgically correcting congenital flaws in that vessel.[22]

In 1944–45, pediatric cardiologist Helen Taussig (1898–1986) and surgeon Alfred Blalock (1899–1964) of the Johns Hopkins University devised and successfully pioneered a procedure to ease the symptoms of tetralogy of Fallot, a set of four congenital malformations of the heart that annually stunted and killed thousands of children by reducing the normal flow of blood from the heart to the lungs, where the blood picks up oxygen to carry to the body. Taussig had suggested that the surgeon could create an artificial duct, or shunt, from one of the arteries near the heart to a pulmonary artery. By boosting the flow of blood into the lungs, the shunt would ease a child's severe deficiency of oxygenated blood.[23] But the Blalock-Taussig procedure was a palliative that did not really correct the congenital defects within the heart. To close a hole in the partition between the right and left ventricles (one of the four congenital defects), a surgeon would have to open a ventricle and disrupt the circulation for many minutes.[24]

World War II forced one surgeon to enter the heart. Dwight Harken was stationed in England at the 160th General Hospital, a huge center set up to receive casualties of the Allied invasion of Europe. There he learned to remove shell fragments and other foreign bodies from the hearts of soldiers. Harken used the most advanced devices and methods available at that time: fluoroscopy to pinpoint the precise location of the fragments, endotracheal intubation, whole-blood transfusion, penicillin to fight off infection. The actual surgical procedure was often brutal because it had to be quick. Harken was prepared to cut a beating heart wide open. By transfusing blood under pressure, his surgical team could keep the patient alive long enough for Harken to remove the missile and stitch up the heart. Harken was a self-dramatizing man and for the rest of his life he enjoyed telling the story: "The first fourteen I operated on . . . all died. The second [group], seven of the fourteen died. The third, two of the fourteen died. I was getting better. But the fourth fourteen, all survived. And the only difference between these first three groups and the fourth group [was that] these were all animals. The final fourteen were all American soldiers, and they all survived." The soldiers were out of bed within days, "living proof that the heart was not such a fragile organ." Every one of Harken's 134 patients survived with a normally functioning heart.[25]

Surgeons still could not actually enter the heart for more than a few seconds — a minute at most. How could they gain more time to work within the heart? Slowing the patient's metabolism by lowering the body temperature seemed a promising approach. In 1950, surgeon Wilfred G. Bigelow (b. 1913) at the Banting Institute in Toronto reported that he and his associate, John C. Callaghan, had cooled anesthetized dogs in ice water to core temperatures as low as 68 degrees Fahrenheit, stopped the circulation of their blood for 15 minutes by tying off the vena cava, and then rewarmed the animals until they resumed breathing. Although 85 percent died, the survivors had suffered no damage to their brains or hearts.

As one of Bigelow's listeners later commented, "It wasn't the 85% mortality rate [that mattered]; the fact was that 15% of the hypothermic animals had survived total cessation of circulation for 15 minutes, a period of time which would be universally fatal at a normal body temperature." If surgeons could slow a surgical patient's metabolism as Bigelow and Callaghan had done with animals, they could greatly prolong the time available to work within the heart. Hypothermia was "the piece necessary to complete the puzzle" of open-heart surgery. Surgical researchers in North America and Europe immediately began to follow up on the Canadian team's work.[26] Within two years, they had brought down the mortality rate in laboratory animals and prepared to use hypothermia with human beings. In the first case, at the University of Minnesota Medical School in Minneapolis, in September 1952, surgeons repaired an atrial septal defect (a large hole in the partition between the right and left atria) in a five-year-old girl whose body temperature had been lowered to 82 degrees Fahrenheit with refrigeration blankets. The child recovered without difficulty. Though the heart had stopped for only five and a half minutes, this was the first operation within the heart in which surgeons had been able to view their work in a dry operating field.[27] They were now equipped to begin lengthier and more complex open-heart operations, though most still hoped for an effective heart-lung machine that would provide a circulation of oxygenated blood to the body while bypassing the heart.

### THE GOLDEN AGE OF CARDIOVASCULAR MEDICINE

When physiologist Julius Comroe included cardiac resuscitation as one of the "top ten clinical advances" in heart medicine since World War II and listed the research findings that had prepared the way, he had a political purpose: writing in the 1970s, he wanted to gain public support for biomedical research on the argument that breakthroughs in treatment were founded on basic scientific knowledge. Comroe cited papers by Stokes, His, Erlanger, Wiggers, and Parkinson. He was perfectly correct that Paul Zoll and other physician-inventors of the 1950s had read these papers or knew of their findings through textbooks, and he was surely correct that these medical scientists had done their research in the faith that it might someday contribute to the betterment of the human condition, but without having a clear treatment in mind at the time they published. Comroe argued that lags between scientific discovery and clinical application have been commonplace in the history of medicine — thus it would be naive for the public and its representatives in the U.S. Congress to expect overnight returns on its investment in biomedical research.[28]

As a historical account, Comroe's discussion was somewhat limited. It did not explain why these scattered papers rather suddenly came together to inform the thinking of Paul Zoll and other innovators after World War II — why the 1950s, rather than some other postwar decade, saw the invention of cardiac pacing. The fact that medical researchers' understanding of Stokes-Adams disease and ventricular fibrillation had grown more sophisticated does not explain why Zoll's external pacemaker, announced in 1952, came into general use and led on to a succession of further inventions in electrostimulation. For this, we must move beyond scientific discoveries and scattered clinical applications to inquire about broader changes in U.S. medicine between the 1930s and the 1950s.

Scientific and organizational developments during World War II contributed powerfully to the emergence of a more technology-oriented medical culture and a more activist approach toward cardiac resuscitation in the postwar years. Between 1941 and 1945, the Committee on Medical Research, an arm of the wartime Office of Scientific Research and Development, distributed about $25 million to universities, hospitals, and business firms for contract medical research on a variety of problems. The war that gave birth to radar, jet airplanes, guided missiles, and the atomic bomb also heralded new treatments for wounds and burns, the mass production of penicillin, and improved knowledge of malaria and other infectious diseases.[29] As many historians have observed, the wartime experience of the United States encouraged a belief that through large organizations such as government and corporate labs and academic medical centers, teams of researchers could develop new technologies and treatments that would drastically improve people's health.[30]

The successes of targeted wartime research inspired broad public support for more of the same after 1945. The public began to think of American medical research as a national asset and often interpreted achievements in the laboratory as harbingers of the new treatments that they could expect in the postwar era. Organizations interested in boosting support for medical research, loosely allied under the leadership of Mary Lasker and Florence Mahoney of the American Cancer Society, sought to channel public enthusiasm into political support for federal funding of programs in medical training, hospital construction, and research on specific diseases. The "disease lobbies" used mass advertising to mobilize public interest in medical research and conducted opinion polls demonstrating the popularity of their cause. At the same time, national commissions such as the President's Commission on the Health Needs of the Nation (1952) repeatedly called for a stepping-up of federal efforts.[31]

The postwar campaigns at times recalled the emergency atmosphere of wartime. Just as they had during World War II, Americans turned to giant agencies like the National Institutes of Health (NIH), organized under the general guidance of a benign federal government and staffed by experts, to put technology to work and force the enemy into unconditional surrender. As Paul Starr has pointed out, members of Congress opposed to national health insurance could demonstrate "their deep concern for health" by their votes in favor of large appropriations for medical research. From 1948 on, Congress repeatedly created new disease-based institutes within NIH, beginning with the National Heart Institute (NHI).[32]

The new enemies of the postwar era were no longer primarily acute diseases such as diphtheria and typhoid but the chronic maladies that tended to afflict Americans in middle age, particularly cancer and heart disease. Epidemiologists had recognized by 1945 that chronic disease, however defined, afflicted one-fifth or more of the population and, with the decline of mortality from acute infectious diseases, now accounted for a growing proportion of the deaths in the United States. Chronic diseases raised a host of problems for the health-care system — problems to do with financing health services for people with complex illnesses who might require care over months or years. Since no consensus on these issues existed in the 1940s, the chronic-disease policy of the United States tended to narrow by default to a focus on research. But the charitable organizations and medical specialty groups promoting research were interested in attacking *particular* chronic diseases rather than in broadly addressing the health problems of the middle-aged and elderly population. And because the events that had legitimated medical research had had to do with finding preventive measures and cures for acute diseases, postwar medical scientists continued to view their role in this way. They should discover new ways to treat chronic diseases and to rehabilitate patients, if not cure them outright.[33]

Most of the research on cardiac pacing in the United States in the 1950s took place in large hospitals affiliated with medical schools and relied in part on funding from the federal government through extramural research grants from NIH. Before World War II, research had not played a prominent role in U.S. hospitals and medical schools. Such research as occurred received most of its support from private foundations or benefactors. But during the war, several medical schools had become involved in government-sponsored research projects, often in collaboration with private industry. These efforts led to such advances as the mass production of penicillin and the discovery of additional antibacterials, the development of dried blood plasma and the gamma globulins (immune agents useful against contagious diseases that often swept through military camps), and cortisone. After 1945, NIH support for research continued to rise, reaching $20.5 million in 1953. NIH training fellowships and research grants supported hundreds of investigators, many of them "brimming with ideas" that they had not previously been able to pursue. For research on heart disease specifically, grants from the agency totaled $3.8 million in fiscal 1950; this had risen to $5.15 million by 1953.[34]

The federal government also funded the construction of modern research and teaching facilities at medical schools and hospitals across the country. W. Bruce Fye notes that between 1949 and 1951, more than two-thirds of the nation's medical schools applied for construction grants from the NHI to build cardiovascular research centers or catheterization laboratories. Success built on success.[35] New facilities and equipment, ongoing research projects with published results, and aggressive department heads made it easier for a research group to win approval for its proposal in the next round of grants. It "would depend to a large extent on where [a research proposal] came from and who the chief of the department was, what kind of clout he had," Paul Zoll recalled.[36]

## THE POSTWAR HOSPITAL

Between Hyman's work at the beginning of the 1930s and the second invention of pacing about 20 years later, the institutions and settings of medical care changed significantly. Joel D. Howell has shown that already by 1940, with the widespread use of X-ray machines and the growing acceptance of the ECG in routine diagnostic work, hospitals had become "diagnostic centers." During and after World War II, large hospitals added advanced therapeutic technologies to their existing technological base. Even in the postwar period, hospitals remained nonprofit institutions for the most part — "community services, with lingering religious, humanitarian, and egalitarian goals." Yet the hospital was also a "supply-driven, technological system, analogous to an electrification system or a system for producing military aircraft." In the case of the hospital, what was produced was "surgery, procedures, X rays, expertise, even babies."[37]

Thanks to the sulfa drugs and penicillin, many patients with bacterial infections could recover at home. The new medications, along with blood plasma and modern techniques for treating shock, also made possible the rapid growth of surgery and postsurgical acute care. Surgeons and other specialists came to view the hospital as an indispensable technological center catering to their needs for diagnostic services, specialized equipment, operating rooms, and skilled nursing care. With the passage of the Hospital Survey and Construction Act (the Hill-Burton Act) in 1946, hospitals were firmly set on the road toward modern high-technology, acute-care medicine.[38]

Among the subspecialties of internal medicine, cardiology developed a particularly close association with hospital-based technology. This became quite clear with the emergence of hospital coronary intensive-care units in the 1960s, but had begun earlier as hospitals became centers for the taking of ECGs, catheterization, and angiography. These procedures were usually carried out in hospitals because the technology was too expensive for a physician's office or because they required special facilities, support staff, and recovery time for the patient.[39]

Conditions in postwar U.S. culture and medicine were propitious for rapid advances in cardiology and heart surgery. In their entrepreneurial enthusiasm for new technologies and procedures in medicine, men like Dwight Harken, Claude Beck, and André Cournand expressed the technological optimism of American culture in the 1940s and 1950s. By 1952, several research groups had reported experiments in which they accelerated the hearts of laboratory animals from extremely slow rates or resuscitated animals from standstill, but, in every case, researchers had stimulated the atrium rather than the ventricle.[40] It remained for someone to make the intellectual connection between electrostimulation of the heart and the disease known as complete heart block.

## PAUL M. ZOLL AND THE EXTERNAL PACEMAKER

Paul M. Zoll, of Beth Israel Hospital in Boston, was already an experienced laboratory scientist when he began his work on cardiac electrostimulation. Born in 1911 into a middle-class Jewish family in Boston and prepared at Boston Latin School, he attended Harvard College, where he graduated summa cum laude in 1932 with a concentration in psychology. His family discouraged an early interest in an academic career, apparently because an older brother had been unable to find a job after majoring in English literature at Harvard. At his mother's insistence, Paul took summer courses in the sciences and went on to Harvard Medical School; he received his M.D. degree in 1936. He did his internship at the Beth Israel and had a year's residency at Bellevue Hospital in New York City. But uncertain of his ability to handle patients and fascinated by research, he returned to the Beth Israel in 1939 as a junior member of a research group led by two experienced laboratory scientists, physiologist Herrman L. Blumgart and pathologist Monroe J. Schlesinger.[41.]

Schlesinger had developed a technique "for injecting hearts [with dye] so that you could really see the pathology in the coronary [arterial] tree. And with this technique, which he had already started, we did every case that came to autopsy; my job was to do the injections . . . and to go through the records and make the clinical-pathologic correlations. It was marvelous. We developed knowledge about the anatomy and the pathology of the coronary arterial tree . . . and we learned all kinds of special things about how prevalent coronary disease, significant multiple obstructions, were, in the absence of clinical symptoms." Zoll's work in the pathology lab contributed to a series of classic papers on coronary artery disease published between 1941 and 1952.[42]

During the war, Zoll was assigned to the 160th U.S. Army Station Hospital in England, where he served as the cardiologist on Dwight Harken's surgical team. As he watched Harken remove shrapnel from soldiers' hearts, he was struck by "how easily excitable the myocardium is. You just touch it and it gives you a run of extra beats — so why should the heart, that is so sensitive to any kind of manipulation, die because there's nothing there to stimulate the chest? It wasn't sensible."[43] Zoll resumed his laboratory research after the war, but also developed a private practice in Boston. From 1947 to 1958, he served as chief of the cardiac clinic at his hospital. Many years later, Zoll would tell of a clinical case around 1947 or 1948 that focused his attention and skill on the problem of stimulating the heart from standstill: he lost a private patient to Stokes-Adams disease. "She suddenly developed complete heart block [with] repeated Stokes-Adams attacks. I was very frustrated. She was my private patient. . . . There's nothing so important to a doctor as having something go wrong with a private patient of his, so that he'll be responsible, at least in his own mind, as a physician, and can't do anything about it. It was awful. . . . My patient lived three weeks after her first episode."[44]

Zoll was slightly acquainted with Albert S. Hyman. He certainly knew of Hyman's invention of a pacemaker in the 1930s and knew that Hyman had intended the device to revive people with healthy hearts brought to standstill by some trauma.[45] He now began to consider a similar device to treat Stokes-Adams attacks. He recognized quickly that Hyman's pacemaker could not serve as a direct model. Zoll suspected that the needle electrode on the earlier pacemaker would tend to become displaced and injure the heart, and he knew that he, and doubtless many other physicians, would be reluctant to stab the electrode into the heart of a patient in extremis. Then, too, pacing the atria, as Hyman had advocated, would be pointless since in heart block the atria were dissociated from the ventricles. A new pacemaker would have to stimulate the ventricles directly — though Zoll was not sure, at first, how to accomplish this. Finally, Hyman's device had relied on cranking by hand and had to be rewound every few minutes, whereas Zoll thought that it might be necessary to pace the heart for an hour or more; this would require an electrical power source.[46]

Zoll's "poor background in electronics" created a major problem for him: "I did not know what type of stimulus to use nor how to provide it." Then in October 1950, at a meeting of the American College of Surgeons in Boston, he attended a presentation by John C. Callaghan, a member of Wilfred Bigelow's research group in Toronto. As we have seen, Bigelow and Callaghan were interested in learning how to lower the body temperature of human beings so as to slow down physiological processes within the body. Their experimental animals would often go into VF or come to ventricular standstill as core temperature dropped. The Toronto researchers dealt with VF by defibrillating the animals' hearts and sometimes succeeded in restoring a spontaneous heartbeat. But for standstill of the heart, a suitable piece of laboratory equipment did not exist, so Callaghan, Bigelow, and engineer Jack Hopps invented one: a laboratory pacemaker consisting of an external pulse generator and a catheter electrode introduced to a dog's right atrium by way of the right external jugular vein.[47] Within a week of hearing Callaghan's report at the surgical meeting, Zoll wrote Bigelow and Callaghan saying that he was "most eager to get more information about the details of the stimulating apparatus." They sent him information about the stimulator Hopps had designed and built, including a circuit diagram for the device.[48]

Working alone and with his weak background in electronics, Zoll could not have used the information sent him from Toronto. Instead he borrowed a standard physiologic stimulator manufactured by the Grass Equipment Company that medical students and laboratory researchers used in their studies of the electrical responsiveness of nerve and muscle tissues. But how best to deliver the stimulus to the heart? An internist rather than a surgeon, Zoll wanted something that would be "clinically easy" to apply when a heart patient suddenly went into profound ventricular bradycardia or standstill. He knew that the esophagus passes just behind the heart; in his initial design, he advanced one electrode down the esophagus until it lay behind the left ventricle and inserted a needle electrode beneath the skin of the chest. With this setup, Zoll was quickly able to accelerate the heartbeat in anesthetized dogs. Recognizing that in a clinical emergency it would be far easier to apply two electrodes to the chest than try to run one down the esophagus, he soon dispensed with the esophageal electrode. "It just took a large current, that's all. And it didn't make any difference whether it was esophageal or totally external."[49]

Nearly two years of studies in the dog lab followed the initial success. Zoll experimented with pulse widths and amplitudes; he paced dogs for days at a time to determine whether a pacemaker stimulus might by chance send an animal into ventricular fibrillation. Using the pacemaker, he successfully revived dogs whose hearts had been stopped for several minutes. He was now reasonably certain that the external pacemaker would be safe to use with human beings. Sometime in 1952, he let it be known at his hospital that he would like to be informed when a Stokes-Adams patient was admitted. Weeks went by — few victims of complete heart block with Stokes-Adams attacks survived long enough to reach the hospital. Finally, a case appeared.

> The first patient that they called me about was a little old man . . . who had heart disease but then developed Stokes-Adams disease, and then had a very bad series of seizures. [He] came in the hospital with recurring seizures, arrests, and so on, intermittently, as happens in the very severe stages. They tried stimulating his heart by injecting adrenaline, epinephrine, into the heart itself with needle punctures through the chest wall, which was one of the desperate efforts that were being tried. And they kept him alive for three or four hours, and then somebody remembered that I had asked to be called when a case like this came along. I went up there and stimulated him for twenty minutes — enough to show that it worked. I could really tell that it worked. And then he died. And he died because one of the [epinephrine] needles had punctured a coronary vein — he died of tamponade, failure of the circulation caused by the blood collection in the pericardial sac.[50]

Zoll's "Case 2" entered the hospital about four weeks later. "R.A." was a 65-year-old man, perhaps less desperately ill than the earlier patient, but still terribly sick. He suffered from congestive heart failure and severe coronary artery disease. He had had a heart attack six years earlier and had been in complete heart block for five months. On his sixth day in the hospital, R.A. began to suffer frequent and prolonged periods of ventricular standstill alternating with bursts of rapid, irregular ventricular beats — classic Stokes-Adams attacks. "I had time to get up there before he had been seriously damaged by the methods of treatment. He had not had his chest punctured."[51] Zoll brought in the artificial pacemaker and positioned the electrodes. He turned on the pacemaker and it immediately began to maintain the patient's ventricular beat. Periodically, Zoll turned off the pacemaker in hopes that R.A.'s heart would proceed on its own; the heart would pick up the beat for a time but then stop. After a moment or two of standstill, Zoll would switch the pacemaker back on. This continued for three days.

Just after noon on October 7, Zoll turned on the pacemaker yet again. For the next 52 hours, R.A.'s ventricles would not beat spontaneously at all; the pacemaker, pulsing away at about 85 beats per minute, kept him alive. Every few hours, Zoll would turn off the pacemaker, then turn it back on because the heart still refused to beat. Through it all, R.A. himself was able to eat, converse, and follow a World Series game on the radio. Late in the afternoon on October 9, the refractory heart began to give out a few spontaneous beats. By 8 P.M., it had established a persistent spontaneous rate of 44 beats per minute, which was adequate to keep R.A. conscious and alert. He had no further Stokes-Adams attacks and, two days later, Zoll removed the electrodes. R.A. later went home. He survived for ten months before another Stokes-Adams attack killed him.[52]

Zoll reported his case in the *New England Journal of Medicine* within a few weeks of the 52-hour episode. Never before had an electrical device, a machine, managed the human heartbeat. While dramatically new, Zoll's approach to pacing the heart was also impressive for its simplicity and directness: the pacemaker consisted of off-the-shelf components with which most doctors were already familiar. (Zoll soon simplified his invention further by discarding the needle electrodes and substituting standard electrocardiograph electrodes that were strapped to the patient's chest.) External pacing carried little danger of serious complications other than the risk of a short-circuit or power surge in the hospital electrical system. R.A.'s apparent recovery was the satisfying closing scene in a ritual that Americans and their doctors had come to associate with advanced, hospital-based medicine: a patient enters the hospital gravely ill; the new treatment is applied; the patient departs "cured."[53]

But the case also demonstrated the problems that would plague external cardiac pacing and drive Zoll and others to invent new forms of pacing within a few years. Because the electrodes were not in direct contact with heart tissue, a stimulus of 30 to 150 volts was required to drive the ventricles. This caused involuntary contraction of chest muscles and, for most patients, proved too painful for extended use. In some cases, stimulation also caused skin ulcers at the sites of the electrodes. R.A. had tolerated external pacing at 130 volts — Zoll even claimed that the patient had been able to sleep while the pacemaker was managing his heartbeat — but most patients found it an unpleasant sensation, at best.[54] Seymour Furman, later an important innovator in cardiac pacing, has told a revealing anecdote about Zoll's invention: "We'd had a patient that I recall, a patient while I was an intern [1955–56], who had been on an external Zoll pacemaker for a long period of time, and had finally committed suicide by turning off the switch just after a pep talk which the house staff, myself included, had given him about the wonders of the future to come, which we didn't believe and he equally didn't believe." As a partial solution, Zoll and his associates developed the practice of intravenously administering epinephrine and iso-proterenol to help manage severely unstable hearts and thereby reduce the patient's time on the pacemaker.[55]

**PAUL ZOLL AND THE FIRST CYCLE OF INNOVATION IN PACING**
Those who have written on the development of the cardiac pacemaker customarily mention the contributions of the research group at Toronto's Banting Institute. Callaghan, Bigelow, and Hopps had devised a remarkable technological device, an electric-powered external pulse generator with a transvenous atrial lead—an insulated wire threaded down through a dog's jugular vein into the right atrium of the heart. One medical historian even credits them with opening "the contemporary era of cardiac pacing." This judgment reflects the fact that nearly 20 years after the work of the Toronto group, transvenous leads gained universal acceptance among doctors who implanted pacemakers. The experimental work in Toronto holds a prominent place in the history of pacing largely because of this later popularity of a technique that Bigelow and Callaghan had pioneered. Some have even suggested that the work of the Toronto group stands in the main line of invention and, indeed, that in 1949–51, Bigelow and Callaghan invented the pacemaker.[56]

Paul Zoll did not invent any one component of the external cardiac pacemaker; what he did was bring together existing knowledge and equipment to solve a clinical problem. Zoll single-mindedly set out to resuscitate victims of Stokes-Adams attacks in the hospital and bent all his efforts to designing and testing equipment that hospital staff could put to practical use in this emergency. Although his external pacemaker had serious draw-backs, "it really worked," as Zoll repeatedly asserted—meaning that it kept people alive by stimulating the ventricles to contract. Zoll's most important insight at this stage of the field was his recognition that he should pace the ventricle, not the atrium.

From an early point in his laboratory research, Zoll also recognized that pacemaker stimuli might be able to restore a normal heart rhythm in circumstances other than Stokes-Adams attacks. Under anesthesia, one of his early experimental animals developed a rapid heartbeat. With his two electrodes, one in the animal's esophagus and the other on its chest, Zoll interrupted the tachycardia by throwing in an extra beat that disrupted the rhythm and set the heart to beating normally again. "It was immediately clear to me," he later wrote, "that the problem of cardiac arrest was solved, at least in theory. There remained but to work out all of the details implicit in this observation."[57] After 1952, Zoll worked out the details: in collaboration with cardiologist Arthur Linenthal and engineer Alan Belgard of the Electrodyne Corporation, a small firm near Boston, he developed an external AC defibrillator and a cardiac monitor that would enable the attending physician to diagnose instantly the onset and nature of a dangerous arrhythmia. Together with the external pacemaker, these machines provided the initial technological basis for the cardiac intensive-care units that mushroomed in U.S. hospitals during the 1960s. Zoll and his associates were fathers not only of pacing for standstill of the heart but of closed-chest countershock for VF and of anti-tachycardia pacing. Their equipment rescued victims of many kinds of heart-rhythm emergencies in the hospital.[58]

In the course of their work on coronary standstill from hypothermia, Bigelow and Callaghan had also become interested in pacing for other forms of severe bradycardia and standstill. Their laboratory device, like Zoll's, consisted of parts or components of a pacemaker, but the group lacked a clear rationale for pacing the atrium rather than the ventricle. At some point before October 1950, Callaghan tried the device on five hospitalized patients who were close to death in cardiac arrest or severe bradycardia following major surgery for lung cancer. As he had done in the animal lab, he introduced the catheter to a vein and advanced it into the right atrium close to the sinus node, the source of the heartbeat. These clinical experiments were not successful: "We were unable to take over . . . complete cardiac action." Callaghan did not report on them in print. Years later, he commented that if he had pushed the catheter just two inches farther, down into the right ventricle, he might have driven these hearts. Perhaps so, but so little is known about the condition of these dying patients that no one can say what pacing arrangement, if any, could have been effective.[59] The Toronto equipment was in some respects more sophisticated than Zoll's, and the group clearly had a better understanding of electricity. The experiments with the five dying patients suggest, though, that, as of 1950–51, Callaghan and Bigelow lacked the very thing that Zoll created, a set of doctrines for the use of their pacemaker. What kinds of patients, what kinds of heart emergencies, might the device treat? How should a hospital worker go about setting up the pacemaker on a convulsing patient on the ward or in an operating room? Most important, what was the rationale for atrial pacing?

Wilfred Bigelow has suggested that Zoll borrowed the design of the Toronto pulse generator without giving proper credit and that Zoll did little more than achieve "the first application of this [Toronto] pacemaker to humans using our . . . circuit."[60] But in the paper that Zoll heard in October 1950, Callaghan mentioned two stimulators. The first, a "physiologic stimulator" manufactured by the Grass Instrument Company, "was found to be satisfactory" for the experimental work of the Toronto group, but "the desired electrical features were developed in a simplified circuit" by engineer Jack Hopps. Zoll may have secured information about both stimulators, but his limited knowledge of electrical equipment apparently steered him toward the commercially available Grass stimulator.[61]

Cardiac pacing as a medical procedure consists of both hardware and a body of technical knowledge concerning when and how to use it. Without that "software," a pacemaker is a useless, inert device. Zoll not only marshaled existing equipment, he addressed a clinical need. By defining first Stokes-Adams and later other forms of cardiac arrest as the problem and showing how external pacing could minister to this problem, he created the first body of knowledge, the first "software." In this sense, he invented cardiac pacing.

In 1954, Electrodyne brought out a commercial version of the external pacemaker. External pacing came into use in U.S. hospitals during the decade of the 1950s, especially in the teaching hospitals where the revolution in heart surgery was under way. This was not cardiac pacing as most people know it today, for Zoll's invention carried with it a set of assumptions and practices quite different from those now associated with implanted cardiac pacemakers. This first version of pacing meant emergency resuscitation in the hospital from ventricular standstill. A pulse generator the size of a breadbox that plugged into the AC electrical system implied a bedridden patient. The high voltage required to capture the heartbeat implied very short bouts of pacing—from minutes to hours—and usually patients who were unconscious or sedated.

Zoll's work inaugurated the first cycle of innovation in cardiac pacing, a cycle based on vacuum-tube technology and reliance on the electrical system of the hospital. The external pacemaker gained a place in the hospitals of the 1950s partly because it was a somewhat conservative piece of hardware. Doctors were already familiar with its components; setting it up was a simple matter; and it did not invade the body. The results of external pacing were rapid, dramatic, easy to interpret. Many patients found its stimuli painful — but the pacemaker kept them alive. General trends in the treatment of heart disease also encouraged the used of hospital-based cardiac pacing; it was one element in a broadly advancing front in hospital coronary care. In retrospect, Mark Lidwill and Alfred S. Hyman can be likened to soldiers in the unenviable position of having gotten a considerable distance out in front of the main line of advance. External pacing is still used in hospitals throughout the United States and elsewhere, but it is clearly a niche technology. Developments in open-heart surgery during the mid-1950s led to the invention of new kinds of pacemakers for a different set of clinical problems.[62]

## CHAPTER 3

# HEART SURGEONS REDEFINE CARDIAC PACING

Reports of Paul Zoll's success at resuscitating patients from standstill of the heart both encouraged others to experiment with cardiac pacing and freed them to do so. Within a few years, many physicians reported in print on their own experience either with the Electrodyne pacemaker or similar devices built for them. By the end of the decade, pioneer open-heart surgeons and a handful of tiny manufacturing firms had also interested themselves in cardiac pacing. These new participants began to reshape the pacemaker and redefine its uses. In the process, they drove pacing a considerable distance from what Paul Zoll had originally had in mind. Even at this early point in its history, "cardiac pacing" had a quicksilver quality. In the mid-1950s, the meaning of terms like *pacing* and *pacemaker* began to undergo a process of redefinition that is still under way.

## THE VARIETY CLUB HEART HOSPITAL

Around 1956–58, the first open-heart surgeons decided that cardiac pacing might solve a hitherto unknown complication that they were encountering in their surgical cases. In adapting Zoll's original idea to their needs, the surgeons invented a radically new version of acute cardiac pacing. The idea of implanting part of the pacemaker emerged first in the Department of Surgery at the University of Minnesota Medical School in Minneapolis.

Events at Minnesota between 1944 to 1960 were typical of the kinds of changes underway in many academic medical centers.[1] During the war, several surgeons at Minnesota had become interested in heart surgery and had learned new procedures developed elsewhere. By 1944, it was apparent that if Minnesota was to become a center for research in heart surgery and postoperative care, the medical school must build a larger and more modern hospital. A development campaign got under way in January 1945 when the Minneapolis chapter of the Variety Club, a service organization for people in the entertainment industry, offered to raise at least $150,000 toward a new heart hospital. With the announcement of the Blalock-Taussig operation for improved blood oxygenation at Johns Hopkins in mid-1945, the campaign in Minnesota gained momentum. The Variety Club eventually raised more than $500,000, the university added $400,000, and the federal government provided $600,000 under the 1946 Hill-Burton Act for hospital construction. The University of Minnesota opened its Variety Club Heart Hospital in 1951 — the first hospital in the United States set up to focus on acute care for heart disease.[2]

Owen Wangensteen, the chief of surgery at Minnesota from 1930 until 1967, trained not just surgeons but surgical researchers: he insisted that every academic surgeon pursue some program of research and that every surgical resident contribute as an apprentice investigator. One resident recalled that "everyone, all residents, had some kind of research project that they either were interested in or gave lip service to because of Dr. Wangensteen's influence. I mean, if you were going to get anywhere around that department, you had to have . . . some kind of research interest. And then you could get tremendous distances." Norman Shumway, who later developed the heart-transplant program at Stanford University, remembered his residency at Minnesota in the same way: "We used to say . . . [that] you had to invent an operation to get on the operating schedule! And of course, that's just what many of the staff members did."[3]

With the new hospital ready for patient care and with funding available for research from the American Heart Association and NIH, Wangensteen pressed his surgeons and trainees to transform the department into an important center for innovation in heart surgery. He understood that new surgical procedures would be possible only with the support of new technology. Surgery within the heart seemed tantalizingly close, but a major roadblock remained: finding some way to stop the heart from beating and clear it of blood while still keeping the patient alive. Efforts had been under way in the United States and elsewhere since the 1920s to build an artificial blood oxygenator, a machine that could maintain the circulation of blood through the surgical patient's lungs to the organs of the body while bypassing the heart itself, but investigators had not reported much success.[4]

Looking for a way to put Minnesota's program in heart surgery on the map, Wangensteen first threw his support behind a large project to build a practical pump-oxygenator.[5] When the leader of this effort left for another university in 1951, taking the experimental oxygenator with him, surgeons at Minnesota turned to hypothermia. Building on the work of Bigelow and Callaghan in Toronto, they practiced operating on dogs that had been anesthetized and then chilled in ice baths or refrigeration blankets. By slowing the animals' metabolism, they gained enough time to repair atrial septal defects in the animals' slowly beating hearts. Minnesota surgeon F. John Lewis had successfully used hypothermia in correcting an atrial defect in a five-year-old girl in September 1952. More complex repairs, however, would require more time for the surgeons to work than hypothermia provided.[6]

## THE DISCOVERY OF POSTSURGICAL HEART BLOCK

While Lewis focused on hypothermia, another surgeon at Minnesota continued the search for a way to oxygenate the blood during surgery while bypassing the heart. C. Walton Lillehei (1918–99), born and raised in Minneapolis, was 33 years old in 1951 and had recently completed his surgical residency as well as a Ph.D. in surgery. During the war, just a year out of medical school, Lillehei had served in North Africa, Sicily, and Italy as commander of one of the army's first M.A.S.H. (Mobile Army Surgical Hospital) units. He returned to Minnesota for his surgical residency under Wangensteen. In line with Wangensteen's belief that a surgeon should have a solid background in basic science and in laboratory research, Lillehei spent months in the physiology lab, published several papers, and in 1951 won a national award for "outstanding research contributions to medical science." Lillehei proved to be Wangensteen's greatest protégé. He emerged during the mid-1950s as a bold and creative heart surgeon and a great teacher of younger surgeons.[7]

Lillehei had not been an enthusiast for the experimental heart-lung machines of the late 1940s because they seemed overly complex and difficult to sterilize. When a group in England published a report on a daring new practice called controlled cross-circulation in 1953, he seized on it as a way to maintain a circulation of oxygenated blood for a patient throughout a lengthy surgical procedure. In controlled cross-circulation, an artery and a vein from the surgical patient, usually a child, were connected to the circulatory system of a donor, usually a parent, lying on an adjoining table. The donor's heart then pumped blood for both bodies, leaving the child's heart inactive and available for surgical repair. Lillehei put this technique into clinical use in 1954 for the repair of ventricular septal defects. Despite the frail condition of the children, the procedure had a high success rate. But many surgeons believed that cross-circulation was too risky for everyday use; one visitor told Lillehei that he was the first surgeon who had ever developed an operation with a potential for a 200 percent mortality rate.[8]

On August 31, 1954, after performing eight operations to close ventricular septal defects, with six children surviving, Lillehei used cross-circulation to operate successfully on an 11-year-old boy suffering from the set of four heart defects known as tetralogy of Fallot. The technique gave him many minutes to make repairs in the heart. By February 1955, his success rate in open-heart operations using cross-circulation had reached 25 survivors with 7 deaths. But one infant had died suddenly and mysteriously 12 hours after surgery, leaving Lillehei to wonder if his sutures in the child's septum had caused heart block.[9]

Both hypothermia and cross-circulation underscored the radical nature of the new surgery, and Lillehei wanted to put them behind him. In 1954, he gave one of his surgical residents, Richard A. DeWall, the assignment of designing a practical bubble oxygenator, a simpler kind of heart-lung machine than the complex pieces of equipment then being tested. Working closely with Lillehei, DeWall succeeded with astonishing speed. By May 1955, he had a prototype ready to try with human beings, an "elegantly simple" device that was safe, easy to assemble, and cheap enough to discard after one use. Over the next year, DeWall and Lillehei made various improvements and Lillehei began to rely on the machine more and more.[10]

By early 1957, Lillehei and other surgeons at the University of Minnesota had carried out 305 open-heart operations, most of them on children with congenital defects. Of these patients, 204 had survived—a good record considering the poor condition of many of the patients and the surgeons' lack of experience with the procedures. Lillehei's fame had spread, and hundreds of surgeons from all over the world flowed through the Department of Surgery at Minnesota to observe his techniques and study cross-circulation and the bubble oxygenator. His surgical residents moved on to appointments at leading hospitals. Lillehei's connections to the larger world of heart surgery would prove important in facilitating the spread of cardiac pacing over the next few years.[11]

Lillehei now turned from cardiopulmonary bypass to another puzzle that threatened the long-term acceptability of surgery in the ventricles. Beginning with the infant who had died 12 hours after surgery in the early months of Lillehei's work, it had gradually become clear that when a surgeon repaired ventricular septal defects, about 1 patient in 10 developed complete heart block because the surgical stitches occasionally damaged the conduction cells that passed the heart's electrical signal from the atria to the ventricles. Typically, the child would have an adequate heartbeat just after surgery, but gradually the ventricular rate would drop and, hours later, the child would die. When it appeared, postsurgical heart block was invariably fatal—"a tremendously significant problem" for the surgeons. But Lillehei believed that if he could keep a child with heart block alive for two or three weeks, the conduction system might heal naturally. External pacing proved a flat failure: "Getting a shock like that fifty, sixty times a minute

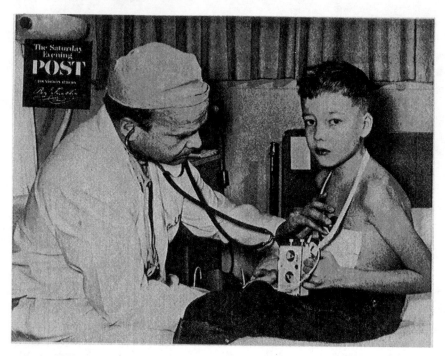

C. Walton Lillehei with a Young Patient Recuperating from Open-heart Surgery, 1961. This photograph appeared in the *Saturday Evening Post*, March 4, 1961. The boy, David Williams, is wearing a Medtronic 5800 external pulse generator. One of the dials on the front of the device controls electrical output; the other, the pacing rate. To prevent children from altering their heart rate the dials were recessed, and it required a small screwdriver to turn them. A wire lead runs from a terminal atop the pulse generator through the boy's skin into the wall of his heart. (Courtesy of Medtronic, Inc.)

is torture. With some of the infants, we were able to restrain them so they wouldn't tear [the chest electrodes] off, but they would develop blisters and ulcers [beneath the electrodes] in four to five days. So that was totally inadequate."[12]

As one of Lillehei's trainees later pointed out, open-heart surgery was all new and "the learning process was one of trial and error." When external pacing proved a disappointment, the group tried several drugs that stimulated the heart to beat. They used epinephrine and similar stimulants in seven cases in 1954–55, and every child died. In mid-1955, Lillehei switched to Isuprel (isoproterenol), a newly available drug for the treatment of asthma that caused the heart to speed up. Now his success rate rose: 9 of 17 children treated with Isuprel survived long enough to revert to normal heart rhythm, 5 remained in heart block but survived, and 5 died. "For our

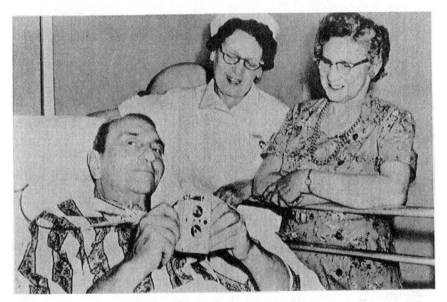

**Warren Mauston with His External Pulse Generator, 1959.** Mauston lived for six and a half years with an external generator and a lead that entered his body through a port in the skin. His case was one of the first to demonstrate that pacemakers could manage aberrant heartbeats over the long term. (Courtesy of Medtronic, Inc.)

purposes, that was a nice improvement over the epinephrine-type drugs," Lillehei later recalled, but "while an improvement, [it] was obviously not satisfactory."[13]

During a morbidity and mortality conference in 1956, a physiologist suggested pacing the children through an electrode that actually touched the surface of the heart. By delivering the electrical stimuli directly to the excitable tissue rather than firing them through the body from outside, they might capture the heartbeat at a much lower voltage. Vincent Gott, one of Lillehei's trainees, borrowed a Grass physiological stimulator (the same piece of equipment that Paul Zoll had worked with in Boston) and took it to the dog lab. Using a standard surgical technique, he created heart block in a dog by throwing a stitch around the bundle of His. He inserted a wire into the animal's heart wall, connected it to the external stimulator, and found that "it picked the rate right up." Gott began this work in the summer or early fall of 1956. Working independently, a second resident, William Weirich, used much the same technique with similar results. Lillehei was immensely encouraged: "Lo and behold, one or two volts, five to ten milli-amps—that's a thousandth of an amp—drove the heart beautifully! Any rate that you'd set. And obviously, one to two volts was totally impercepti-ble to the animal. . . . We ran some animals—I don't know, ten, fifteen—and it just worked beautifully."[14]

"There was no FDA [Food and Drug Administration] back then," Lille-hei added—meaning that he could go ahead and use the new kind of pacing with a human patient whenever he judged that it had been adequately tested with animals. The wire concerned him. The heart was moving all the time; would the wire dig a hole in the *myocardium*, the muscular wall of the heart? Then there was the question of leaving a metallic object inside the body for several days. From their work with dogs, Gott and Weirich had settled on a silver-plated, braided copper wire in a Teflon sleeve. The last two inches of the wire were bare. The surgeon would insert the wire into the myocardial tissue at a shallow angle so that the tip emerged again. The wire was tied down at the points of entrance and emergence. That was enough to keep the electrode in place for a while.

After trials with some 50 dogs, Lillehei decided to use the "myocardial wire" the next time a surgical patient showed signs of heart block. An opportunity soon arose: on January 30, 1957, he performed surgery on "a little girl, about six years old, who had a large ventricular defect; and com-ing off the pump oxygenator it was obvious that we had complete heart block." Lillehei and his assistants inserted the wire in the wall of the child's left ventricle and tied it down. They brought the wire out through the surgical incision, buried a second wire under her skin as an indifferent electrode, and connected both to an external pulse generator. The girl's ventricular rate jumped from 30 to 85 beats per minute. Soon Lillehei was using the wire whenever a patient showed signs of block during an open-heart operation. The surgeons found that when the heart seemed able to resume control of its own beating, they could pull gently on the wire to dislodge it from the myocardium, then draw it from the body without reopening the chest.[15]

## "KIND OF AN INTERESTING POINT IN HISTORY"

The myocardial pacing wire was the first electrical device ever to be im-planted in the human body and left there for a period of time. Surgeons at Minnesota were now able to pace children for days or weeks after heart surgery. By October 1957, they had used the technique with 18 patients. But Lillehei now grew uneasy about the Grass stimulator because it was bulky and plugged into the electrical system. The surgeon wanted to get his heart patients out of bed and moving around, but the stimulator had to accompany them on a wheeled cart. The electrical cord was a further nuisance. "Many of these [patients] were kids. They wanted to wander around and get active. Well, they *were* active. They couldn't go any further than the cord. We had to string wires down the hall. . . . And then, if they needed an X ray or something that couldn't be done in the room, you couldn't get on the elevator so you had to string them down the stairwells. It seemed that almost everything you wanted was on a different floor. We needed something battery-operated."[16]

The plug-in stimulator was more than an inconvenience, for by introducing the myocardial pacing wire, Lillehei and his associates had connected the hearts of their surgical patients to the 110-volt electrical system of the hospital. Everyone in the program knew that an electrical surge might send patients into ventricular fibrillation or that a power outage could leave them without pacemaker support. On October 31, 1957, an equipment failure at a large Twin Cities power plant caused an outage lasting nearly three hours in Minneapolis.[17] The University hospital had auxiliary power in its surgical suites and recovery area, but not in patients' rooms. None of his heart patients died — but Lillehei viewed the event as a warning. Shortly after the power failure, he asked a graduate student in physics who worked part-time in the Department of Surgery's experimental lab to build a battery-powered stimulator. Weeks passed, but the student never got around to this project. Lillehei next turned to Earl Bakken (b. 1924), a young engineer who owned a small medical electronics business called Medtronic and repaired and serviced equipment for the Department of Surgery.[18]

Bakken first thought of "an automobile battery with an inverter to convert the six volts to 115 volts to run the AC pacemaker on its wheeled stand. That, however, seemed like an awfully inefficient way to do the job, since we needed only a 10-volt direct-current pulse to stimulate the heart." Powering the stimulator from a car battery would have eliminated the need for electrical cords and plugs, but would not have done away with the wheeled cart. Bakken then realized that he could simply build a stimulator that used transistors and small batteries. "It was kind of an interesting point in history," he recalled — "a joining of several technologies." In constructing the external pulse generator, Bakken borrowed a circuit design for a metronome that he had noticed a few months earlier in an electronics magazine for hobbyists. It included two transistors. Invented a decade earlier, the transistor was just beginning to spread into general use in the mid-1950s. Hardly anyone had explored its applications in medical devices. Bakken used a nine-volt battery, housed the assemblage in an aluminum circuit box, and provided an on-off switch and control knobs for stimulus rate and amplitude.[19]

At the electronics repair shop that he had founded with his brother-in-law in 1949, Bakken had customized many instruments for researchers at the University of Minnesota Medical School and the nearby campus of the College of Agriculture. Investigators often "wanted special attachments or special amplifiers" added to some of the standard recording and measuring equipment. "So we began to manufacture special components to go with the recording equipment. And that led us into just doing specials of many kinds. . . . We developed . . . animal respirators, semen impedance meters for the farm campus, just a whole spectrum of devices." Usually the business would sell just a few of these items. When Bakken delivered the battery-powered external pulse generator to Walt Lillehei in January 1958, it seemed to the inventor another special order, nothing more.[20] The pulse generator was hardly an aesthetic triumph, but it was small enough to hold in the hand and severed all connection between the patient's heart and the hospital power system. Bakken's business had no animal-testing facility, so he assumed that the surgeons would test the device by pacing laboratory dogs. They did "a few dogs," then Lillehei put the pacemaker into clinical use. When Bakken next visited the university, he was surprised to find that his crude prototype was managing the heartbeat of a child recovering from open-heart surgery.[21]

### FROM THE DOG MODEL TO THE 5800

Lillehei's myocardial wire in the heart and Bakken's transistorized pulse generator were offshoots of two revolutions under way in the 1950s — the one in heart surgery, the other in microelectronics. In certain respects, the new pacing system still resembled Zoll's original version of pacing. In both external and myocardial pacing, the patient was assumed to be gravely ill, confined to the hospital, and pacemaker-dependent. Both systems treated acute crises, whether Stokes-Adams attacks or postsurgical heart block. In both, the pacemaker was defined as a piece of hospital equipment; its transformation into a more or less permanent addition to the patient's own body was still a year away. But the new pulse generator opened up new possibilities for cardiac pacing in ways that Zoll's original invention never could have.

As a regional distributor for the Sanborn Company, a leading U.S. manufacturer of ECG machines, Bakken regularly visited surgical departments throughout the upper Midwest. He recognized that there might be a market for the transistorized pulse generator. In spring 1958, he and his half-dozen employees at Medtronic did what they could to redesign the device as an attractive product for hospitals that were setting up programs in heart surgery. The first commercial model had recessed knobs to prevent the children from changing their own heart rates. It also sported two little handles—cannibalized from an old ECG machine—so that doctors could secure the device to a child's chest with straps: the new product was not only portable but wearable. The housing "was a kind of carvable Bakelite paneling that was available at the time. It was layered, white inside, black on the back, and then it was carved" to bring out the white lettering that formed the words *Medtronic Pacemaker.* The generator also had a red neon light that blinked reassuringly with each pulse. Bakken decided to call it Model 5800 "because we made it in 1958." A few days later, he added one further design change: by reversing the Bakelite housing, he transformed the 5800 from a black box to a white one because "it appeared to me that white was more appropriate with the whiteness and cleanliness of a hospital."[22]

The battery-powered pulse generator reduced the enormous complexity and uncertainty that surgeons faced in performing open-heart surgery, freeing them to focus on other aspects of the procedure and on the care of their patients. Like the Zoll pacemaker, the system invented in Minnesota clearly had flaws. By running a wire out of the patient's body, the surgeons were creating a source of infection that required daily monitoring. Battery drain was not a problem—it was easy to replace the battery in an external pulse generator—but sometimes the doctors would discover that the myocardial wire had broken within the body. Most of the early pacemakers that were partly or completely implanted were plagued with problems of this sort. Doctors soon learned that the heart's beating would cause the wires to flex back and forth. This might not cause trouble over a few days of pacing, but as some patients began to remain on pacemakers for weeks and months, it became a major source of failure.

For all surgeons who tried the myocardial wire, the most puzzling problem and the greatest source of pacemaker failure involved rising thresholds of stimulation: the voltage had to be turned higher and higher as days passed or the pacemaker would fail to "capture the heartbeat." After a few boosts in the voltage, the patient's chest muscles might begin to twitch. In many cases, the pacemaker would simply quit. But Lillehei did not believe at first that the typical patient would need a pacemaker for more than about two weeks, so he did not define these as major problems.

One might be tempted to conclude, on learning that Bakken built the first transistor pacemaker in just days using an article in *Popular Electronics* as his principal engineering source, that this was a simple invention of no great technical interest or cultural significance. Certainly, when compared with the implantable pacemakers of the 1990s, the Medtronic 5800 looks primitive. Today pacemakers have hundreds of thousands of transistors; the 5800 had just two. But the device should be assessed within the context of its own time.

Ten years earlier, the idea of connecting a pulse generator to a wire sewn into the wall of the heart itself would have been unacceptable to physicians anywhere in the world. Permitting a person so unfortunate as to be dependent on such a device to get out of bed, walk the hospital corridors, and perhaps even go home would have been inconceivable. By the late 1950s, surgeons were accumulating knowledge and developing a sense of confidence about going through the pericardium, working around the exterior of the heart, and even cutting into the heart itself. Certainly, Lillehei's myocardial approach to pacing was a child of the revolution in heart surgery and of a more aggressive, manipulative attitude about the heart itself. At the same time, the friendly little 5800, with its blinking red light, offered reassurance. Ordinary Americans were no longer in a position to understand the technical details of many important inventions because they lacked the necessary knowledge of the underlying science; but they could in some cases assess a machine or artifact on its outer appearance and its effects. The wearable pulse generator, like the kitchen blender, the portable radio, and other "transistorized" products of the 1950s, for some years received a surprising amount of attention in the national press. Perhaps one reason is that, after all, it was external: you could see it and photograph it. What it did was also relatively easy to comprehend.

The feature that most intrigued people who observed the 5800 was its blinking red light. According to Bakken, the flashing light reassured both physician and patient that the device was really stimulating the heart. "We went to putting a screw switch [on the pacemaker housing] so that people could turn the light off because it would double the length of the battery. Nobody would do it." Bakken told an interviewer that "one night we had a lot of these [pulse generators] going at the University of Minnesota, and the nurses started saying, well, sometimes at night these pacemakers quit. I couldn't figure that out—why should they quit at night? And so I finally took them in a darkroom and, sure enough, at times it appeared that the neon wasn't flashing. I didn't know whether it was an optical illusion or what it was." Bakken tested one of the devices and found that "the output didn't quit"—the pacemaker was still stimulating the heart—"it was just that we had these adjusted so we would just barely trigger the neon light in an illuminated condition because we didn't want to wake [the children]. It was drawing half the current anyway, was going to flashing the bulb. We said, well, it's just at the threshold, and it took the ambient light to put it over the threshold so it would flash. We told them just to shine their flashlight on it. If it's flashing, everything is O.K. So that's all [it took]."[23]

Open-heart surgery provided hope for children hanging on to life with congenitally malformed hearts. But as with polio victims in their iron lungs, it must have seemed to some families and other onlookers that the technology of rescue was nearly as dismaying as the illness itself. In the case of open-heart surgery, the child had to undergo hypothermia (under anesthesia, to be sure) or coronary bypass, through cross-circulation or the heart-lung machine, before the surgical procedure could even begin. Photographs of open-heart surgery published in national magazines depicted teams of ten or twelve doctors, nurses, and technicians surrounded by the technologies of the modern operating room: monitors, hoses and tubes, intense lights, instruments by the dozen. A cover story in *Time* included a half-page photograph of a patient immersed in an ice bath before surgery; another photograph, in color, showed her chest sliced open from one side to the other.[24] The surgeons themselves, invariably males, cultivated their image as intense and decisive wielders of these instruments and techniques. In the days before the Model 5800, some children were left dependent for weeks on a wire sticking through their skin and attached to an electronic box plugged into a wall socket. The whole technological array, including the surgeon, probably frightened many.

Arriving in early 1958 Bakken's external pulse generator contrasted sharply with the early equipment and procedures of open-heart surgery. Like the Apple computer of the 1980s, the Medtronic 5800 had reassuring, "friendly" features. It was white, small enough to hold in the hand, battery-powered, comprehensible to ordinary people. And it had that blinking light. The toylike 5800 bridged the gulf between the idealized world of childhood and the somewhat frightening world of open-heart surgery. It was a reassuring technology, a token of a better future. The decision to reverse the color of the housing from black to white reinforced this reassuring quality.[25]

The surgical program at Minnesota in effect marketed the 5800. A few hospitals in the United States and western Europe that were setting up programs in open-heart surgery sent in orders for the device. Then in 1960, Lillehei and others at the university, together with Bakken, published a paper in the *Journal of the American Medical Association* that described the device and discussed its clinical uses. Two illustrations showed the 5800 and the Medtronic name up close; one of them pointed out nine important features of the pacemaker, such as its handles, neon flasher, control knobs that "cannot be accidentally changed," and white case that "allow[s] for damp scrubbing with alcohol." No ad agency could have prepared a more effective promotion.[26]

Viewed from long afterward, the larger significance of these events of 1957–58 seems clear. The invention of the myocardial pacing wire at the University of Minnesota and Bakken's small external pulse generator encouraged other inventors to push on toward fully implantable pacemakers that managed heart block, not for days or weeks but for years. But at the time, the implications of Lillehei's and Bakken's work were not at all clear. The new field of pacing was not moving toward some fixed, universally desired consummation.

## "THE GROPING STAGE OF PACEMAKER DEVELOPMENT"
The lack of a clear vision became obvious in September 1958, when the Radio Corporation of America (RCA) and the Medical Electronics Center at the Rockefeller Institute (today, Rockefeller University) held a one-day conference on future developments in cardiac pacing. The occasion brought together about two dozen engineers, surgeons, and cardiologists from several of the leading U.S. research groups for an informal exchange of ideas. Cardiac pacing stood on the brink of a major transition from a technology for emergency resuscitation to a treatment for long-term management of chronic illness, yet the experts did not see the change coming and in fact were unable to agree on anything. The field was seven months away from the first case of continuous pacing over many months to maintain an adult patient suffering from complete heart block. But in September 1958, a consensus did not yet exist as to whether long-term pacing would be possible or even desirable.[27]

Despite the surge of research activity in pacing after 1950, Victor Parsonnet and Alan D. Bernstein are certainly correct that, as of mid-1958, "no one had yet demonstrated that cardiac stimulation was possible as a routine lifesaving measure."[28] Only two groups, Zoll's in Boston and Lillehei's in Minneapolis, had accumulated a significant amount of clinical experience with any form of cardiac pacing. By September 1958, Zoll and his associates had supervised the use of the external pacemaker in more than 100 cases and the group at Minnesota had paced 57 patients. Few patients had relied on a pacemaker for longer than two weeks. At the conference, Lillehei's assistant William Weirich remarked that the Minnesota group had paced one patient for 21 days, another for 57. This last case, if correctly reported (the reporter may have confused the number of days with the number of patients), certainly represented the longest survival on a pacemaker.

The moderator, an engineer, opened the conference by asking participants to address "the kinds of problems one might get into if one had to do very long time cardiac stimulation for months, perhaps years, or indefinite periods." Discussion jumped from subject to subject but kept coming back to this question — and it genuinely presented itself as a question for most of those present. Several conferees thought it the wrong question for the time; they wanted to address the known shortcomings of short-term pacing rather than turn to an entirely new set of problems. The most outspoken was Lillehei's assistant Weirich, who had helped develop myocardial pacing for postsurgical heart block. Weirich characterized long-term pacing as "an impractical idea." He seemed preoccupied with the challenges of managing patients following open-heart surgery and was little inclined to refocus his thinking.

A second group argued that long-term pacing would require rate variability — in other words, that the ventricles would have to speed up or slow down with the patient's level of activity. They believed that no single rate — whether 60 beats per minute, or 70, or any other — could maintain a human being through the myriad activities of the day and the hours of sleep at night. In a normally functioning heart, the sinus node varies its rate of impulse formation based on cues about the status and activity level of the body that reach it via the nervous system.[29] The impulses emanating from the sinus node spread through the atria and cause them to contract. If an artificial pacemaker could be designed to deliver its stimulus into the ventricle when it sensed the natural atrial signal — the P wave of the ECG — rather than firing at some fixed, preset rate, this arrangement would achieve the goal of a variable heart rate.

Several investigators had built laboratory equipment able to trigger ventricular contractions off sensed P waves, but on the whole these efforts had left them discouraged. They warned that it would take years to design and build a practical pacemaker capable of synchronizing the atria and ventricles. Of the physicians who had investigated "atrial coordinated" pacing, cardiologist Herman K. Hellerstein, of Western Reserve University, was probably the most experienced.[30] Several years earlier, Hellerstein had built a remarkable experimental apparatus that he had used with laboratory animals. A pickup catheter electrode in the right atrium sensed the P wave; this signal was amplified and, after an appropriate delay, triggered a pulse from a ventricular catheter electrode. Built in 1949–50, near the end of the vacuum-tube era of electronics, this device weighed some 50 pounds and had to be suspended on a trolley atop the cage of the animal. In experiments, Hellerstein's group had used the setup to emulate a rare rhythm disorder known as Wolff-Parkinson-White syndrome by varying the interval between atrial and ventricular contractions.

In 1950, Hellerstein saw the clinical implications of this laboratory device and had suggested "a possible application in human patients with sudden complete A-V block [heart block], where other therapy has failed."[31] At the Rockefeller meeting, he argued that pacemakers "should be made to simulate what nature does." But he was deeply pessimistic about actually building equipment that would embody this principle. The sinus node was "quite intelligent"; emulating its action would require a degree of complexity too great for an electronic device. For example, a ventricular pacer triggered by an atrial signal would have to "flip off" in the presence of atrial fibrillation; otherwise it would risk driving the ventricles into fibrillation, too.

Stanley Briller, of New York University, was equally dubious about achieving rate variability in a pacemaker. The group at NYU, in the course of their studies of electrical activity within the heart, had built a sensing apparatus that could pick up the P wave. As Briller described it at the Rockefeller conference, this device had "about 70 vacuum tubes in it. It is about two rackfulls of equipment." After 15 minutes of fine tuning, it would sense the atrial signal, but "then the patient jiggles or moves and you have to resynchronize." Briller argued that "really what is needed" was not long-term pacing but "an emergency type of apparatus" for postoperative heart block or recurrent Stokes-Adams attacks. "The important bill of goods [is] to ask the engineers for help with . . . some device which will be available to the patient continually without failures for long periods of time, [and] which will have the electronic intelligence to turn itself on when needed." The device "must be portable. It must be simple. It must be ever present." It must take over "very quickly, within a few moments' time."[32]

These discouraging reports are hardly surprising: in order to coordinate the atria and the ventricles, pacing required a level of electronic complexity well beyond the state of the art in pacemaker engineering in 1958. Such a pacemaker can be likened to a jumper-wire bridging the atrium and the ventricle. It has two electrodes on or in the heart—one to sense atrial activity and the other to deliver stimuli to the ventricle. The sensing electrode must discriminate between the atrial signal and the much larger ventricular signal. It must ignore electrical interference and muscle potentials. As Hellerstein noted, the pacemaker must be capable of shifting into 2:1 or higher conduction ratios if the atrial rate speeds up, so as to avoid passing an atrial tachycardia on to the ventricle. At the opposite extreme, the device must also maintain a minimum pacing rate for the ventricle in the event of atrial fibrillation, a slow sinus rhythm, or loss of atrial sensing. Hellerstein and Briller were telling the Rockefeller group that long-term pacing would require all this—and warning that it would be many years before pacemaker engineering would be able to deliver.

They were correct in believing that reliably synchronizing the upper and lower chambers lay years in the future. But was synchrony truly a requirement for all long-term pacing? Did the heart require rate variability? Two physicians at the Rockefeller Conference argued that rate variability was a frill and that continuous pacing over the long term was a practical goal for the immediate future. As a surgical resident at Johns Hopkins in 1956–57, Stanley Brockman had briefly paced four postsurgical patients via a myocardial wire; this was at about the same time similar work was going forward at Minnesota. Two patients had survived the first stormy days after their heart surgery. One converted to normal sinus rhythm after eight days, but the other remained in block and died suddenly five months after the operation, probably from cardiac standstill. Brockman recognized that this fourth surgical patient had developed a chronic block for which short-term pacing was not the answer. Such patients would "die of Stokes-Adams, all of them, eventually." Brockman did not believe that it was necessary to synchronize the atria with the ventricles: "I think nature can be improved upon," he remarked—otherwise, why hold a conference on artificial pacemakers? The atria were "just dilated veins" whose principal function was to "maintain a filling pressure for the ventricle." The ventricles could pump blood to the body adequately even without being filled efficiently by the atria. This opinion sparked a spirited exchange—which the moderator cut off with the remark that "we are getting a little bit off the subject here."

Of all the physicians at the Rockefeller Conference, Paul M. Zoll seemed most confident about making the transition to long-term pacing for chronic disease. Zoll, who had first reported the clinical use of short-term pacing, had subsequently lost many patients to repeated Stokes-Adams episodes. "After the initial excitement of saving the patient from the initial episode of standstill, everybody relaxes and you come back later . . . and find the patient had another episode. . . . You can resuscitate a patient . . . if you are ready all the time for the rest of the patient's life, and that is a big order. . . . What we need is a reliable pacemaker that will drive the heart indefinitely for the rest of the patient's life."

Acute pacing was a settled issue; now the field must move on to chronic stimulation: "It does not make much difference in most [elderly] patients how fast one drives the heart," Zoll argued. "We have found that usually if we drive the heart in the fifties or sixties this is fast enough to maintain an adequate circulation and keep the patient well." The patient might suffer "some degree of congestive heart failure," but "this minor degree of inefficiency . . . is not ordinarily very important from the clinical point of view."

Challenged by the advocates of "atrial coordination," Zoll conceded that his patients led lives as semi-invalids. But he dissented "very violently" from the goal of rate variability. "There are many other factors involved in changing cardiac output besides cardiac rate. . . . One does not have to worry particularly about what the cardiac rate is as long as [it lies] between 60 and 100." Toward the end of the day, after listening to several hours of rather confused discussion, Zoll elaborated on his objections:

> There is no question in my mind that it might very well be desirable to have an artificial ventricular rate follow the rate of the atrial beat. This would be ideal perhaps, and it would be very desirable also to turn this on when necessary. . . . [But] these are accessory, secondary considerations, it seems to me. . . . They are secondary problems. The problem at present is to find something that we can use over a long period of time. . . . It seems to me that the block, I still have not heard anybody get around it, is that we cannot stimulate the heart directly for more than a month or two.

For some time, Zoll's group had been investigating the requirements for a fully implantable, permanent pacemaker. They had run into "trouble and that is the reason I am here." Zoll described the problem of rising thresholds for pacing with eventual loss of the ability to stimulate the ventricle. The difficulty, he thought, seemed to lie in tissue reactions to the electrode. But others at the meeting did not pursue the problem, and Zoll went away disappointed.

The Rockefeller Conference ended inconclusively with neither the moderator nor anyone else able to sum up the day's discussion. Without exception, participants recall it as a chaotic meeting. The most significant innovations and issues of the next few years received hardly a mention. Participants' frustration with the conference can be ascribed to the fact that there was little settled doctrine in the field of pacing—knowledge that everyone held in common and took for granted. "So much was unknown at that time," said a participant, "that it wasn't even clear what the crucial questions were." Virtually every aspect of pacing seemed an open issue. Working from a very limited published literature, the conferees could not agree on a list of the key problems and needs that they ought to discuss.

This situation was compounded by the fact that they came from many different research centers, did not know one another personally, and often seemed unfamiliar with one another's work.[33]

All of the physicians at the meeting understood that pacing for chronic heart block or other rhythm disorders would require that they rethink the experience of acute pacing between 1952 and 1958. Implicitly, they recognized that chronic pacing would be a radical or discontinuous development, not just an improvement on existing practice. Perhaps this recognition was the major achievement of the meeting. Yet because chronic pacing raised so many new problems, participants could not agree on whether it was a practical or desirable next step. "It was the 'groping stage' of pacemaker development," said one Rockefeller participant many years later.[34]

### "ALL OF A SUDDEN IT STARTED TO PUMP"

Paul Zoll had said, "you can resuscitate a patient . . . if you are ready all the time for the rest of the patient's life, and that is a big order." Just such a situation presented itself to Dr. Samuel W. Hunter in April 1959. After completing his residency in cardiothoracic surgery under Walt Lillehei, Sam Hunter (b. 1921) had begun a surgical practice in St. Paul, Minnesota. One day a call came in from an internist he knew who reported that a 72-year-old man had rapidly developed complete heart block following a heart attack. The patient's ventricular rate varied between 16 and 36 beats per minute and he was suffering dozens of Stokes-Adams attacks daily. The house staff at Bethesda Hospital were keeping him alive with injections of epinephrine and Isuprel and by pounding on his chest. Was there any chance that Hunter could give this man a pacemaker?[35]

The case of Warren Mauston came to Hunter because some doctors in St. Paul knew that Hunter had been working with a young engineer from Medtronic, Norman Roth, on animal tests involving a new pacemaker electrode. Roth had realized in 1958 that it might be possible to use Earl Bakken's transistorized pulse generator not only to pace children recovering from open-heart surgery, but to manage the heartbeat in elderly people who suffered from fixed complete heart block.[36] No one knew how many cases of chronic heart block there might be, but finding a new use for the Model 5800 pulse generator would perhaps boost its sales modestly and provide some much-needed income for Bakken's business. Roth visualized a system rather similar to what surgeons were using at the University of Minnesota: external generator with a myocardial wire touching the heart and poking out of the chest. It was obvious that the wire was the unstable part of this arrangement. Animal studies had shown that if you relied on the wire for more than two or three weeks, "one of two things [would happen]: either the wire would break, or there was a tendency for fibrotic tissue [scar tissue] to build up around the wire, and the resistance go up, and they would get to a point where you couldn't drive [the heart]." For either sort of failure, the only remedy would be to reopen the chest, remove the malfunctioning wire, and implant another. "It was not a happy thought to have to reimplant an electrode" — particularly if the replacement electrode might also fail within weeks.[37]

Roth "came up with the idea that there probably were several things involved. One might well be the stability of the electrode itself. With just a wire in there, you were bound to get a lot of movement, stressing, flexing, and so on, which could likely irritate tissue and cause the fibrotic buildup to become larger. Along the same line, it was pretty well understood that current density was the primary requirement of stimulation. And so that's when I started to play with the bipolar idea." Lillehei's myocardial wire had been a unipolar configuration: one electrode in the heart, the other buried under the skin several inches away. Roth suspected that unipolar pacing caused the electrical stimulus to dissipate instead of concentrating it in the excitable tissue. This made it more difficult to pace the heart, especially when fibrosis set in around the myocardial electrode. It was possible to raise the amplitude of the stimulus to a point, but sometimes pacing had failed even with the amplitude at its highest setting.[38]

Roth designed a new myocardial lead terminating in a small "platform" of silicone rubber from which protruded two stainless-steel pins. Maintaining a high current density was uppermost in his mind: he hoped that when stitched down against the surface of the heart, the platform would stabilize the electrodes, thereby reducing the size of the lesion and the amount of scar tissue. And with anode and cathode now just half an inch apart, the stimulus would be highly concentrated.[39]

In mid-1958, Medtronic's entire business was located in two garages in north Minneapolis. The company had never manufactured an implantable device, it had no animal laboratory, and it obviously could not conduct trials involving dogs. Roth needed a surgeon who would collaborate on developing the platform electrode. When he approached the surgeons at the University of Minnesota, he was "kind of rudely handled": Lillehei's people were preoccupied with surgery on children and uninterested in a pacing lead that they couldn't remove from a patient's body simply by pulling on it. At the suggestion of his personal doctor, Roth then approached Sam Hunter. Hunter recalls that he, too, was skeptical about the platform electrode: "[Roth] was adamant that there were people out there, adults, in [heart] block, who could benefit. But I said, 'You can't get it out! What are you going to do — leave it in there?'" But Hunter had a small animal-research lab and was "more or less casting around for things to do." The same day that he met Roth, Hunter installed the platform electrode on the heart of a dog. When they connected it to an external pulse generator, it proved able to overdrive the dog's natural heart rate. Impressed, Hunter agreed to work with Roth on the animal studies.[40]

During the next few months, Roth made up electrodes at the Medtronic garage and Hunter implanted them in dogs. They would pace the dog for a few days or weeks using the experimental electrode, then sacrifice the dog and examine its heart for evidence of infection or fibrotic buildup around the site of the electrode. On advice from a Dow Corning scientist whom he met on an airplane flight, Roth chose a silicone rubber product called Silastic to insulate the wire lead and to create the platform sewn down on the heart. The Silastic regularly caused huge infections until Roth flew down to Dow Corning's research labs at Midland, Michigan, to learn how to cure the rubber properly. After several months of trials, "we got to the point that we were having extremely good, long-term (as things were at that time) results with the dogs." The threshold of stimulation for the animals' hearts would rise, then stabilize at six or seven milliamps.[41]

Roth continued, "I'll never forget one night . . . Sam called up and said, 'How soon can you have me an electrode to go into a person?' I said, 'You've got to be kidding.' He says, 'Nope, we've got a fellow down here and he needs it, not going to last without it; he wants to go [with the pacemaker].' So I told him about how many cure-hours and so forth it would have to have. He says, 'Get to it.' So I went down to the plant and put one together, cured it up, gave it a good cure [and] wash-out."[42]

According to Hunter, Mauston was "essentially dead when we brought him to the operating table" the next day, April 14, 1959.

> We kept his heart going by pounding his chest. . . . It was a funny-looking heart when I got in there, sort of blue. . . . I stitched the electrode on his right ventricle and, as it worked out, his ventricle didn't rupture. It's a won-der it didn't, because I'm sure I had to do it as quickly as I could, holding the electrode, pushing it against the heart, stitching it, and tying down on that poorly perfused right ventricular musculature. We got it on, and I imme-diately threw the leads out to the engineer . . . and when it started I just couldn't believe my eyes. Because a nice little compact heart in a child is one thing; but this was a 72-year-old man with a big bulbous heart that was like a jellyfish; . . . and all of a sudden it started to pump, vigorously and according

to the rate that we wanted, and we could control it, and all of a sudden he started to wake up! So we had to put him to sleep and finish the operation.

The full procedure required more than four hours.[43]

Hunter's second case, some two months after Mauston, did not have such a happy outcome. When the surgeon exposed the patient's right ventricle, he discovered a large area where the wall of the heart was dead and "mushy" from a recent heart attack. "I had to put the [electrode] where it would fit between coronary arteries. . . . So I fitted it where I could, and I think I was on the second or third stitch, when suddenly the heart split at that point. The myocardium just sort of disintegrated as I pulled down on the stitch and tied down, even though I was going as gently as possible, and [the stitch] cut right through the myocardium, and the next thing I knew, the ventricle beat and . . . pulled apart like opening doors. . . . We had no way of controlling the hemorrhage." Within moments the patient, a woman, bled to death on the operating table. Afterward, Hunter reflected that the damage to the patient's right ventricular wall had been only partly responsible for the disaster: "The right ventricular wall is very thin. It's only about one-fourth or one-fifth [the thickness of] the left ventricular wall. Now, would the pacemaker lead work on the left ventricular wall?" Further laboratory studies with dogs showed that a pacing electrode on the left ventricle worked perfectly well. "So never again did I go to the right!"[44]

## "NOTHING TO WORRY ABOUT"

In 1959, there were at most a few dozen surgeons who used myocardial cardiac pacing on human beings in the United States. Some of them, such as Hunter, not only used but participated in the invention of pacing components and systems. These surgeons and a few cardiologists like Paul Zoll were clearly the dominant influences on the nascent field of pacing. By necessity, pacing at first remained largely confined to large teaching hospitals, though it would later spread to community hospitals as part of a technological package that included cardiothoracic surgery and acute coronary care. With its operating rooms, catheterization labs, and skilled nursing care, and with procedures such as electrocardiography, AC defibrillation, and cardiac catheterization, the large hospital had already emerged by the mid-1950s as the appropriate locus for the practice of acute-care medicine relating to the heart. Pacing was not only nurtured in the hospital, it promised to reinforce the hospital's role in the acute care of heart disease.[45]

Although postsurgical heart block had added thousands of new patients to the number who might be assisted by pacing, the total population with

Stokes-Adams disease or postsurgical block still seemed quite small to most clinicians. When representatives from manufacturing firms began to inquire about the market for pacemakers in the late 1950s, physicians in the field gave them estimates on the order of five hundred units per year for the United States. Such figures were probably based on the assumption that one external pulse generator, whether plug-in or battery-powered, could pace dozens of patients over a few years because the pacemaker was a piece of hospital equipment, not (yet) a part of the patient's own body.

Mauston, Hunter's first patient, lived for six and a half years, dependent on his external pulse generator the entire time. A journalist wrote that "although he occasionally frets at being unable to go out on the golf course as he used to, he putts on the living-room rug . . . , gets up and down stairs and walks around the neighborhood." In 1961, Mauston survived major cancer surgery, pneumonia, and an automobile accident. He remained active until the very end. On the day before his death in October 1966, he drove 60 miles to Lake City, Minnesota, to visit his son and watch a World Series game on television.[46]

Mauston always declined an implanted pulse generator, confiding to Hunter that his grandmother had rejected indoor plumbing for her farmhouse because "some things just don't belong inside." Because of this choice, Mauston always had an open wound through which the pacing lead protruded. Hunter gave him antibiotics for a time, but stopped when he grew concerned that this would encourage the growth of drug-resistant bacteria. So the surgeon or an assistant would drop by Mauston's house in St. Paul once or twice a week to clean the point of emergence of the catheter with alcohol and change the dressing. About once a month, Mauston's pulse generator would get a new battery. Roth had "modified the [generator] to have a large capacitor in it so that you could take the battery out and you'd still get eight or ten acceptable pulses — give you plenty of time to take the old one out and put the new one in." The Hunter-Roth electrode served Mauston for four years, long past the time when Medtronic had withdrawn it as a commercial product. In 1963, Hunter abandoned the original lead and gave Mauston a transvenous lead that delivered the electrical stimulus within the right ventricle and emerged from the external jugular vein at the base of the patient's neck.[47]

Mauston had an ideal personality for the role of pioneer patient: he thrived in the limelight and did not appear unduly anxious about the external pulse generator or the transcutaneous wire. According to Roth, "he liked to see the light blink" with each stimulus from his generator. Hunter recalled that "a lot of people wanted to see him and take pictures of him and talk to him. He was in *Reader's Digest*, he was in *Saturday Evening Post*, he was in newspaper articles. Throughout the country, people were picking up the story." When cardiologists visited from out of town, Mauston would occasionally allow Hunter to turn the pacemaker off to demonstrate the effect. "If I set him at 60, and then turned [it] off—bang—he would be O.K. . . . for four seconds. And then he would start to slide quickly and go unconscious. . . . And he always said he was falling back, down a well or down a big barrel. And he said it wasn't unpleasant. Then I'd snap it on again, and he'd come right out of it. . . . I had a lot of [ECG] tracings . . . all over the laboratory: Mr. Mauston sliding toward eternity because I'd turned off his pacemaker."[48]

The technology available for long-term pacing in 1959 was crude by later standards; it had not been thoroughly tested because cases usually arose as clinical emergencies. Often critically ill with multiple problems, the patients were hardly the best candidates for clinical trials involving open-chest surgery lasting four hours. Typically, the doctor had little opportunity to try to stabilize and strengthen the patient before proceeding. The choice was stark indeed with a patient like Warren Mauston: try to manage the heartbeat with the equipment then available, imperfect as it was, or let the patient die. In a medical culture that rewarded boldness and believed in leaps forward, surgeons and patients often elected to go with the new technology.

As they tried cardiac pacing for postsurgical heart block or used it with elderly patients like Mauston, surgeons began to accumulate a body of practical experience. When they had managed a few cases, the early implanters of pacemakers would present the results at medical conventions or in print, thus beginning the process of converting their experience into more formal doctrine. Roth traveled extensively to introduce the Medtronic 5800 and the platform electrode at hospitals. Hunter gave a paper on the Mauston case at a national heart meeting. About 18 months after he operated on Mauston, a surgeon on the West Coast called Hunter. "He used some very uncomplimentary words. He said, 'This is the most jackass pacing equipment. I couldn't get the lead to fit into the external pacemaker. . . . This thing just is designed so poorly I can't believe it. I can't believe you would talk about it.'" Hunter eventually realized that his caller had forced the two spikes of the electrode into the sockets atop the external pulse generator and stitched the other end down on a patient's heart. "He put it on backwards. And the funny part of it is, it worked." Hunter added, "I immediately wrote a paper with drawings [and] descriptions, showing how to put it on."[49]

Among the early pacemaker patients, Mauston stands out for his cheerful willingness to serve as an object of medical study and his determination to stay with an external pulse generator. He contributed to the shaping of his own care and helped spread the message that it was possible to manage the heartbeat over months and years. The fact that many pacemaker recipients rose from their beds to do light housework, go bowling and dancing, or work in their gardens was big news around 1959–61 and stories appeared in national newspapers and magazines. "I used to dread going outdoors alone," Louise Kreher, of Buffalo, New York, told a reporter, "because I was always afraid I'd faint and fall. I'd go across the street to the store and then wonder why I had come [because her heart block made her dizzy and forgetful] and would feel like crying. Now I go out often; I can shop for an hour or two and only feel a little tired, nothing to worry about."[50] Testimony like this reinforced Americans' tendency to place their faith in the miracle cures of modern high-tech medicine.

The Mauston case, along with a few others from the United States and England, had demonstrated by 1960 that long-term cardiac pacing was possible. Until superseded by implanted pulse generators two or three years later, the Medtronic 5800 paced the hearts of at least a hundred older men and women in chronic complete heart block. Bakken particularly recalled a patient who resided at the Veterans Administration hospital in south Minneapolis. "They would release him every weekend to go home to Bemidji [in northern Minnesota], and he was an avid square dancer, that was his big love of life. He'd go dancing . . . and he'd break his wire and then he'd retract to a slow heart rate, would have to sit down, get back to the VA, and I'd be called invariably on Mondays to come out and solder his wires back together." Cases like this convinced him that the pacemaker was benefiting "the whole personality of the person," effecting a mental and spiritual as well as a physical restoration. Out of such cases came the earliest Medtronic slogan — "Toward Man's Full Life" — and the image at the company's web site of a human figure arising from a sickbed to a standing position.[51]

The existence of a treatment encouraged physicians to redefine chronic heart block, essentially to reconceive the disease. Physiologists had understood what happens in heart block since the early twentieth century, but their research had attracted little attention from clinicians because it yielded no practical clues as to how to treat the condition. Once Bakken, Roth, and Hunter had pioneered a plausible treatment, research into heart block picked up again and clinicians acquainted themselves with the symptoms of the disorder. As they did so, they discovered still other disorders of the heartbeat for which a pacemaker seemed the appropriate treatment. The most recent set of formal guidelines for pacemaker implantation lists dozens of rhythm disorders, many of which had not been carefully studied or even noticed until after the invention of cardiac pacing.[52]

Pacemaker patients since the early 1960s have had no way to participate meaningfully in shaping the development of cardiac pacing. But in the early days, when pacing was a revolutionary and untried therapy, the patients did play a key role. The children and elderly men and women made clear to physicians, by their behavior, that they wanted to look and feel like healthy people even though their hearts refused to beat on schedule. As surgeon Seymour Furman points out, "a surprisingly large number did survive and even prospered," sustained by the primitive pacing technology of the day and by their own determination.[53] By accepting or refusing pacemaker therapy and by trying to restore elements of normal life for themselves, the early patients helped the doctors and engineers to grasp the idea of permanent pacing of the heart. No one yet knew whether permanent pacing would prove feasible in more than a few cases; no one knew how to do it. But there was one point on which all investigators agreed: There should be no opening in the patient's skin — no path of infection leading from outside directly to the heart. Could a pacing system be designed that would stimulate the heart on schedule, over months or years, yet permit the surgeon to close up the patient's body?

## CHAPTER 4

THE MULTIPLE INVENTION OF
IMPLANTABLE PACEMAKERS

Despite the confused discussion at the Rockefeller Conference in October 1958, pacing the heart for weeks or months at a time loomed as a desirable next step in part because heart specialists had begun to realize that quite a few adults developed heart block or chronically slow heartbeats in late middle age. Between 1957 and 1960, at least eight research groups designed and tested pacemakers that were fully or partially implantable and intended to work for the remaining lifetime of the patient; all eight groups used their invention with human beings. Today a few men and women are alive who have been pacemaker-dependent for thirty years or more, but in the late 1950s and early 1960s, "permanent" pacing had a more limited meaning: physicians who implanted pacemakers probably hoped that their patients would survive for two or three years.

Complete heart block in adults seemed a rare malady, yet there were suggestions in the medical literature that clinicians might be overlooking many cases. But this was almost beside the point: inventing an apparatus to stimulate the heart electrically for months and years would be a superb technological achievement and would open the way for other biomedical inventions using implanted circuitry, batteries, and wires. David Sarnoff, chairman of the board at RCA and a noted technological sage of the 1950s, explained what was at stake when he predicted that "miniaturized electronic substitutes will be developed to serve as long-term replacements for organs that have become defective through injury or age. . . . It is not too far-fetched to imagine a man leading a normal life with one or more vital organs replaced by the refined substitutes of the future." Sarnoff added, "One day artificial kidneys, lungs, and even hearts may be no more remarkable than artificial teeth."[1] Would the cardiac pacemaker prove to be the doorway to this electromedical utopia?

Medical and engineering innovators like C. Walton Lillehei, Paul Zoll, and Earl Bakken wanted to contribute to the relief of human suffering. One cannot talk to these men without being impressed with their empathy for

TABLE 2. THE SEARCH FOR A LONG-TERM PACING SYSTEM, 1956–1960

| Group | Work Begun | Clinical Use | Description |
|---|---|---|---|
| Zoll, Belgard, Frank Boston | 1956? | 1960 | Implantable generator, myocardial lead |
| Senning, Elmqvist Stockholm | 1957 | 1959 | Implantable rechargeable generator, myocardial lead |
| Chardack, Greatbatch Buffalo, N.Y. | 1958 | 1960 | Implantable generator, myocardial lead |
| Hunter, Roth Twin Cities, Minn. | 1958 | 1959 | External generator, Hunter-Roth lead |
| Furman, Schwedel Bronx, N.Y. | 1958 | 1958 | External generator, endocardial lead |
| Glenn, Mauro New Haven, Conn. | 1958 | 1959 | RF pacemaker |
| Abrams et al. Birmingham, U.K. | 1958? | 1960 | Inductive coupled pacemaker |
| Cammilli, Pozzi Florence, Italy | 1959? | 1961 | RF pacemaker |

the stricken people they were trying to restore to some degree of health. But they were also well aware that electrostimulation of the human heart was a "deep and consequential problem" that would exercise their talents to the fullest and might establish their reputations in the medical and engineering communities of which they were members. Their employers and professional associates valued innovation in medical devices, admired it, and in some cases stood to profit by it. As Stuart S. Blume puts it, innovation was normatively sanctioned. Thus encouraged, the key innovators took it for granted that inventing an implantable cardiac pacemaker for long-term use was a project well worth the effort.[2] By the late 1950s, eight groups from four countries were designing pacemakers for long-term use (see table 2).

**THE TECHNICAL CHALLENGE**
Physiologists and clinicians had often stimulated the human heart with electricity, but they had never placed a machine inside the body, closed the

skin, and left the thing to function. Fully implantable pacemakers seemed practical by 1958 because two inventions of the 1940s, small batteries and silicon transistors that were tiny and drew little current, were becoming available by the mid-1950s. But numerous problems of design remained.[3] The requirements for long-term pacing in the late 1950s were:

— No ports in the skin through which infections could enter the body.

— A small battery with high energy density and a probable worklife of two to five years, or a practical rechargeable battery.

— A pulse generator small enough to implant in the patient's abdomen.

— A wire lead able to withstand tens of millions of flexions each year for several years.

— A stable electrode in contact with heart tissue that would not cause the threshold of stimulation to rise to unacceptable levels.

— A biocompatible encapsulant for the pulse generator that would allow the discharge of hydrogen gas, a byproduct of the chemical reaction in zinc-mercury batteries.

— Circuitry shielded from both battery discharge and incursion of body fluids.

— A transistorized circuit yielding pulses of about 2 msec. duration and amplitudes of 15 milliamperes across the load resistance of the heart, at a steady rate of about 70 impulses per minute.

Without exception, physicians wanted to avoid a break in the patient's skin. This was a fundamental requirement. Lillehei's myocardial pacing wire (1957) was the first implanted electromedical device in medical history, but it was obviously a transitional technology. No one regarded the wire as appropriate for more than a few days of pacing: the wire was unaesthetic, disturbing to patients and their families, and a source of infection. Most patients who lived with partly implanted, partly external pacemakers had recurrent infections. Cases like Warren Mauston's (1959–66) would show that a patient could survive for months or years with an opening in the skin, but some patients died of infections.

The decision to implant the entire apparatus and aim for continuous, long-term heart stimulation meant that the generator must be quite small, it must include a tiny battery that was either rechargeable from outside the body or very long-lived, and it must be encapsulated in some biologically inert material that would protect the circuitry from the warm, salty environment of the body.

Designing a stable, implanted wire lead that terminated in a bare metallic electrode in direct contact with the heart would cause bigger problems than the generator. Between 1956 and 1962, that contact point preoccupied the inventors of pacemakers and caused much of their frustration. At 70 contractions per minute, the heart beats 36,792,000 times in a year. What metal, what configuration of a strand of wire, could stand up to tens of millions of flexions without breaking? Lillehei's group did not report on broken wires, but others experimenting with myocardial pacing in laboratory animals found that breaks in the wire did occur. When this happened, pacing became intermittent or ceased altogether. In a patient who depended on the pacemaker for every heartbeat, a broken lead would be a major emergency requiring immediate open-chest surgery.

Leads had other limitations besides their tendency to break. Scar tissue invariably developed at the point where the lead entered the heart. This tissue did not respond to the pacemaker stimulus and it shielded the responsive heart tissue from the stimulus. A stimulus powerful enough to trigger a heartbeat during the first few days of pacing might thus prove too weak once the scar tissue had formed, and the pacemaker would "lose capture" of the heartbeat. Designers would try a myriad of solutions in the 1960s, but for the earliest implantable pacemakers the only realistic way to manage this problem was to design the pacer to deliver a stimulus several times more powerful than what was thought necessary — even though this would reduce the longevity of the battery.

Some "problems" discussed at the Rockefeller Conference did not greatly concern the research groups that invented pacemakers in the late 1950s. Many patients in complete heart block will have occasional normal beats in which the atrial signal is not blocked but reaches the ventricles. As everyone understood, a pacemaker that fired a stimulus into the ventricle could cause patients to experience "competition" between the pacemaker impulses and these normal heartbeats. To avoid competition, a pacemaker must sense the onset of ventricular depolarization — the QRS complex in the ECG — and inhibit itself from firing. By 1959, most pacemaker inventors had decided to ignore this problem and simply to pace the ventricle steadily. This simplest way to stimulate the heart came to be known as fixed-rate, or asynchronous, pacing.

As the term *fixed-rate pacing* implies, the earliest implantable pacemakers always fired their stimuli at a rate set by the designer's choice of circuit components; as a later generation would say, they were hard-wired. The devices could neither slow the rate down when the patient went to bed at night nor pick up the pace when he or she walked upstairs or strolled around the block. Designers accepted fixed-rate pacing for the same reason that they accepted pacemaker competition: anything more sophisticated would require that the device sense electrical activity within the heart. Sensing would require more complex circuitry and a second electrode on the atrium to detect the P wave. Prudence dictated that they had better master fixed-rate pacing before attempting anything more subtle.[4]

## RADIOFREQUENCY AND INDUCTION PACING: A PATH NOT TAKEN

Several groups designed pacemakers that in effect circumvented one or another of the requirements for an implantable pulse generator while holding to the basic idea of avoiding permanent breaks in the patient's skin. A group at Yale University Medical School led by surgeon William Glenn (b. 1914) introduced a radiofrequency (RF) pacing system. Glenn had long been interested in developing new techniques for repair of the heart and the major blood vessels and in the problem of resuscitating victims of cardiac arrest. He quickly recognized the drawback in Lillehei's myocardial pacing technique. As he later recalled, it was unsuitable for long-term pacing "because of the development of infection at the site of incision, proceeding down the wire to the heart." Yet implanting the pulse generator seemed equally impractical to Glenn because "the battery sources . . . that were available to us . . . were either very bulky or very short-lived, and would obviously require reoperation in a very short time for replacement."[5]

In 1958, Glenn contacted Yale physiologist Alexander Mauro, a specialist in stimulating neural tissue by radiofrequency induction. "I asked him if it was possible to broadcast a signal to the heart in a way similar to that that had been done to the brain; and he said of course it was, and he thought it would be a fairly simple matter to construct such an electronic device." Together the men designed a small external transmitter and a buried receiver with a myocardial electrode. They would implant only part of the pulse generator and leave the rest outside the body. The implanted part was a receiver circuit enclosed in a polyethylene jacket, to which was attached the myocardial lead. The transmitter, taped to the wearer's chest, sent signals from a coil through the skin to the implanted receiver directly beneath. The initial version plugged into the AC electrical system, but Glenn and Mauro later developed a battery-powered model. This technology seemed to provide the best of both worlds. It retained the advantages of

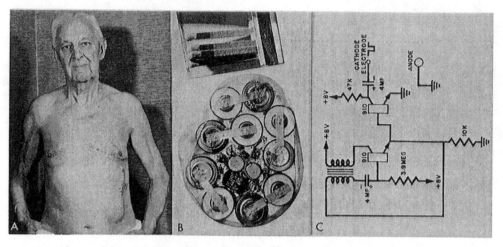

**Frank Henefelt and the First Chardack-Greatbatch Pacemaker, 1960.** Henefelt, age 77, was the first U.S. patient to receive a fully implanted cardiac pacemaker. Surgeon William Chardack initially implanted a pacing lead on the outer surface of the patient's heart and paced him from an external pulse generator. Seven weeks later, Chardack connected the lead to the implantable generator that engineer Wilson Greatbatch had constructed (shown in cutaway, center) and implanted it in Henefelt's abdomen. The circuit diagram of the Chardack-Greatbatch implantable pacemaker appears at right. Note the 10 mercury cells that constituted the battery. (Courtesy of Dr. William M. Chardack)

the external pulse generator by permitting the physician to replace batteries and adjust pacing rate and output without performing surgery, yet avoided a permanently open wound in the patient's body.[6]

A clinical case came Glenn's way in October 1958—an elderly man experiencing severe heart block and Stokes-Adams attacks. In January 1959, Glenn gave this patient an RF pacemaker. The device functioned for three weeks before failing because of a cracked wire; the patient declined a replacement and died soon afterward in another Stokes-Adams attack. Through the end of 1963, Glenn's group paced 17 patients with the RF system. Several were still being paced in mid-1964, with one patient having gone 26 months on the pacemaker—long enough to demonstrate that it could manage chronic heart block. Most failures had come from broken wire leads within the body and were unrelated to the radiofrequency apparatus itself. Glenn's group was justified in claiming that inductive pacing was "a controllable, reliable, and simple technique for long-term remote control of cardiac stimulation."[7]

About the time of Glenn's first clinical case in 1959, a surgical group in Florence, Italy, led by surgeon Leonardo Cammilli and engineer Renato

Pozzi, invented an RF pacemaker that was similar in concept and in some respects technically superior to the Yale design. Cammilli's group had already tried the Lillehei technique—external pulse generator with a wire through the chest and into the heart—but, like others, found it unsatisfactory if used for more than a few weeks. To avoid Glenn's problem with implanted wires breaking, the group came up with a unique design for the implanted receiver. Instead of locating it just beneath the patient's skin and connecting it to the heart by a wire lead, they constructed a receiver the size of two stacked U.S. quarters with two projecting pins and sutured it down directly on the ventricular surface. Despite the constant motion of the receiver with respect to the external transmitter, Cammilli reported no loss of stimulation.[8]

Leon D. Abrams, a surgeon at the University of Birmingham in the United Kingdom, led a group that invented an even simpler device (hence offering fewer possible points of failure) than the Yale and Florence designs. In the Abrams pacer, the patient carried the pulse generator in a small case with a shoulder strap. A wire lead from the generator terminated in a coil that was taped down on the patient's chest. The secondary or receiving coil was located just beneath the skin and consisted of a thousand turns of fine copper wire wound to form a ring and covered with silicone rubber. From this implanted coil, two wires were attached to the heart wall. Abrams summarized the merits of this system: "The implant is of the simplest possible construction consisting essentially of a length of wire. . . . There are no electronic components to fail. The pacemaker [pulse generator] itself is external so there is no difficulty in changing it for any reason. . . . Moreover pacemaking is voluntary and can be stopped if the patient reverts to stable sinus rhythm."[9]

Clearly, there were advantages to keeping the power source and controls for pacing rate and amplitude outside the patient's body. For this reason, groups in Colombia, Japan, China, the Soviet Union, and Israel developed and implanted devices of similar design in the early 1960s. Yet none made the transition from experimental design to successful commercial product. With RF pacing, the physician's control was compromised because patient and family members could still see and touch part of the pacing system. As the British surgeon Harold Siddons observed in 1963, "Psychologically the arrangement is not ideal, as patients find it hard to forget their condition and are apt to keep on fingering the small transmitter and belt." It was crucial that the transmitter remain in position on the patient's torso. If it moved even an inch, the implanted receiver might lose the signal and fail to pace. An adhesive held the transmitter in place, but the device needed frequent battery replacements or, in the case of the Italian unit, a recharge. Once a week, the patient was supposed to pull the transmitter aside and clean the skin, then apply new adhesive and reposition the transmitter. This maintenance routine risked loss of capture of the heartbeat, a serious drawback in patients with complete heart block. Cammilli remarked that the patient was always "obliged to remember that his life depends on this device." Under the circumstances, the patient could hardly put this thought out of his mind for long.[10]

## ELMQVIST, SENNING, AND THE RECHARGEABLE PACEMAKER

The first successful pacemaker that was fully implanted in the patient's body came out of the Karolinska Institute in Stockholm, Sweden, in 1958. The key figures were Rune Elmqvist (1906–96), the director of research at the electronics firm Elema-Schönander in Stockholm, and heart surgeon Åke Senning (b. 1915), of the institute. Like several of the early pacemaker engineers, Elmqvist had already accumulated a great deal of experience in designing electrical devices with medical applications. He also resembled his U.S. counterparts in retaining many of the attributes of an amateur: he was curious, playful, and uncomfortable as manager of a corporate department. Rejected for the Royal Institute of Technology as a young man, Elmqvist had worked as a hospital lab assistant and studied for his medical degree. But he never practiced medicine; instead, at Lund University Hospital, in 1931 he built several original pieces of equipment, among them a multichannel ECG machine. He worked for Jerns Electrical in Stockholm, later joining Elema-Schönander. After the war he built the machine he considered his best accomplishment—the mingograph, an inkjet printer for use with ECG machines. Elmqvist had met Senning in the 1930s when Senning was a medical student, and during and after the war they collaborated on an early defibrillator. By the 1950s, Elmqvist's technical preparation for further design work on machines that would treat heart arrhythmias was second to no one's.[11]

Åke Senning was trained as a general surgeon and after World War II involved himself in several of the major projects that prepared the way for open-heart surgery: open-chest defibrillation, external pacing, hypothermia, and attempts to develop a heart-lung machine. Senning traveled widely and kept abreast of new machines and surgical techniques all over western Europe and North America. A protégé of Clarence Crafoord at the Karolinska Institute during these years, Senning assisted in Crafoord's first operations to repair heart malformations in 1954. Three years later, Senning became chief surgeon of the Department of Thoracic Surgery at

Karolinska. He visited Minneapolis on more than one occasion and observed Lillehei's use of the myocardial pacing wire in 1957. Back in Stockholm, Senning experimented with the Lillehei pacing technique; Elmqvist built an external pulse generator for him. Senning found, however, that almost every patient developed serious infections because of the open wound through which the pacing lead entered the body.[12]

Work on a fully implantable pacer began in 1957 with animal studies to gather information about appropriate waveforms, amplitudes, and rates. Elmqvist considered using small batteries based on zinc-mercuric oxide, but feared that the implanted unit would be too large and heavy. Mercury cells had other drawbacks, too. "I had already had experience with the mercury cells," he later wrote. He knew that the cells gave off hydrogen gas as a chemical byproduct and he did not know how it would affect the generator or how it could be vented. He chose, instead, a rechargeable nickel-cadmium battery of two cells. The pulse generator as a result was smaller than other implantable generators of that era, occupying a volume of 51.4 cc. For the leads, he chose suture wires of stainless steel and encapsulated them in polyethylene sleeves.[13]

Senning implanted the pacemaker in a human being before its design had matured, Elmqvist believed. The recipient was a 43-year-old man named Arne Larsson who was himself an electronics engineer and owner of a small firm that produced measuring instruments. Larsson had developed a severe infection that attacked his liver and heart muscle, damaging the conduction system and leaving him hospitalized with complete heart block and recurrent Stokes-Adams attacks.[14] Larsson's wife called Elmqvist and Senning daily to press them to complete their animal studies and give Larsson the implanted pacer. She got her wish on September 8, 1958. Elmqvist put together two pulse generators, molding them in epoxy resin in a shoe-polish can. Senning implanted the electrodes and generator, but the first unit failed after six hours. The next morning, when he implanted the second, it paced Larsson's heart effectively. Recharging was simple but required several hours, overnight, about every two weeks. A light coil of wire was taped to Larsson's abdomen and connected to a vacuum-tube RF generator. A current passed through the coil was transmitted by induction to the implanted pacer and would recharge the battery. Eventually, though, rising thresholds of stimulation, probably the result of fibrotic buildup around the electrode, made it impossible to capture the heartbeat. Senning removed the device after seven weeks. Larsson's condition had improved "not much, but a little," and he survived without a pacemaker until November 1961, when he underwent another procedure to receive a redesigned

Elmqvist pacemaker. In mid-2000 Arne Larsson was still living—on his twenty-sixth pacemaker.[15]

At a professional meeting for surgeons in Los Angeles in April 1959, Senning briefly described the rechargeable pacemaker but conveyed a profound sense of pessimism about it. "We have had the same experience as all other people have," he said. "After a while the resistance in the heart, or whatever it is, increases more and more, so that a higher and higher voltage is required, and when one reaches the neighborhood of 20 volts the whole patient jumps, but the heart goes [back] to its own rhythm [because the pacemaker stimuli no longer trigger heartbeats]. This takes about four, five, or six weeks. So I don't think this is the way to tackle this problem. I think we have to seek other ways to treat an A-V block for long periods." But that spring and summer, he and Elmqvist, assisted by Arne Larsson and engineers at a Swedish telephone company, redesigned the pacemaker lead and developed a small disc electrode of pure platinum.[16]

At least three and perhaps five of the redesigned, rechargeable Elema 135 pacemakers were implanted in human beings after the Larsson case. Senning has said that two were implanted in Stockholm in 1959; and an Elema unit was implanted in Montevideo, Uruguay, on February 3, 1960, in a 40-year-old woman who survived for 9 months, then died of an infection from the surgical wound. One of the surgeons in this case, Dr. Orestes Fiandra, had been able to acquire the pacemaker because he had worked in Stockholm under Senning on a training fellowship. On March 31, 1960, surgeon Harold Siddons, at St. George's Hospital, London, implanted a unit that functioned for 10 months. These remarkable cases went almost unnoticed in the United States, perhaps because Elmqvist believed that the pacemaker was still too early in its development for a formal announcement. Efforts to improve the rechargeable pacemaker from 1959 on drew Elema-Schönander further into the field of cardiac pacing. In 1960, the company introduced an implantable pacer with a nonrechargeable mercury battery, despite Elmqvist's dislike for this power source. U.S. clinical investigators knew nothing of the work going on in Sweden. Their own efforts were entirely separate, but in some respects ran along parallel lines.[17]

## PAUL ZOLL'S SEARCH FOR AN IMPLANTABLE PACEMAKER
Two research groups in the United States set out to develop completely implanted pacing systems. This design decision eventually yielded pacemakers acceptable to the relevant community of users, the cardiothoracic surgeons. From the earliest announcements of success in 1960, the implantable pacemaker gave off an aura of "inevitable next step," even though

the inevitability of this step had not been obvious a year earlier and the devices had flaws that required more than a decade to eliminate.

Designing an implantable pacemaker was not such a straightforward proposition as it may seem in retrospect, a point that the experience of the Beth Israel group in Boston richly illustrates. Paul Zoll's team began work earlier than any other U.S. research group, probably in 1956. Zoll's work on external pacing had given him unequaled practical experience with short-term stimulation; now he wanted to find out how the ventricles would respond to artificial electrical signals over weeks and months. He decided to begin by studying the effects of long-term chronic electrostimulation on heart tissue in laboratory animals. For all his inventiveness, Zoll was a cautious man and a member of a conservative department of cardiology. As his Beth Israel colleague Howard A. Frank recalled, "He would wait, wait and study, and so forth; and the environment here was not for easy entry into testing of untested things in humans."[18]

Many years later, Zoll enjoyed telling the story of visiting the Simplex Wire Company in Cambridge, Massachusetts: "When I told them we needed wires that would flex 36,792,000 times a year for 50 years they informed me that their wires would withstand only a few thousand flexions. When I hesitated in response to their question of how much wire I might need, they informed me that their wires were used in building bridges and were usually sold in multiple-carload lots."[19]

Zoll's subsequent studies of the properties of wire leads were less amusing. He knew that the pacemaker stimulus must deliver enough energy to depolarize the heart cells around the electrode. These cells would trigger their neighbors to depolarize, the impulse would quickly spread throughout the walls of the ventricles, and the chambers would contract. The minimum stimulus that will reliably induce ventricular contraction is called *the threshold of stimulation.* This threshold is not invariant for all hearts in all circumstances; the responsiveness of heart tissue to electrical impulses varies depending on a variety of factors such as the surface area of the electrode, the condition of the heart muscle, and medications that the patient may be taking. Zoll and his colleagues had to determine how the threshold of stimulation would change as electrostimulation continued week after week. Would the chronic threshold settle at a stable level, and would it be higher than the acute threshold? If possible, they had to design an electrode and a pulse generator that would be able to control the heartbeat over the long term without depleting the battery too quickly.[20] Results for the Beth Israel group were discouraging at the outset and remained so. In Howard Frank's words, they "did a great number of dogs . . . and in a high proportion of them, the initially low stimulus threshold didn't stay low. They could go back to those electrodes in the animal, and retest them and retest them, and the thresholds rose and rose, and sometimes rose to a level requiring a strength of stimulus that you couldn't deliver with the battery-powered systems."[21]

A myriad of possible causes might have contributed to the rise in thresholds. Having no theoretical reason to suspect one more than others, the group was forced to investigate numerous possibilities. They placed their experimental electrodes in several locations on or in the myocardium; they experimented with a variety of conductive materials for the lead — tantalum, stainless steel, vitallium (a conductive plastic), copper, gold, platinum. They considered electrodes of various sizes and shapes and even varied the sutures used to hold down the electrodes on the heart surface: silk, Dacron, fine Teflon-coated stainless steel wire. Almost always, they found that within one week they had to raise the amplitude of the stimulus. "At times," Zoll reported, "three- to 10-fold rises within a month were followed by stabilization below 6 milliamperes or 6 volts, which were considered the practicable limit. In other experiments, thresholds rose excessively, usually within a few weeks but occasionally not for a few months."[22] There seemed no pattern.

Zoll considered a variety of more remote hypotheses. Did the electrical current itself bring about the changes in threshold? To test this idea, the group implanted electrodes and left them in place without sending electricity through them; later, when they tried pacing through the electrodes, they found that the threshold of stimulation had risen. Would bipolar electrodes induce lower threshold levels than unipolar? Trials of both produced no discernible difference. After many months of work with animals, the group decided that inflammation of the heart tissue at the site of the electrode was the probable cause of the increases in stimulation threshold. But what was causing the inflammation?[23]

Zoll decided that the electrodes, although sterile, were "dirty. . . . There were microscopic particles on them and there was a foreign body reaction." He recalled reading a paper in a chemical journal warning that "to use this kind of material, a metal material . . . , in the body, they have to be not only sterile, but absolutely microscopically clean. [The author] said what to do. So what we did was, we developed a ritual in which we took electrodes . . . and we'd boil them in Ivory Flakes . . . and not touch them thereafter. . . . And at the operating table they would come out with a great ceremony, . . . and the surgeon would wash the dust off his gloves, you know, the talcum powder."[24] Zoll later smilingly compared this procedure to a religious ritual. It proved efficacious — or rather, a particular configuration of materials and procedures proved able to keep the threshold of chronic stimulation within acceptable limits. The lead that the group settled on consisted of two multistranded braided wires of stainless steel manufactured by the American Cyanamid Company and enclosed in a multilayered Teflon sleeve. The electrodes, two short pieces of uninsulated pure platinum wire, were not located at the tips of the two wires but were silver-soldered into the lead about four inches back. It remained unclear which factor was the crucial one in keeping thresholds within practical limits.[25]

Howard Frank was reminded of this mystery years later when he visited a surgical center that had set up elaborate sterilization procedures for all staff members in order to halt a series of infections that had run through the facility.

> You went through a series of chambers. . . . You took off all your clothes. . . . You walked through a shower and you bathed very thoroughly. And then you put on freshly sterilized stuff, and masks and whatnot. . . . And apparently the occurrence of infections went sharply down — and then they were left, they were stuck, with this technique, this technology, probably nine-tenths of which was irrelevant. That's why I think of it in connection with [implanting the pacemaker electrode]. But they couldn't stop, because they didn't know what they should give up.[26]

Along with the work on electrodes, Zoll's group was also considering the problem of pulse-generator design and the overall configuration of the pacing system for long-term stimulation of the heart. They developed a battery-powered radiofrequency transmitter and a subcutaneous receiver, but Zoll decided that there was too much risk that the transmitter would unpredictably lose contact with the implanted receiver and leave a pacemaker-dependent patient without artificial support. In fall 1958, Zoll also observed a patient at Montefiore Hospital in the Bronx who was being paced by means of a catheter electrode positioned within the right ventricle. This technique, too, struck him as unsafe; as he later wrote, it ran "risks of infection, thrombosis and embolism, hemorrhage, cardiac injury and arrhythmia, and loss of electric contact with slight movement of the wire." He also rejected implanted rechargeable batteries as "too short-lived to offer any advantage."[27]

Zoll (and other investigators) had been assured that a pacemaker battery consisting of several zinc-mercury cells would power a pacemaker for five years. On this basis, they believed that mercury would definitely free the patient "of the continuous concern with cardiac action and electrical equipment inescapable in systems that depend upon external units." He left the design of the circuitry and choice of a suitable encapsulant to his electronics expert, Alan Belgard. The result was a pulse generator that used a simple blocking oscillator circuit to transform a current flow into 70 impulses per minute. Circuitry and the six mercury cells were embedded in epoxy resin and enclosed in a steel jacket coated with Teflon. The unit weighed 170 g and occupied a volume of 66.3 cc in the abdominal cavity.[28]

The overriding theme in the work of the Beth Israel research group from 1956 to 1960 was the search for safety. Despite his accomplishments as a clinical investigator and an inventor of electronic devices, Zoll fully shared the conservative outlook of his department at the Beth Israel. An internist rather than a heart surgeon, he never displayed the bravado of his Harvard classmate and wartime associate, the flamboyant surgeon Dwight Harken. To the end of his career, Zoll maintained a small general practice. He performed annual checkups and prescribed medications for ear infections — he even made house calls. In pacing, Zoll was acutely conscious of how little he and others really understood, and he determined not to design a device whose complexity would pose a threat to the patient. Thus the four-year quest for a suitable myocardial lead. Thus, too, his single-minded focus on pacing the ventricles without trying to design in a sensing function so that the pacemaker could sense atrial depolarization and then stimulate the ventricle after an appropriate delay — an idea much discussed in the late 1950s and thought by some to be a requirement for cardiac pacing. By the mid-1960s, Zoll's conservatism would appear extreme, though it had, after all, yielded an implantable pacemaker.[29]

### A NEW RESEARCH GROUP TAKES SHAPE

A third team, this one in Buffalo, New York, was wrestling with many of the same questions as Zoll's group in Boston and Senning and Elmqvist in Stockholm. Here the key figures were William M. Chardack (b. 1914), chief of thoracic surgery at the Veterans Administration Hospital, and electrical engineer Wilson Greatbatch.

The initial impetus came from Greatbatch. Born in 1919, he had grown up outside Buffalo and served in the Pacific during World War II. Afterward, he worked as an installer-repairman with the New York Telephone Company, then earned his bachelor's degree in electrical engineering at Cornell in 1950. Greatbatch worked on the electronic instrumentation of animals at Cornell's Animal Behavior Farm and later at a private company in Buffalo. As an assistant professor of electrical engineering at the University of Buffalo at mid-decade, he was involved with several projects in the field of medical electronics and became a well-known figure in the local engineering community. He organized a chapter of the Institute of Radio Engineers (IRE) Professional Group in Medical Electronics to bring together physicians and engineers and encourage collaborative work. After completing a master's degree at the University of Buffalo in 1957, he joined Taber Instruments, where he worked with transistor modules and aerospace medical amplifiers in early NASA projects to send animals into space atop Mercury and Atlas rockets.[30]

Greatbatch probably began to think about an implantable pacemaker in 1956. At that time he was working with the first generation of silicon transistors to become commercially available; he also happened to construct a blocking oscillator circuit in the course of building a piece of equipment at the Chronic Disease Research Institute in Buffalo. Greatbatch loved to tell the story of reaching into his resistor box but misreading the color-coding and choosing the wrong resistor, with the result that his circuit "started to 'squeg' with a 1.8 ms pulse, followed by a one second quiescent interval. During the interval, the transistor was cut off and drew practically no current."[31] Later he realized that these characteristics — the brief pulse coming about 60 times per minute, and with a low current drain — would make the circuit highly appropriate for an implantable device that stimulated the heart. In his conversations with cardiologists at IRE meetings, Greatbatch probably heard about complete heart block and raised the idea of long-term pacing.[32]

He did not meet William Chardack until 1958. Chardack held his medical degree from Downstate Medical Center of the State University of New York; since 1952 he had headed the Department of Surgery at the VA Hospital in Buffalo. He had also set up "a modest surgical research laboratory" using space originally planned for the hospital laundry. Over the years, he pursued numerous research interests, including a new surgical technique for the removal of a hitherto inoperable tumor of the lung and synthetic substitutes for skin in the treatment of burns. Since 1952–53, Chardack and colleagues had investigated the flow of blood in the coronary arterial system with an eye to devising improved surgical procedures to alleviate coronary artery disease.[33] As their work proceeded, the surgeons decided that they needed a two-channel electromagnetic flowmeter to measure blood flow simultaneously in two coronary arteries and suitable for use in an operating room; a transistorized, battery-powered version would be best. Because no such piece of equipment was available commercially,

Chardack looked for an engineer to build one. He was eventually directed to Greatbatch, who was said to be unusually knowledgeable about transistorized circuits.[34]

Chardack visited Greatbatch in early 1958 and described his flowmeter problem. The two met several times, both at Taber and at meetings of the IRE group. On one of these occasions, Greatbatch asked "whether there would be some interest in the medical profession in a cardiac pacemaker, implantable, powered by the Ruben mercury cell and the circuitry being transistorized." Chardack was interested but wary: "Greatbatch's firm and persuasive conviction that the state of the art permitted the construction of a low current drain transistorized pulse generator" certainly impressed him. He took some time to consider the idea. "I was chief of a large surgical service, director of two surgical training programs, involved in teaching students, and committed to several other research projects." Although he had read about surgically induced heart block in children, Chardack had not performed open-heart surgery. As for chronic heart block in older adults, he had never seen a case. When he inquired among cardiologists in Buffalo, he discovered that no one else had seen a case, either.[35]

But when he reviewed the published literature on complete heart block, Chardack found a recent paper analyzing 224 cases of the disease. Heart block was associated with a variety of common conditions, including coronary artery disease, high blood pressure, and previous heart attacks. Of the 224 patients, 127 were known to have died; 39 of these had died suddenly, probably in Stokes-Adams attacks.[36] The numbers suggested that heart block was more common than it appeared. "The disorder with its ominous prognosis had taken its toll well before the patient was referred to the specialist. Since no effective therapy was available, the level of diagnostic suspicion was low." After thinking it over, Chardack decided to set aside some of his time and resources to work with Greatbatch on the pacemaker. From the start, the project was a joint venture. The initial idea and enthusiasm definitely had come from the engineer, but the surgeon recognized the possibilities of this idea and agreed to go ahead in the face of considerable skepticism among local heart specialists.[37]

Once he had decided to accept Greatbatch's invitation, Chardack spent several days mastering the established surgical technique for creating heart block in laboratory dogs by cutting the bundle of His or squeezing it in an encircling suture. He and his colleague Andrew Gage purchased a General Electric stimulator and an oscilloscope. Feeling that he was "a fairly inexperienced investigator," Chardack did not organize a set of formal studies of stimulus strength and duration as Zoll had insisted on doing. Instead, using stainless-steel wires coated with Teflon, like the ones Lillehei had been using, he and Gage essentially replicated earlier work on stimulation thresholds as published by the physiologists Brian Hoffman and Paul Cranefield in 1950. In April, hoping to benefit from the physiologists' experience with electrodes for electrostimulation of the heart, Chardack visited Hoffman and Cranefield at the Long Island College of Medicine. They had an electrode that resembled "a little plastic button," but it wasn't suitable for an implanted pacemaker. "They were only concerned with short-term experiments, and then the thresholds deteriorated." Chardack returned to Buffalo and his myocardial wires.[38]

Greatbatch, for his part, began to split his time between Chardack's crowded laboratory at the VA Hospital and the small workshop in the barn behind his house. In the lab, Greatbatch designed a simple circuit connecting the oscilloscope to the stimulator and electrodes. On April 16, 1958, he assembled his first pacemaker breadboard, using a germanium transistor, and tested the electrical properties of the blocking oscillator circuit with a mercury-cell battery.[39] He never doubted that the best power source for the pacemaker would be several of the mercury cells that Samuel Ruben had developed during World War II to meet the demands of the U.S. military for a small, reliable battery to power "walkie-talkies." Mallory & Company, of Indianapolis, had brought the mercury cell to the civilian market in 1947.[40] Greatbatch was impressed at that time with its small size and high energy density, but was concerned that it would be subject to significant internal self-discharge at body temperature. Still, no other battery seemed remotely suitable for implantation. The only possible alternative would have been a rechargeable nickel-cadmium (NiCad) battery, but Greatbatch believed that, even allowing for recharging of the NiCads, the nonrechargeable mercury cell would actually outlast nickel-cadmium.[41]

April 1958 was just six months after the launch of the Soviet Union's Sputnik I satellite, and Greatbatch fell to calling these early pacemaker models his Tikniks. He bench-tested each model in a warm environment to emulate conditions within the body. At first, the units would fail within a few hours, and Greatbatch would track down the problem and substitute different components. When model 4 (April 20) pulsed for 14 hours and model 5 (April 25) for 24 hours, Greatbatch and Chardack felt ready to try pacing a dog. This brought to the fore the problem of how to seal the pulse generator from the hostile environment of the body. For their earliest animal experiments, the inventors wrapped the circuitry and mercury cells in electrical tape and dipped the assembly in epoxy.[42]

On May 5, Chardack created heart block in a dog, implanted a myocardial electrode, and attached it to Tiknik 6; but because of a component failure, the unit quickly ceased functioning. Two days later, a slightly rebuilt pulse generator (model 7) was implanted and it paced the same dog for 24 hours. An elated Greatbatch wrote in his lab notebook, "Dr. Chardack suggests similar excitation of phrenic nerve to get polio patients out of [iron] lungs!"[43] In 10 weeks, the inventors had identified the key problems that they must address and—although it was probably not yet completely clear to them—had found a workable circuit and a power source and had made some progress on other roadblocks. They had developed a prototype of their implantable device that functioned effectively for hours within the body of an experimental animal. But they were still a long way from having a pacemaker appropriate for a human trial. They must now prove that their device could pace dogs not just for hours, but for months.

## CRITICAL PROBLEMS: ENCAPSULATION AND A PACING LEAD

From May 1958 through January 1959, Greatbatch continued to experiment with various circuit components. But as they worked their pacemakers' longevity up from 24 hours to several weeks in dogs, he and Chardack began to redefine the critical problems facing them. They turned from circuitry and the battery to encapsulation and the baffling problem of rising thresholds of stimulation.

Until the 1970s, implantable pulse generators could not be hermetically sealed: they had to be gas-permeable to allow for the escape of hydrogen, a byproduct of the chemical reaction in the mercury cell. The inventors had to select materials that would permit the hydrogen to exit the pulse generator while still shielding the electronic components both from the corrosive electrolyte of the battery and from body fluid. As Greatbatch explained:

> We found that gas (including water vapor) would readily diffuse through both the epoxy and the silicone sheath around it, raising the interior to an eventual 100% humidity. Thus our electronics were essentially operating under water, but distilled water. After all, the anodes of powerful radio transmitting tubes operate at thousands of volts with cooling water in direct contact with the anodes. This is safe, but should a pinch of salt get into the cooling system, the transmitter will go through the roof! This meant that we had to keep all salt ions out.[44]

A sealant would have to flow around the circuitry and then harden. Chardack and Greatbatch chose Scotchcast V, an epoxy from Minnesota

Mining & Manufacturing (3M) that was inert and body compatible and would firmly keep the components in place. Unable to mount their components on a circuit board because the epoxy would not bond to it, they fixed the "floating" circuit in an epoxy block. They used epoxy to glue the electronic components together, hand-soldering the circuit and shaping it to fit between the ten cells of the battery, positioning it, then encasing the entire assembly in still more epoxy. After curing overnight at body temperature, the generator was given two outer coats of Dow Corning silicone and the drips were trimmed off with a scalpel.

According to Greatbatch, "We studied various silicone coatings. . . . We looked at [Dow Corning] RTV 501 . . . which handled a little better than RTV 502, but which had mercury in its accelerator. We settled on the RTV 502 after deciding that its tin accelerator was safer. Many such judgments had to be made off the seat-of-our-pants because there just wasn't much data around on the long-term behavior of materials in the body." By early 1959, the partners had agreed on a solution to the encapsulation problem, and they did not deviate from it in their first human implantations. They continued to experiment with square and circular configurations for the pulse generator before settling on an egg shape in which the 10 mercury cells formed a lopsided ring, the circuitry being in the center.[45]

Designing an electrode for chronic pacing proved as frustrating for the group in Buffalo as for Zoll's group in Boston. Indeed, by 1959 it was clear that the entire effort hinged on this question of finding a stable means of delivering the electrical stimulus to the myocardium. Chardack and Greatbatch began with a wire like the one in use at Minnesota. Like Zoll, they tried numerous metals and configurations in an open-ended effort to come up with something that would eliminate the problem of rising thresholds of stimulation. "We tried stranded wire, solid wire, silver wire, stainless steel, orthodontic gold, and platinum and its alloys." This went on for about a year.[46]

"If I had been an experienced investigator," Chardack reflected, "I would have doubled the amplitude and let it be." In other words, he and Greatbatch could have chosen to set the stimulus strength well above the minimal requirements to capture control of the heartbeat; this would have overwhelmed any rise in threshold. But this seemed an inelegant solution and would deplete the battery too quickly. Instead, "what we did was improvise electrodes." Threshold testing required that the investigator gradually reduce the amplitude of the stimulus until it lost capture of the heartbeat, so it made sense to use an external pulse generator with an amplitude control.

The Buffalo group bought several of the Medtronic 5800s, the battery-powered external pulse generator that engineer Earl Bakken had invented for Lillehei early in 1958.[47]

Sometime in the second half of 1959, Medtronic announced to a few clinical research groups that the Hunter-Roth lead with its platform electrode was available. Norman Roth and a Medtronic colleague visited Paul Zoll in Boston, then stopped in Buffalo to introduce Chardack to the new lead. Zoll was interested but cautious. Zoll twice sent his associate Arthur Linenthal to St. Paul to examine Hunter's patient Warren Mauston, but never used the Hunter-Roth lead. But Chardack and Greatbatch accepted as reliable the results that Hunter and Roth had gotten from months of tests in animals as well as Hunter's first few patients: thresholds of stimulation rose during the first three weeks of pacing, then stabilized at amplitudes between 4 and 14 milliamperes.

When the Buffalo team conducted a controlled trial on a small number of animals, pacing some with the myocardial wire and others with the Hunter-Roth lead, the results confirmed what Hunter and Roth had reported. "This represented a drastic and significant reduction in threshold requirements over our previous unipolar work," they wrote. "In none of our bipolar experiments have we seen the threshold rise above 4 or 5 ma [milliamps] and in some animals we have seen thresholds of 0.5 ma after as long as five weeks." They believed that the bipolar configuration of the electrode was the key factor, but it is more likely that the silicone rubber insulation on the Hunter-Roth lead lowered the risk of infection and the silicone rubber platform reduced the buildup of scar tissue by stabilizing the electrode pins against the constantly moving heart wall.[48]

By late summer 1959, Greatbatch and Chardack had solved what they understood to be the critical problems standing in the way of permanent pacing of the heart with a fully implanted pacemaker. The device they had invented used a blocking oscillator circuit that triggered a transistor switch to open and close. Parameters were still somewhat fluid: Greatbatch continued his experiments with the circuitry and mercury cells; hence, the number of impulses per minute and the pulse width and amplitude changed from one model to the next. He finally settled on a battery of 10 cells. Potted in epoxy and coated with silicone rubber, the pulse generator occupied a volume of 57.3 cc and weighed 159 g. The all-important connection for the wires of the Hunter-Roth lead consisted of two protruding sockets in the side of the pulse generator, with set screws to hold the wires in place.[49]

With a stable design for the pulse generator, an acceptable lead, and experience in using the system to pace dogs for up to four months, the moment for a clinical trial loomed before the partners. Chardack had decided on a two-stage surgical procedure: "Operate on the patient, place the electrode, bring it out through the skin, make repeated threshold measurements, and then after four to six weeks, attach the implanted unit, soldering the stainless steel filament, and cover it with a silicone preparation known as medical adhesive, let it cure on the outside for a few days since a curing process produces acetic acid. Then we would sterilize this as best we could, and implant it."[50]

## AN IMPLANTABLE PACEMAKER: 1960 AND AFTER

In September 1959, less than 18 months after Greatbatch's first breadboard assembly, Chardack carried out stage one of the planned procedure by implanting a Hunter-Roth electrode on the right ventricle of a 65-year-old man in complete heart block with a slow ventricular rate and Stokes-Adams attacks. The surgeon brought the lead through the patient's skin and attached it to an external pulse generator. The patient wore this unit for five weeks — then, a day or two before he was to receive the implantable generator, the patient had a heart attack and died. If he had died a day or two after the implantation rather than before, the death would probably have been ascribed to some defect of the pacemaker.[51]

Chardack and Greatbatch did not have another opportunity for six months. Finally, in April 1960, they were introduced to Frank Henefelt, another sufferer from heart block, age 77, who wore a football helmet when walking around his house to protect himself when he blacked out in Stokes-Adams attacks. On April 18, Chardack carried out stage one and on June 6 he implanted the pulse generator. After the months of difficulties with pacing thresholds in dogs and the sudden death of the first patient, Henefelt's case was entirely uneventful. He recovered normally and lived for 30 months with the implanted pacemaker.[52]

After the implantation, the inventors immediately began to spread the news of their clinical success at medical meetings and in the engineering community. In July, Greatbatch filed for a patent on the implantable pulse generator and it was duly awarded.[53] In October, Chardack reported on the pacemaker in the lead article in the journal *Surgery*. Physicians and engineers quickly came to think of William Chardack and Wilson Greatbatch as the inventors of "the" pacemaker, as if there had been no others. In 1984, the National Society of Professional Engineers selected the implantable pacemaker as one of the 10 most important engineering contributions to society over the preceding half-century and honored Greatbatch as its inventor.[54] Medtronic's dominant position in the world pacemaker industry dates from the 1960s, when the company manufactured a succession of devices vetted by Chardack and Greatbatch and bearing the name Chardack-Greatbatch Implantable Cardiac Pacemaker. If theirs was not the only pacemaker of the 1950s, it appears to be the only one that survives today in the collective memory of the community of the physicians, engineers, and businesspeople whose careers are tied to the pacemaker.

The rechargeable pacemaker invented in Stockholm had been clinically effective in at least two cases and had certainly preceded the model developed in Buffalo. The Chardack-Greatbatch pacemaker stood out from other prototype implantables of the late 1950s not because it was first or clearly a better design, but because it succeeded in the U.S. market as did no other device. In October 1960, the inventors licensed their creation to Medtronic and at the end of the year the company began to manufacture it in small numbers. From the beginning of 1961 through August 1963, a shakedown period in which implanted pacemakers often failed and major improvements were rapidly introduced, Medtronic sold 2,178 Chardack-Greatbatch pacemakers; the price was more than $600 per unit.[55] Models commercialized by Electrodyne and General Electric around the same time shared the basic features of the Medtronic pacemaker; notably, the mercury battery. In Stockholm, Rune Elmqvist also decided at the beginning of 1960 to switch from nickel-cadmium to mercury; hence, the Elema model 137, released later that year, also resembled the Medtronic device. All manufacturers in fact were converging on the essential design features that Chardack and Greatbatch had pioneered, but Medtronic was best positioned to dominate the new market for implantable pacemakers.

Both Zoll's pacemaker and the Chardack-Greatbatch unit paced the heart to the standards of reliability possible at that time. Zoll's group in Boston had begun with far more experience in designing and using devices that delivered electrical impulses to the heart, yet took four years (compared with the two years that Chardack and Greatbatch required) from inception to clinical success. Explaining why one course of events unfolded rather than another is always treacherous; it is easy, after the fact, to list any number of reasons. Still, it is striking that the Buffalo team seemed more trusting of data that other groups had reported, and more inclined to use components invented elsewhere; Zoll, on the other hand, spent much laboratory time essentially replicating the work of others to be certain that it was right. Zoll developed an explanation for the threshold problem that burdened him with implanting only the cleanest and purest of electrodes. Chardack and Greatbatch, in contrast, tested the Hunter-Roth electrode, found that it gave them stable thresholds, formulated an ad hoc explanation for this success, and accepted the Hunter-Roth for their pacemaker. The Buffalo partners wanted to complete the pacemaker and were prepared to let the definitive understanding of the threshold problem emerge later.

By standards of safety and effectiveness that prevail today, neither the Chardack-Greatbatch pacemaker, Zoll's unit, nor any other could be considered a reliable invention at the time of first clinical use in 1960. The FDA, had it regulated implantable devices at that time, would have required years of additional testing with animals and human subjects before permitting manufacturers to release the pacemaker for general use. Of the first 16 clinically implanted Chardack-Greatbatch pacemakers, 10 had wire breakages, 5 had premature battery failures, 1 had a failure in an electronic component, and 3 caused infections. After these 16 procedures, Chardack, in April 1961, abandoned the Hunter-Roth lead in favor of a new design that proved more reliable (see chapter 5). In Zoll's first 14 cases, only 4 were uneventful: 3 pacemakers failed because of broken or ineffective electrodes; 1 pulse generator eroded through the patient's skin four months after implantation; several patients had severe inflammation around the generator or the electrode; and several units failed after a short while because tissue fluid leaked into the generators and caused battery failures or short circuits. These were inventions in the early stage of product development. The only justification for using them at all with human beings — a compelling one — was that the patients had little chance of survival without some electrical assist for their heartbeats.[56]

The implantable pacemaker emerged out of the new and dramatic practice of heart surgery. Thoracic surgeons devised the standard implantation procedure and held a near-monopoly on the implantation of pacemakers for many years. The dominance of surgeons was inevitable given the design of the pacemaker, a device that required the surgical opening of both the abdomen and the chest. Some years later, smaller pulse generators and leads introduced to the inside of the ventricle through a vein would make it possible for cardiologists to implant pacemakers in catheterization labs, but surgeons retained their dominant position in pacemaker work until the 1980s. As late as 1993, when the catheter lead was almost universally used, an estimated 38 percent of pacemaker implanters were surgeons.[57]

More broadly, the pacemaker and the practice of implanting pacemakers were products of the 1950s and of the immense prestige that interventional medicine enjoyed. Established companies played a secondary role in the invention of the implantable pacemaker. Electronic engineers such as Bakken, Elmqvist, and Greatbatch actually built the pacemakers of that era, but they designed to the requirements set by physicians. The standard pacemaker design that established itself in the market during the early 1960s embodied physicians' choices on three crucial points: first, the device would be fully implanted inside the patient's body so as to reduce the risk of infection and the risk that patients or others might mess something up by manipulating the external part of the device; second, the battery would not be rechargeable because this would require the patient to present himself every few days at a charging station, and because the longevity of nickel-cadmium batteries had not been shown to surpass that of mercury batteries; third, the pacemaker would abandon the idea of AV synchrony by ignoring the atrium and pacing the ventricle only. All of these fundamental choices were intended to reduce the number of ways the device could fail, to minimize the doctor's or the hospital's responsibility for managing the case after implantation, and to foil the patient's interest, if any, in playing doctor.

## CHAPTER 5

MAKING THE PACEMAKER
SAFE AND RELIABLE

The earliest pacemaker implanters were often stunned at their patients' sudden and dramatic improvement. At a medical meeting around 1963, a well-known cardiologist told surgeon Victor Parsonnet that he doubted whether pacing was an effective treatment. Parsonnet "tried to explain that once having seen a patient in Stokes-Adams seizures immediately regaining consciousness when a paced rhythm returned, turning from a cadaveric white to a normal pink," no one could question the effectiveness of the new technology.[1] Parsonnet was confident that implantable pacemakers would eventually be able to prolong many patients' lives. But in 1961, the invention was still at an early stage of development. Its chief appeal lay in its promise of a definitive solution, a "cure," to the patient's rhythm disorder. Before it could realize this promise, the pacemaker would require some improvements.

By the standards of later years, the earliest implantable pacemakers appear large and ungainly. They delivered stimuli at a preset rate, regardless of any intrinsic electrical activity in the heart. The pacemaker pulse was typically several times stronger than required. "The objective was simply to drive the heart," Wilson Greatbatch later wrote, "without much regard for economy of battery life" or other refinements.[2] Virtually all of the early patients were pacemaker-dependent and faced a serious risk of sudden death if the implanted devices failed. And often they did fail.

Clinical experience in 1961–63 revealed that all pacemaker models still had serious design flaws — a common experience with new medical technologies, which often go into regular clinical use at an early stage of development. Extensive real-world experience may be necessary before physicians and manufacturers can pinpoint flaws and decide how best to correct them. Physicians had hoped that the battery-powered pulse generators would function for three to five years, but to their dismay the generators often failed suddenly within a few months.[3] The wire leads would sometimes break within the patient's body, and the pulse generators occasionally

TABLE 3. IMPLANTATIONS REPORTED, JANUARY 1961–AUGUST 1963

| Manufacturer | Number |
|---|---|
| Medtronic, various models | 2,178 |
| Electrodyne TR-14 | about 1,000 |
| General Electric, various models | about 500 |
| Cordis Atricor (from June 1962) | about 100 |

Sources: Ann N Y Acad Sci 111 (11 June 1964): 1063, 1068, 1088; Sol Center et al., "Two Years of Clinical Experience with the Synchronous Pacer," J Thorac Cardiovasc Surg 48 (October 1964): 513, 524.

stopped firing for no apparent reason. At best, the pacemaker batteries seemed to provide power for only 18 months or so. The unpredictable shutdown of life-sustaining devices implied many emergency replacement procedures. One surgeon recalled that "it was always Saturday, my wife was always in the midst of serving dinner to twelve guests, and Mr. Jones was convulsing in the emergency room because his pacemaker had broken." Every pacemaker implanter faced a succession of broken wires, generator failures, and other forms of sudden failure. Yet few lost confidence in the basic Chardack-Greatbatch design: that issue was settled. The agenda that investigators now set was not to invent new pacing systems, but to improve the existing one.[4]

A mere handful of physicians and engineers defined the critical problems in pacing, invented new components and procedures, and spread the word about what would work and what would not. As these pioneers talked and wrote about their experiences, more doctors took up pacing. Many of the first generation of pacemaker implanters were strongly oriented to research and publication, but over time a growing number of ordinary practitioners also entered the field. Out of the intense sharing of experience came a strong sense of membership in a special group, a community of physicians who were collectively trying to understand and perfect a dramatic but still flawed technology (see table 3).[5]

Once they had addressed the most severe pacemaker problems, the pioneers turned to more subtle ones. A consensus began to grow that opening the chest and exposing the surface of the heart to install the myocardial electrode put the patient at unnecessary risk. By the late 1960s, an alternative route to the heart had come into use. It also became apparent that routine checks of the pacemaker's functioning could warn of impending failure and enable the physician to perform a scheduled, rather than an emergency, replacement. Physicians also reached a consensus that the metronomic pacemaker stimuli could in certain circumstances kill by sending the patient into ventricular fibrillation, and engineers came up with a solution to the problem.

## BROKEN WIRES AND THE COILED-SPRING LEAD

Chardack and Greatbatch had chosen the Hunter-Roth platform electrode for their early pacemaker patients on the basis of animal experiments and news of the Mauston case. Then, in 10 of Chardack's first 15 patients after Frank Henefelt, the wire lead broke. Chardack found that it was often easy to diagnose a broken wire because pacing became intermittent, resuming when the two broken ends happened to come into contact. By having the patient change posture, he could demonstrate cessation and resumption of stimulation. But while it was easy to spot, a broken wire was obviously a grave problem because it both deprived the patient of pacemaker support and entailed major surgery to remove the failed lead and implant another.[6]

The wire broke because it was always in motion — always flexing as the heart beat. Chardack recalled that "we made some [filmed] radiological studies. . . . The motion was almost invisible; it was just a little motion at crucial points, like where the wire was soldered to the pin." He thought that a wire lead would be more reliable if it were coiled like a spring and lacked any welded or soldered junctions of dissimilar metals. He needed a strong but nonirritating metal. "Platinum [was] the least irritating metal that you could put into the body. But it is also a very soft metal." A small company near his lab manufactured wire products. When Chardack paid a visit, a salesman suggested to him, " 'Look, why don't you use our orthodontic spring? It's the proper caliber, it's used by dentists all over the country, and it's well tolerated in the oral cavity.' I asked about the composition; it was nickel, cobalt, copper . . . and lead." Chardack decided instead on an "80–20 mixture of platinum and iridium to provide some flex resistance."[7]

A friend on the engineering faculty at SUNY-Buffalo "constructed a wheel with little spokes and a jig, and we would test out the flex resistance by simply spinning the wheel. We could very easily simulate . . . years of exposure within a day or two." The coiled-spring lead proved highly resistant to breakage. Borrowing a design feature of the Hunter-Roth electrode, Chardack added a silicone platform, or plate, that stabilized the tip of the lead against the outside wall of the ventricle. He allowed the coiled wire to ride freely within its silicone rubber sleeve and learned to leave an extra loop in the body to relieve strain on the wire.[8]

Implanted in a human being in April 1961 and released to the market almost immediately as Medtronic's model 5814, the coiled-spring myocardial lead was the first reliable pacemaker lead. After two and a half years, Chardack was able to report that there had been only 17 failures, and in those 17 only a single documented case of breakage, in 2,100 implanted leads. Other manufacturers soon imitated the coiled-spring configuration, so that Chardack's design became the starting-point for later lead designs in pacemakers. He had been able to come up with the new lead quickly because he could take certain aspects of the problem for granted and focus intensely on others. He relied on existing knowledge about resistance to electrolysis and corrosion in electrodes made of various metals and concentrated his work on mechanical breakdown as the central issue. The coiled-spring lead embodied all that the pacing community had learned about leads to that point. It opened the way for cardiac pacing to move from an experimental to a widely accepted medical treatment. As Earl Bakken later said, "what finally made pacemaking practical was that spring-coil electrode."[9]

## SEARCHING FOR A BETTER POWER SOURCE

At Newark Beth Israel Hospital, "the Beth," in New Jersey, a pacemaker research group took shape during the early 1960s. The prime mover was Victor Parsonnet (b. 1924), a vascular surgeon whose great-uncle, Aaron Parsonnet, had coauthored a textbook on electrocardiography with Albert S. Hyman in the 1920s.[10] In 1960, Victor Parsonnet was already an established figure at Newark Beth Israel. "When pacing came along," he later said, "it kind of intrigued me. I'd always been interested in cardiology." He read the early papers, including Chardack's announcement of the implantable pacemaker, but the report that "really made an impression was the one that [Arthur] Linenthal and Howard Frank and Paul Zoll wrote." Parsonnet knew Frank, the surgeon who implanted the Zoll model in July 1960, because Frank had been one of his preceptors during his internship year in Boston in 1947-48. Parsonnet arranged to spend a few days under Frank's tutelage late in 1960 "to watch the implants and learn their technique." Back in Newark, Parsonnet carried out his first implantation at the Beth in January 1961.[11]

By the time he had done four or five procedures, Parsonnet was considered an expert. A few months after his initiation into pacemaker work, he had a call from Dryden Morse, at Passaic General Hospital. Morse asked Parsonnet to "send somebody over" to help him implant his first pacemaker, but nobody was available from Newark Beth Israel. So Parsonnet explained the technique to Morse over the telephone. Morse recalled that after the call "we put wires on the skin of a patient, an old man, and paced him through the skin while we opened the chest and put wires into the heart and put a pacemaker in, according to the Parsonnet technique."[12]

Finding that others at the hospital considered him a cardiac surgeon because he did pacemakers, Parsonnet arranged in 1962 to go to Baylor College of Medicine in Houston on a training fellowship and there apprenticed in heart surgery with Michael E. DeBakey and Denton A. Cooley. In the 1960s, Parsonnet carried out hundreds of pacemaker procedures and gained experience with every kind of pacing device then available. His hospital soon earned a reputation as a cardiac-pacing research center of international importance.

Troubleshooting pacemakers and trying to improve their performance required some expertise in electronics. Parsonnet brought in engineer George H. Myers from the nearby Bell Telephone Laboratories: the two men saw themselves as helping to introduce electronics into the practice of medicine. To their older arrays of instruments and machines, hospitals in the 1950s had added modern ones such as flowmeters and monitoring systems. Computers, automated laboratory analysis, and the pacemaker were harbingers of more exotic technologies to come. Surgeons at the Beth since 1957 had routinely monitored the heartbeat electronically on every patient undergoing general anesthesia. In 1963, with funding from NIH, the hospital installed a sophisticated instrumentation system for heart surgery. The new instruments displayed numerical values for blood pressure, heart rate, and other physiological indicators every three seconds throughout an operation. Meanwhile, an oscilloscope displayed analog data such as the electrocardiogram and pulse-pressure waveforms.[13]

For Parsonnet, "the big problem" hindering the widespread acceptance of pacing was the pacemaker power source. One firm, the T. R. Mallory Company, produced the mercury-zinc cells and supplied all the manufacturers. Doctors had been impressed with the high energy density of the Mallory cells and believed that a mercury battery would function for five years or more, but in actuality the typical pacemaker lasted about 18 months — if that. The cells lost nearly as much of their charge through self-discharge as from pacing, but what concerned doctors more was that many failed unexpectedly because of internal short circuits or damage to the circuitry of the pulse generator as the corrosive liquid electrolyte (sodium hydroxide) migrated out of the battery. Because the cells emitted hydrogen gas as a byproduct, the pulse generator could not be sealed hermetically against the intrusion of body fluids.[14]

The manufacturer made incremental improvements, but unhappiness

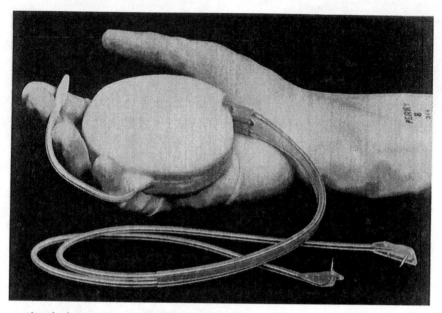

**A Chardack-Greatbatch Implantable Pacemaker of the Early 1960s.** Here is the Medtronic 5850, with the coiled-spring myocardial electrodes invented by William Chardack. The pulse generator was encapsulated in silicone rubber, weighed 140 g, and occupied a volume of 21.4 cc. The "subcutaneous extension" on the left, known affectionately to implanters as the "pigtail," was intended to give the doctor a means of controlling the pacer if it unexpectedly misbehaved. Positioned just beneath the patient's skin, the extension contained three wires. In the event of a problem with the pacemaker, the surgeon could make a small incision and bring the tip of the extension out of the patient's body. Connecting wires A and B raised the pacemaker output; connecting B and C disabled the pacemaker. It was also possible to pace temporarily from outside the body by connecting wires A and C to an external pulse generator. (Courtesy of Medtronic, Inc.)

with the mercury-zinc cells also induced a flurry of inventive activity involving a range of different concepts. Parsonnet's group decided to "look for something that has a more infinite source of energy [than a chemical cell], such as biologic energy. The human replenishes his energy source by eating. And the heart keeps beating because you supply it with essential nutrients." The group looked for "ways we could harness natural energy and convert it to electrical energy, enough to provide power for an electronic circuit. . . . We sort of dreamed up various things we could use."[15]

One possibility was temperature differentials within the body. "The temperature in the liver is something like . . . 102 degrees Fahrenheit; the temperature of the skin on the back of the neck may be closer to room temperature. You get the cold junction of a thermocouple close enough to the

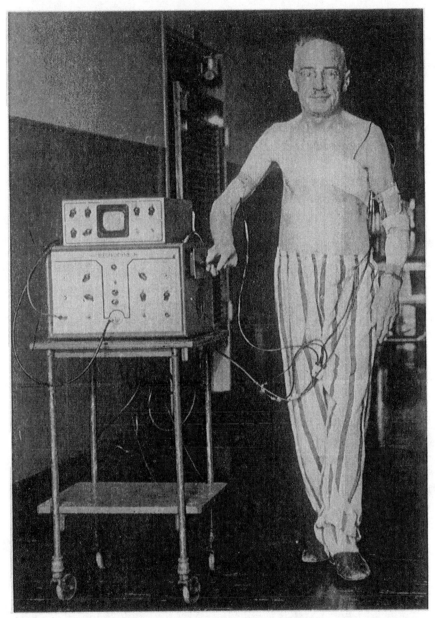

**P.S. Walking the Hallway of Montefiore Hospital, Fall 1958.** P.S. was the second patient ever to receive pacemaker support for the heartbeat via a transvenous lead, the invention of Dr. Seymour Furman while he was in surgical training. The cart holds an Electrodyne PM-65 external pacemaker-defibrillator. Resting atop it is an Electrodyne heart monitor with a small oscilloscope. Both machines were developed by engineer Alan Belgard in collaboration with Paul Zoll and his colleagues. The pacing lead enters a vein at the inside of the patient's left elbow. With the external pulse generator and his transvenous lead, P.S. was paced intermittently for 96 days. (Courtesy of Dr. Seymour Furman)

skin. . . . We entertained that possibility. But the temperature differential wasn't enough to provide power." The group next tried converting some mechanical process within the body to electric power. "What's moving in the body that would produce energy? One thing that is moving, obviously, is the heart." The group wondered whether it might be possible to tape a physical contraption to the heart and convert its motion to electrical energy. But they felt that they did not know enough about how one point on the surface moved in relation to other points, so they looked for other motion within the body.[16]

They chose expansion of the thoracic aorta as a possible power source. One member of the group knew a little about piezoelectric crystals; a second was an amateur jeweler. In their first attempt, the team obtained some matchstick-sized piezoelectric transducers, made of a ceramic material, and soldered the ends of several strips to a copper ring. They tapered the ceramic strips to the diameter of a dog's aorta so that the strips would touch the vessel without constricting it. After bench tests with simulated aortas, they took the device to the animal lab. They severed the aorta on a dog, slipped two of the rings over the aorta, and repaired the vessel. With each heartbeat, the motion of the aorta deformed the ceramic strips, and this motion produced electrical energy. Eventually the device proved capable of energizing a simple pulse generator and pacing the heart in experimental animals. But problems remained, especially the problem of how to seal this apparatus. The group experimented with silicone jackets and other ways to keep body fluids from short-circuiting the invention.[17]

In a second version, Parsonnet's group employed two thin blades of a piezoelectric ceramic that were hinged together at one end like a clothespin. They clamped this apparatus around the descending aorta. The vessel's expansion and contraction moved the ceramic pieces, thus generating electricity. The electricity was transmitted to capacitors encapsulated in the pacemaker, and a transistor circuit released the stored energy in pulses. With this unit, the group paced animals' hearts for as long as 25 days at a time. The group also considered other ideas such as pacing off the motion of the diaphragm using a mechanism like a watchspring (designed and supplied by the Bulova Watch Company). As the spring unwound, it turned a cam, whose teeth snapped against wafers of piezoelectric material to produce output pulses. The researchers found, however, that the diaphragm eventually became weaker and had a diminished excursion. All this meant "years of fun" for the investigators, but by about 1967 they had recognized that the avenue they had chosen was not likely to yield a clinically reliable way to power a pacemaker. The test devices produced only limited amounts of power, and since they involved moving parts, members of the group wondered how reliable they would prove to be over time.[18] But by then, Parsonnet had thought of a new approach to the problem.

## PACING GOES NUCLEAR

In 1965, Parsonnet approached the U.S. Atomic Energy Commission (AEC) with the suggestion that the agency oversee the development of a radioisotope, or "nuclear-powered" pacemaker, and the AEC proved receptive. After a bidding process, the AEC awarded a contract to the Nuclear Materials and Equipment Corporation (Numec), of Apollo, Pennsylvania, to design and build a prototype nuclear-powered pulse generator modeled on an established, commercially available generator, but with a 10-year pacing life. (Numec was acquired in 1969 by Arco Nuclear, a subsidiary of Atlantic-Richfield.) Other companies followed with designs of their own, and eventually seven firms (four of them U.S.; three European) introduced nuclear pacemakers that met safety standards established by the International Atomic Energy Agency.[19]

The Numec/Arco device used a thermoelectric process. A tiny metallic slug of radioactive plutonium-238 was triply encapsulated inside the pulse generator. The plutonium emitted alpha particles (helium ions) that bombarded the walls of the inner capsule, heating it to 200 degrees Centigrade. The design included an ingenious thermocouple arrangement to convert this heat to an electrical current: hundreds of thin thermoelectric wires were woven into cloth tapes, six of which wound around the fuel capsule. The thermocouple stepped up the output voltage to five volts without the need for a DC-DC converter.[20] The entire power-generation unit was enclosed in a corrosion-resistant titanium case. Only the lead from the thermopile passed through the case to supply the electronic components of the pulse generator.

A human subject first received a nuclear pacemaker on April 27, 1970, in Paris. This was not the Numec/Arco design but the somewhat different model jointly developed by a French firm, Alcatel, and Medtronic. Nuclear pacing came to the United States in April 1973, when Parsonnet implanted 15 of the Arco pacemakers over a two-day period. Both events generated intense press coverage.[21]

The 165 milligrams of plutonium in an Arco pacemaker was completely enclosed and shielded. Developers estimated that the danger of radiation poisoning was smaller from an intact nuclear pacemaker than from the radium dial on a wristwatch. But the extreme toxicity and long half-life of plutonium (86.4 years) carried a more serious risk than radiation. Completely aside from its radioactivity, plutonium is one of the most poisonous substances on the face of the earth, and the tiny capsule in a pacemaker would remain toxic for centuries. Would the containers burst and spill the plutonium if the person carrying the implant were in an auto accident or a plane crash? What about the burial or cremation of a deceased patient with an implanted nuclear pacemaker?[22]

The nuclear pacemaker was a straightforward and effective technology, it solved the problems of the mercury cell, and it had been cleared by the AEC. Viewed from a narrowly technological perspective, the nuclear pacemaker did not fail: it was, as Wilson Greatbatch observed, "the most reliable pacing system ever built." The cost of nuclear pacing over the lifetime of the average patient was theoretically lower than with nonnuclear pacing because a nuclear pacer would rarely need replacement. Yet by early 1979, fewer than three thousand had been implanted worldwide.[23]

From the start, the nuclear pacemaker faced some unique problems that it proved unable to overcome. The Nuclear Regulatory Commission in the United States (the NRC, successor to the AEC) and similar agencies abroad established stringent licensing requirements for manufacturers and implanters. Pacemaker centers had to provide extensive documentation about each implantation, about all follow-up checks of the pulse generator, and about any accidents. These requirements discouraged many hospitals from using nuclear pacemakers and slowed the development of improved models. Most countries regulate the transport of nuclear materials, and some extended this practice to nuclear pacemakers. Thus patients planning to travel outside the United States sometimes found that they had to register their pacemakers in advance. These regulations, intended to guarantee the safety of nuclear pacemakers, probably sent just the opposite message to the public. Fear of things nuclear did not abate but grew throughout the 1970s, reaching a climax in the spring of 1979 with the accident at the Three-Mile Island nuclear power plant in Pennsylvania. Anticipating patient resistance, physicians and hospitals grew less willing to suggest nuclear pacemakers.

"If we were still saddled with the two-year life of mercury pacemakers," Greatbatch commented, "all pacemakers would have become nuclear-powered by now." But by the mid-1970s, nuclear was competing with a new kind of pulse generator that combined lithium batteries, low current drain, hermetic encapsulation, and more advanced circuitry. The lithium battery showed promise of functioning for 8 to 10 years. In an era of rapid progress in pacemaker technology, might the nuclear pacemaker deprive the patient of the benefits of later technological advances? The nuclear generator worked to perfection — but progress in other pacemaker components rendered it unnecessary. Public uneasiness about nuclear safety and environmental pollution might have doomed the nuclear generator, anyway, but professionals in pacing had reached their own verdict by the late 1970s.[24]

A pacemaker relying on bioenergy never made it all the way from laboratory bench to clinical practice. Nuclear pacemakers did come into use, but eventually the pacing community showed a preference for other designs. Both rejected systems proved to be incompatible with broader developments in the technology of pacing. In the competition among radically different power sources, both technical and nontechnical considerations proved important. Doctors were of course highly interested in the technical properties of the nuclear battery, but their judgments about future developments in pacing technology and even about public attitudes toward plutonium also came into play.

## THE TRANSVENOUS ROUTE TO THE VENTRICLE

A pacemaker surgeon of 1962 who, Rip Van Winkle–like, had fallen asleep for ten years and awakened in a hospital in 1972, without the benefit of a decade's incremental learning, would still have recognized and understood the hardware and rituals of cardiac pacing — with the exception of one big change. Implanting a pacemaker in 1962 meant putting a sick patient under general anesthesia, opening the abdomen and the chest, and exposing the surface of the heart. Thus surgeons performed pacemaker implantations, and the procedure qualified as major surgery. In the early and mid-1960s, hospital mortality rates in the days after implantation averaged about 7.5 percent. Then the standard procedure changed greatly.[25]

A doctor who implanted pacemakers in 1972 almost never opened the patient's chest to stitch the pacing electrode onto the surface of the heart. Instead, the lead was introduced to the inside of the right ventricle by making a small incision in the upper chest and advancing it down a vein. Once the lead was positioned and tested for its electrical characteristics, the physician plugged it into the pulse generator and buried the generator in the chest between the layers of subcutaneous tissue and muscle at the site of the incision. The patient remained awake, under local anesthesia. An experienced implanter could carry out this procedure in an hour or less, though complex cases would take longer. Within days, fibrous tissue would grow around the electrode and bind it tightly to the *endocardium*, or inner wall of the heart. This procedure, well established by 1972, remains the accepted one today.[26]

The discovery of a transvenous path to the right ventricle actually antedates the earliest fully implantable pacemakers. A second-year surgical resident in New York City first introduced an endocardial pacing lead into a patient's ventricle in 1958.[27] At Montefiore Hospital in the Bronx, Seymour Furman (b. 1931) was helping to set up the open-heart surgery program. The surgeon in charge of organizing the effort asked Furman to study the Minnesota technique of pacing the heart through a wire connected to an external pulse generator. Using dogs, he set to work on this task.

Brooklyn-born, Furman was the son of an immigrant garment worker: his parents had pointed him toward a doctor's career from boyhood. Furman attended the Washington Square campus of New York University, then took his medical training at Downstate Medical Center of the State University of New York, graduating in 1955. For temporary pacing, the young trainee first constructed a myocardial lead by enclosing a length of stainless steel wire in a polyethylene sleeve. The pulse generator presented a greater challenge. Furman did not know of Earl Bakken's transistorized generator and understood that the Electrodyne external pacemaker, with an output of 10 to 150 volts, was not suitable for myocardial pacing. "There was no one available at the hospital who could provide any assistance. I called on all of my limited knowledge of electrical theory and decided that the best way of reducing the output voltage would be to place a resistor across the external generator output, and take the output going to the heart from that resistor. In that way the output voltage could be reduced to about one-tenth" its earlier value. With some help from the firm Electronics-for-Medicine, Furman built his voltage reducer as a small plastic box that attached to an Electrodyne pacemaker.[28]

Furman was also learning the techniques of cardiac catheterization under the tutelage of cardiologist Doris Escher. It occurred to him that "the polyethylene tubes which I had used to insulate the steel wires could equally easily be converted into cardiac catheters." In dogs, the external jugular vein provides ready access to the right ventricle. Why not try putting a pacing lead *inside* the ventricle? He passed a braided copper wire down a polyethylene catheter and wrapped the exposed tip around a small piece of tinfoil to create a crude electrode. On March 12, 1958, he inserted this primitive lead down the external jugular vein of a dog into the animal's right ventricle and attached the other end to an external pulse generator. With this *transvenous* lead, Furman was able to capture the dog's heartbeat and pace the animal artificially.[29]

Furman soon discarded the copper wire and tinfoil in favor of a standard catheter that cardiologists employed to take electrical readings within the chambers of the heart. After pacing several dogs, he showed the new system to his seniors at Montefiore. One of them was the chief of cardiology, John B. Schwedel, a man who "felt that new things should be tried." Furman recalls that "for the first time I was able to excite enthusiasm for the research effort in a person other than my wife." As a cardiologist, Schwedel was particularly receptive to the idea of a catheter lead. He "grasped the importance of the technique instantly, became extremely excited, and assured me that a clinical trial was mandatory and that he would help me." Schwedel also assured Furman that "he would bear responsibility for the disasters, if any, but that if the technique [proved] successful the credit would be mine." In July 1958, Furman successfully paced an elderly man with heart block for two hours during colon surgery by inserting his catheter lead into the median basilic vein at the crook of the patient's arm and threading it through the venous system across the chest and into the heart. The patient tolerated the surgery well but died about 10 days later following a second operation during which pacing was not used. Furman recalls that Schwedel "by this time was beside himself with the possibilities here." Schwedel told Furman that "it was a phenomenal technique." "Push on," he said — "it's yours."[30]

The story of Furman's second patient, P.S., has become a classic in the medical lore on cardiac pacing. P.S. was a 78-year-old man, a private patient of Schwedel's suffering from congestive heart failure. When Furman began on the case, in August 1958, P.S. had developed heart block with occasional Stokes-Adams seizures. The team at Montefiore maintained him on an external pacemaker while they tried drug therapy to stimulate the heart, but his condition deteriorated and he remained under sedation most of the time. On August 18, 1958, Furman pushed the catheter from the patient's left arm into his right ventricle and embedded the second electrode under the skin in his chest. The external pacemaker immediately achieved control of the heartbeat. With a steadier and more rapid heart rhythm, P.S. gradually improved. In October, he was allowed to sit in a chair in his hospital room. Soon he was reading and writing — and now Furman decided to get him up and about. Furman put the pulse generator on a hospital cart that P.S. could push. He had a long extension cord made up so that his patient had "a range of almost 100 feet down the hospital corridor. We'd have cords at one outlet in the corridor, cords at another outlet. He'd walk the full extent of one cord, then I'd come by, the nurse would hold him tightly, and I'd switch plugs. Off he'd go again."[31]

P.S. would remain an invalid for the rest of his life, but he was alert and responsive, his condition had stabilized, and his heart appeared able to shoulder the burden of sustaining its own rhythm. On November 21, Furman withdrew the catheter. The pacemaker had managed this man's heartbeat for 96 days. Furman and Schwedel emphasized that with an endocardial electrode, it was not necessary to raise the voltage as the days passed. The case demonstrated "the maintenance of sensitivity of the endocardium to stimulation after many weeks of application of electric current." P.S.'s pacing threshold on day 96 remained what it had been on day 1—just 1.5 volts.[32]

Furman recognized that transvenous pacing could claim some significant advantages over the more invasive myocardial approach that every other research group favored. Most important, the physician could gain access to a vein and introduce the catheter without subjecting the patient to general anesthesia and major surgery. Furman also found that the catheter also reduced the risk of damage to the heart tissue because it was not stitched down.

At first, Furman did not know what to make of his new way of pacing the heart. "I remember coming home after doing patient number one. . . . Obviously I was excited about what was happening; and . . . I told [my wife] we'd finally put it in a patient and that it worked as advertised. And she said 'Well, frankly Sy, I've never heard of [heart block]. . . . Do you think there'll be many patients?' I told her that I'd barely heard of the condition myself and I certainly didn't think there would be many patients. . . . But that was 1958. By 1960, I was beginning to realize that there were a number of people around and that the number was not wholly inconsequential."[33]

Today Furman's work seems a dramatic case of medical innovation, but for several years the inventor had trouble getting a hearing. His youth and junior status may have been factors initially. The opposition of Paul Zoll definitely hampered Furman and his supporters for years. Zoll was widely considered the world's leading expert on electrostimulation of the heart. In September 1958, while in New York for the Rockefeller Conference, he had visited Montefiore to observe P.S. at the request of the patient's family. In Furman's recollection, Zoll "told the family that the pacing effort was without value and that no benefit would accrue"—an opinion that P.S.'s wife and son had rejected. The following year, Furman and Schwedel submitted a paper to the *New England Journal of Medicine* that described the first two cases in detail. The journal's editor rejected the manuscript after an anonymous referee gave it a scathing review.[34] Furman read the comments and realized that it was probably Zoll who had written them; he felt that the questions Zoll had raised "were the same kind of questions that naysayers can always ask."

I took a deep breath, and . . . I decided to do something which I certainly would not have had the courage to do a couple of months before. Because by this time I was convinced that there was something real here; [transvenous pacing] was no longer just a joke or a trick—there was something real. I wrote a long and impassioned letter . . . to [the editor] . . . and pointed out that the questions that were being asked . . . were legitimate in some regards—but that I couldn't give them all the answers, and that I really shouldn't have to give them all the answers. That the case was described honestly and carefully, and . . . —I think this may have turned the trick—I pointed out to him that one of the associate editors of the *New England Journal* had in fact seen the patient in consultation, and why not ask the editor? Ask him if there was anything in the description of the case which was not true. And that indeed if he wanted the hospital records, they could be provided, but in fact the patient *had* been maintained that long, and the editor had seen the patient.

Some time later, the editor wrote back suggesting minor changes but agreeing to publish the paper.[35]

Furman recognized quite early that he might have invented a technique adaptable to long-term pacing of the heart. He tried to interest medical electronics companies in developing an implantable pulse generator to go with the transvenous lead, but he recalls being "universally turned down, politely but firmly." By mid-1960, his group had used the transvenous lead with a total of 19 patients, of whom 6 were still living (2 of the 6 would survive until the end of the 1970s). After having to reposition the leads time and again when patients moved their arms and caused pacing to stop, Furman switched to the right external jugular vein as the route to the ventricle. He would attach the pacing lead to a battery-powered external pulse generator like the Medtronic 5800; thus, the system was partially external. Case 7 particularly impressed Furman. The patient, H.N., lived at home for 41 months with this arrangement, but died in November 1962 after open-chest surgery to implant a long-term myocardial pacemaker. Furman commented that "his death tended to discourage other thoracotomies for pacemaker implantation."[36]

## THE DIFFUSION OF TRANSVENOUS PACING

Furman disappeared from the scene for two years' service in the navy and a year's surgical residency under Michael DeBakey at Baylor. On returning to Montefiore in 1963, he found the pacemaker program essentially unchanged from 1960. In his absence, there had been about four dozen additional patients, but all had received the kind of pacing system in use when he had left — a catheter lead introduced from the jugular vein and attached to a battery-powered external generator. This arrangement "worked" in the sense that it successfully stabilized and maintained patients in chronic heart block. However, in about one-third of the Montefiore cases, bacterial infections entered the patient's body through the port in the skin. On the whole, it was not a system likely to appeal to surgeons, few of whom saw it as a possible solution to the problems hindering progress in myocardial pacing.[37] More broadly, surgeons did not normally perform cardiac catheterizations, but use of the catheter as a diagnostic tool was perhaps the defining ritual of cardiology. Yet a surgeon's skills were needed to implant a pulse generator, and the medical world had already grown accustomed to the idea of pacemaker implantation as a *surgical* procedure. Transvenous pacing sprawled awkwardly across the specializations in cardiovascular medicine.[38]

The technique spread beyond Montefiore during the early 1960s as a way to stabilize a patient for a few days before implanting a myocardial pacemaker. Experience with temporary transvenous pacing helped accustom surgeons to the techniques of catheterization and opened their minds to the possibility of using the transvenous lead with permanently implanted pacers.[39] After 1965, new voices spoke out in favor of transvenous pacing. Surgeon Hans Lagergren, from Karolinska Hospital in Stockholm, reported in a U.S. journal that his group had used transvenous leads in a hundred consecutive patients since 1962. Lagergren described an extremely flexible pacing lead available from Elema-Schönander and emphasized that the internal electrode had done no damage to the endocardium. At first, Lagergren had accessed the heart through the external jugular vein and used an external pulse generator. However, beginning in May 1962, he brought the lead beneath the skin from the jugular vein across the chest and through the diaphragm to a generator implanted in the abdomen. Lagergren strongly reinforced Furman's point that transvenous pacing was considerably safer than myocardial.[40]

Reports from Parsonnet in Newark and from Aubrey Leatham's group at St. George's Hospital in London provided still more clinical evidence. The St. George's group pioneered a new placement of the pulse generator that would become standard a few years later. In 1963, surgeon Harold Siddons began to locate the generator in the patient's left or right chest just below the collarbone. This placement gave good access to the jugular vein and put the entire system above the diaphragm. Medtronic's introduction of a flexible transvenous lead in 1965 (a variant on Chardack's coiled-spring design) contributed to a shift toward the transvenous technique in the United States. Medtronic also modified two of its pulse generator models for transvenous leads. Although they were still quite large (an inch thick and nearly three inches across), it did prove possible to implant them in the patient's chest or under the arm.[41]

Furman now began to advocate a crucial improvement in transvenous pacing: physicians should abandon the jugular and insert the permanent lead via the left or right cephalic vein. These run horizontally across the upper chest below the collarbones. Furman had become "disenchanted" with the jugular vein "largely because the lead had to cross over the clavicle [to reach an implanted pulse generator] . . . and was, therefore, a nuisance to the patient." Tunneling the lead beneath the collarbone risked damage to the large bundle of nerves that controls the arm. But the cephalic vein provided easy access to the right ventricle. Venous access close to the site where the generator would be buried greatly simplified and shortened the entire implantation procedure. Furman implanted the generator beneath the layers of skin in the chest but above the muscles — a placement that did not require a deep pocket. Patients typically needed only "light barbiturate sedation and local anesthesia" and the procedure could take place in a cardiac catheterization lab rather than an operating room.[42] Furman and his associates also showed how to position the tip of the catheter in the apex of the right ventricle, where it would remain in contact with the inner surface of the heart. Emulating Chardack's papers of the early 1960s on myocardial pacing, Furman published a series of papers that walked the reader through the entire transvenous implantation procedure and explained how to resolve the most common complications based on his group's experience with more than a hundred patients.[43]

Perhaps the clinching factor for transvenous pacing proved to be doctors' growing realization that Furman's procedure was less risky for their elderly patients. Hospital mortality rates from transvenous pacing ranged from 0 to 3 percent. Around 1966, William Chardack himself began to use the transvenous route. By 1970, experienced implanters had switched in large numbers to transvenous pacing, and new entrants to the field were accepting it as the normal path to the ventricle.[44]

Transvenous pacing was a radical innovation: it required that surgeons learn an entirely new technique of implantation. Yet they willingly made the leap from myocardial to transvenous leads because the new route to the ventricle had proved itself safer and because it simplified and speeded up the typical implantation procedure. In theory, the use of transvenous leads opened pacemaker implantation to cardiologists. Impediments remained, however, and few cardiologists implanted pacemakers entirely on their own for another decade. Pulse generators remained bulky and the skills of a surgeon were usually required to implant them in the chest. Sometimes, too, the cephalic vein proved unsuitable for a transvenous lead. In that case, prudence dictated that a surgeon should handle the alternative routes, the external jugular or the subclavian vein.[45]

Victor Parsonnet and Alan Bernstein have pointed out that invention of transvenous pacing "spawned new specialties and industries." As Furman's technique gained acceptance in clinical cardiac pacing, researchers discovered that they could use transvenous electrodes to sense and record the spread of the electrical impulse within the heart itself. These intracardiac recordings, or *electrograms*, yield much more information than an ECG recorded on the surface of the body. The electrogram formed the diagnostic basis for an emerging subspecialty of cardiology called *clinical cardiac electrophysiology*. By the 1990s, electrophysiologists were implanting defibrillators that used transvenous leads to terminate rapid heart rhythms by delivering powerful electrical shocks. Most of their diagnostic and interventional tools, in fact, depended in one way or another on transvenous leads through which they could sense or deliver treatment within the heart.[46] No one could have predicted this chain of events in the 1960s. Innovators adopted transvenous pacing for more proximate reasons. The procedure seemed less stressful for patients and, once implanted, the lead proved more stable and reliable than earlier leads.

## PARSONNET ON PACEMAKER COMPLICATIONS

In the 1960s, Victor Parsonnet and Seymour Furman tried to identify and solve the full range of problems hindering the progress of cardiac pacing, ranging from imperfect scientific knowledge to inadequate hardware and from the organization of services in the hospital or clinic to the training of new implanters. They did not limit their attentions to the clinical realm but treated all these disparate problems as important. They organized pacing as a set of structured activities that would not end with implantation and immediate postoperative care but would continue throughout the remaining lifetime of the patient.

Beginning with his second patient, Parsonnet tried to improve the procedure for implanting a pacemaker. He felt that he could stabilize the heart rhythm, enable the patient to build strength, and prevent Stokes-Adams episodes during surgery by starting the patient on temporary transvenous pacing with an external generator several days before going forward with the implantation. Reported in 1962, this technique was widely adopted and continues in general use today for patients whose overall condition is poor. It gave surgeons their initial experience with the catheter lead and thus prepared the way for acceptance of permanent transvenous pacing a few years later.[47]

Parsonnet had not originally intended to pace permanently via a transvenous lead, but this seemed a logical thing to try. An opportunity arose in October 1962 when an 80-year-old farmer from western New Jersey was admitted with Stokes-Adams attacks. Parsonnet began this man on a temporary transvenous pacemaker, then discovered that he was basically in good health, with no detectable heart problems other than the recent onset of complete heart block. He seemed an ideal candidate for a trial of long-term transvenous pacing. Four days after the first procedure, Parsonnet introduced a new transvenous lead via the cephalic vein in the patient's left chest, then tunneled the lead down to an incision in the upper abdomen where he placed an Electrodyne pulse generator. This system sustained the patient for about two months, but in mid-December — the day after Parsonnet had described the case at a regional medical meeting — pacing became intermittent because of a rising threshold of stimulation. Parsonnet now switched his patient to a conventional myocardial pacemaker and removed the transvenous lead. The patient subsequently required several more pacemaker replacements but lived at home until 1968 and survived until December 1971. The case revealed that Parsonnet would be quick to identify opportunities for innovation in pacing and was more interested in trying new approaches than in endlessly replicating established practices.[48]

Parsonnet's choice of problems was related to his analysis of the cardiac pacemaking community. It seemed to him that, in general, surgeons who implanted pacemakers were highly interested in carrying out the initial procedure, less interested in troubleshooting defective pacers later on, and less interested still in routine oversight of patients whose pacers appeared to be functioning normally. Yet the well-being of the patient demanded that every pacemaker be tracked and every implanter or hospital be ready to solve problems that arose suddenly. Cardiac pacing was an interdisciplinary field — the expert must know surgery, cardiology, and electronics. "Often [the surgeon] cannot decipher all the complications by clinical judgment and 'routine' laboratory work alone but must resort to an analysis of voltage, resistance, impedance and current, all ghostly terms from his dimly remembered past." Since the benefits of pacing were apparent, surgeons would have to master troubleshooting and routine follow-up.[49]

One day in fall 1962, while sitting in the gallery of the operating room at

Baylor, Parsonnet began writing a paper on complications in cardiac pacing. "[Denton] Cooley looked over my shoulder while I was at work one morning and expressed amazement that I should write about complications. With tongue in cheek he said that I should learn never to do that, but only write about the good results." As both men understood, just the opposite was true: doctors learned more from reports of bad results—and there had been many in cardiac pacing. The paper reviewed the common failure modes such as battery exhaustion, dislodged myocardial electrodes, and broken leads. Parsonnet explained how each defect would manifest itself in a chest X ray and an ECG and discussed in detail how to resolve each problem using his own cases as examples. This early paper, perhaps more than his reports on new hardware, in many ways defined Parsonnet as a voice in the physician community. It hinted at one of the important themes in his later career, his concern about sources of complexity in cardiac pacing and his interest in codifying his own experience and that of others so as to enable physicians to deal with the myriad of problems that could arise with implanted electronic devices.[50]

## PACEMAKER FOLLOW-UP AS A CRITICAL PROBLEM

Pacemaker reliability was already improving by the middle years of the 1960s, and patients were living longer on pacers. But all pacers would eventually stop working for some reason or other. As other sources of sudden pacemaker failure became less common, battery exhaustion by 1970 accounted for four-fifths of all failures. The mercury battery would run down with little warning. "The battery voltage and, therefore, the pacemaker rate and output might decline . . . to actual failure to stimulate the heart . . . within several days or, at most, one week."[51] The ailment troubling most patients at that time was complete heart block: without artificial pacing, their hearts would slow down and they might experience Stokes-Adams attacks. Doctors would then have to perform unscheduled replacement procedures.

The case of Wesley Johnson, of St. Paul, Minnesota, provides an instructive example. Johnson received his first implanted pacemaker at the age of 54 in July 1966. He had suffered a "fainting spell" while doing yard work at the family cabin in northern Wisconsin. His wife Helen drove him home to St. Paul that evening, and an ECG at their family doctor's office the next day revealed that he had developed complete heart block. Dr. Sam Hunter gave Johnson a standard implanted fixed-rate pacemaker with a mercury battery and a myocardial lead.[52] To Hunter's surprise and delight, after the implantation Johnson continued to work as a supervisor at a local steel plant and pursued his outdoor activities as vigorously as ever. "He has got to be the most interesting pacemaker [patient] we have," Hunter wrote in 1969. "This man canoes up in the Knife [River] chain and just returned from a seven-day trip. On one day he did five portages but in order to get across the portage he carries the canoe first and then two heavy packs, so he actually is making 15 portages in one day with a cardiac pacemaker. If I ever write up my memoirs I certainly must include this case." The Johnsons also maintained their cabin and enjoyed downhill skiing. Johnson was a classic case of the pacemaker patient whose "electric wiring system is off" but whose heart in other respects remained healthy.[53]

Then on the afternoon of December 24, 1969, as he walked up his driveway, Johnson "suddenly . . . felt kind of let down." He told Helen, "I feel like I've been turned off." Not understanding what had happened, the couple drove to a Christmas Eve service at their church. She recalls that during the service, she "looked at him and he was sitting there taking his pulse and acting funny." Johnson said that he could barely feel his pulse. The pacemaker had stopped and his heart rate had dropped below 30 beats per minute. They slipped out of the pew and called Dr. Hunter from the church office. As Helen Johnson tells the story, "Dr. Hunter said, 'Read me your pulse over the phone,' and Wes did. And he said, 'I want you down in the hospital.' But we were going from church to his sister's for Christmas Eve supper! Dr. Hunter told him he could go. . . . We ate dinner, he opened his packages, and then my son-in-law and I took him down to the hospital."

On Christmas morning, Hunter gave Johnson a local anesthetic, made an abdominal incision, and replaced his failed pulse generator, connecting it to the original myocardial lead. The hospital released him at two in the afternoon and he got home in time for Christmas dinner. This story had a happy ending, but it was the kind of episode that made pacemaker physicians uneasy. With a ventricular rate below 30, Johnson could easily have gotten dizzy and slipped on an icy sidewalk. If his heart rate had dropped further, he might have suffered a Stokes-Adams attack. The point was that doctors had learned how to implant a pacemaker and were learning how to troubleshoot the device if it failed, but they lacked routine procedures for keeping track of patients and managing their implants. During the 1960s, several pacemaker groups recognized follow-up as a critical problem and began to invent new equipment and procedures that would routinize the long-term management of both patient and pacemaker.[54]

Extending the worklife of the battery and anticipating battery exhaustion were the keys to long-term management. Both Furman and Parsonnet looked for ways to reduce battery drain. Parsonnet's group devised an ingenious endocardial pacemaker electrode that made it possible to stimulate the heart at greatly reduced amplitudes. Although it did not come into general use, it alerted the manufacturers to the possibilities of redesigning their electrodes to reduce battery drain. Furman demonstrated that shortening the stimulus duration and using electrodes with a smaller surface area (hence, higher current density), would manage the heartbeat more efficiently.[55]

The other line of attack—organized pacemaker follow-up to spot impending battery failures—required both organizational and hardware innovations. This was particularly evident at Montefiore Hospital, where a rudimentary pacemaker follow-up clinic was taking shape as early as 1959. With Furman's patient P.S., the staff had learned to check frequently for signs of infection at the site where the lead entered the arm. Soon the hospital was managing half a dozen new patients who had been discharged from the hospital with transvenous leads and portable external pulse generators. They "had to come back on a long-term organized basis so that the wounds could be cleaned, the fragile connections resecured, the batteries in the external pulse generators replenished." Already the central features of a pacemaker clinic were emerging. The clinic had a designated location in the hospital and each patient was scheduled for regular visits. A nurse would take an ECG, the doctor or a technician would attend to the (partially external) pacemaker, the doctor would examine the patient and review the situation with him or her. As Furman later commented, "patient support was extremely important . . . because the one guarantee we had for a patient with a pacemaker was the frequent repetition of potentially life-threatening complications."[56]

With the coming of fully implantable pacemakers after 1960, several pacemaker groups in England, Australia, and the United States developed routines for handling a variety of common pacemaker crises. Some larger centers organized outpatient pacemaker clinics within the hospital or nearby. These clinics resembled Montefiore's model. As patient populations rose, pacemaker clinics learned to keep detailed records on each patient and to give routine reminders by telephone or postcard.[57]

Parsonnet had been tracking pacemaker patients since his earliest cases, and Newark Beth Israel set up a formal pacemaker clinic in September 1966. The clinic's doors were open three days a month, with 20 to 30 patients being reviewed each day. The typical patient would visit the clinic every three to four months during the first year after implantation, every two months thereafter. Although the examination was simple, involving a tracing from Lead 1 of the ECG, Parsonnet and his associates organized an elaborate record-keeping system. In 1968 they began to store data on computer, using a regional time-share network. Parsonnet reported in 1970 that 64 percent of the pacemakers monitored at the clinic had been replaced on a planned basis, thereby avoiding high-risk emergency procedures. This compared with nonemergency replacement of only 9.5 percent of the pacemakers not monitored.[58]

By the late 1960s, pacemaker clinics were adding instruments that improved the analysis of the implanted devices. Here Furman played a critical role as an innovator. Montefiore had hired an engineer, Bryan Parker, to manage the growing array of electronic devices in the hospital's coronary-care area and cath lab. Parker and Furman designed and built two devices that a physician could use to assess the electrical functioning of a pacemaker. The first was an output-and-impedance meter that tested the electrical properties of the pulse generator and the leads and could be used at the time of implantation or while troubleshooting a defective unit. The second device simply told the doctor whether the pulse generator was firing stimuli. Surgeon Adrian Kantrowitz had discovered that if he placed a portable AM radio over the implanted generator and tuned it to the lowest possible broadcast frequency, the radio would pick up a ticking sound with each pacemaker pulse. This revealed whether the problem in a malfunctioning pacemaker lay with the generator or the lead. Furman recalled "carrying an AM radio with me as I made rounds as a quick and dirty check of pacemaker function. By taking the patient's pulse and simultaneously 'listening' to the pulse generator . . . I could rapidly determine whether all was well, even without the use of the more cumbersome, albeit more accurate, ECG." But the technique was "inefficient" because the magnetic energy of a pacemaker fell within a portion of the audio spectrum for which a radio receiver was relatively insensitive. Parker designed a more sensitive detector, complete with a pair of earphones for the doctor to wear.[59]

The radiofrequency detection method also permitted the doctor to check the pacing rate by counting the ticks. This was extremely important information because the rate was the most important indicator of the health of the battery. Furman realized that many patients would not visit the pacemaker clinic often enough for someone to catch the signs of impending battery failure, and that some implanters would probably never establish clinics anyway. But if the ticking of a pacemaker could be transmitted over the telephone, a physician might detect impending battery failure in time to replace the generator before it failed entirely. Furman proposed this idea to Parker, probably in 1968.

Parker designed and built a transtelephone monitoring system that utilized a transmitter, kept by the patient at home, and a central receiving terminal, at the hospital. The patient dialed into the receiving terminal and placed the home telephone handset into a cradle on the transmitter. By holding an electrode in each hand, the patient completed a circuit that transmitted each pacemaker impulse over the telephone as an audible click. An automatic interval counter at the receiving end determined the impulse rate. By comparing the rate with previous data for this pacemaker, staff at the clinic could graph the pacing rate over time and detect the slight declines in rate that heralded impending battery failure.[60]

With transtelephone monitoring, a patient could reduce the number of visits to the pacemaker clinic or the doctor's office. The system in theory made it possible to provide a basic level of oversight for the thousands of patients who were then receiving little or no organized follow-up care. Furman emphasized that the transtelephone system "cannot and does not replace electrocardiographic and clinical examination of patients" — but it could warn of impending battery exhaustion. He arranged for a medical-equipment company to manufacture the transtelephone apparatus. At the same time, one of his former patients, Theodore Stern, invented a device for transmitting a one-lead ECG over the telephone and founded Cardiac Datacorp to manufacture it. Other device manufacturers introduced similar equipment, though Furman continued to believe that knowledge of the rate alone (i.e., without the ECG) was sufficient for spotting impending battery exhaustion. To simplify follow-up further for the physician, several commercial services soon sprang up that received and stored transtelephone data for many physicians and alerted the physician when a replacement was needed.[61]

By October 1966, Furman's group had implanted about 150 pacemakers, Parsonnet's about 130; these numbers more than doubled in the next three years.[62] Neither surgeon was the most prolific implanter in the United States, or even in his state. What set them apart was their commitment to clinical research and their ability to learn something useful from every case. Both men invented new devices and components, pioneered new operative techniques, and experimented with the organization of pacing-related activities in their hospitals. The two surgeons had established themselves by 1970 as champions of new technology and new procedures. They had codified existing know-how in pacing and had proved themselves adept at framing the next problems that innovators should tackle. Through publication and personal contacts with manufacturers and other physicians, both tried to spread their knowledge widely.[63]

## PACEMAKER COMPETITION WITH NORMAL HEARTBEATS

Could a pacemaker stimulus trigger ventricular fibrillation? Some patients on pacemakers had occasional, normally conducted beats; others had "ectopic beats" in which an impulse originated from some other source in the heart than the sinus node; still others experienced heart block alternating with periods of normal conduction. In all these situations, the ventricles might receive natural and artificial signals in competition.

It had been clear since Carl Wiggers's experimental work in the 1930s that a stimulus delivered to the ventricles during one brief "vulnerable period" of the heartbeat could induce fibrillation. If that happened, the patient could die. Few pacemaker patients were attached to a heart monitor at the moment of death, so researchers lacked conclusive proof. But studies of patients who had occasional conducted beats or intermittent block with long episodes of normal rhythm did begin to create a body of presumptive evidence. Edgar Sowton, a cardiologist at St. George's Hospital in London, showed in 1965 that patients with unusual rhythms received pacemaker impulses during the vulnerable period of the heart cycle about twice a minute on average, or three thousand times a day. He also reported that the death rate for these patients was five times higher than among patients with fixed complete heart block.[64]

Paul Zoll and William Chardack remained skeptical that pacing could set the heart into fibrillation; they believed that the pacemaker stimulus, even when it fell into the vulnerable period of a conducted heartbeat, was too small to trigger this disaster. Zoll, a stubborn man, declared categorically, "I think that a fixed rate pacemaker should be implanted in every patient if a pacemaker is to be used at all."[65] Others replied that the irritability of the heart muscle and hence the danger of VF increased significantly when the pacemaker fired into tissue damaged by an earlier heart attack or when drug therapy lowered the threshold of stimulation.[66]

Many implanting physicians were also dissatisfied with asynchronous pacing because patients sometimes found it uncomfortable. In heart block, the upper chambers often continue to contract normally and at varying rates. Some patients had sizable fluctuations in cardiac output and blood pressure because sometimes the paced ventricles happened to be in proper synchrony with the atria but at other times were not. When the heart produced normally conducted beats, many patients, particularly younger ones, could feel the competition between the heart and the pacemaker impulse.[67] Interest in the question intensified when an experimental pacemaker that avoided competition was announced at the American Heart Association meeting in August 1964. The new pacing concept did not come from a physician-led research group but from a biomedical engineer at a device-manufacturing company — an inventor who maintained close ties to physicians.[68]

Barouh V. Berkovits (b. 1924), an engineer at the American Optical Company, was already well known as the inventor of the DC defibrillator and the cardioverter, a device that interrupts a rapid heart rate (tachycardia) with low-energy shocks. He knew that when the cardioverter discharged randomly into the tachycardia, it would "occasionally not only not stop the tachyarrhythmia . . . but would produce ventricular fibrillation." Cardioversion had to be synchronized to fall within the QRS complex and avoid the vulnerable period of the heartbeat.[69] In 1963, Berkovits applied this principle to cardiac pacing.

To solve the problem of competition, Berkovits in 1963 designed a sensing capability into the pacemaker. His invention behaved exactly like an asynchronous pacer until it detected a naturally occurring R wave, the indication of a ventricular contraction. This event would reset the timing circuit of the pacemaker, and the countdown to the next stimulus would begin anew. Thus the pacer stimulated the heart only when the ventricles failed to contract. It worked only "on demand." As an added benefit, noncompetitive pacing extended the life of the battery. (Below, I will use the term *noncompetitive pacing*, rather than American Optical's proprietary word *demand*.)[70]

The notion of an automatic on-off function for the pacemaker was not an entirely new one. In 1956, during the vacuum-tube era of electrostimulation, engineer Geoff Davies of St. George's Hospital in London had built an external pacemaker that was able to detect the R wave and shut itself off, then resume pacing if the patient's heart defaulted for six seconds. There was no thought of avoiding pacemaker competition; Davies had intended only to spare the patient as much of the painful external stimulation as he could. Then, in the early 1960s, a group in Düsseldorf, West Germany, put into clinical use a complicated apparatus designed by the French engineer Fred Zacouto, who combined a conventional implanted pacemaker with an optional external sensing unit. On sensing the R wave, the external unit would send a radiofrequency signal to deactivate the implanted pulse generator; if the heart beat too slowly, the sensor would signal the pacemaker to begin firing again. The physician could select the lower rate limit below which the sensor would activate the pacemaker.[71]

For physicians, these experimental pacemakers embodied a continuing tension between two ideals: *simple devices* and *physiological devices*. The goal of having a straightforward, easily comprehended, easily managed device shimmered before physicians like a desert mirage. Later, when noncompetitive pacemakers came into use, the principal argument that Chardack and others would mount against them was that they were too complex, with more things that could go wrong. This was a powerful point. But although doctors valued devices that were simple, they were also learning to value inventions that emulated what they understood to be nature's own way of doing things. The term *physiological* became a shorthand way to refer to this somewhat utopian idea of emulating nature. The word took on connotations extending beyond its dictionary definition, "characteristic of or appropriate to an organism's healthy or normal functioning."[72]

Berkovits championed his "demand" pacemaker in the 1960s on the ground not only that it avoided pacemaker competition but that it "actually delivers its stimulation in the same fashion as an autonomic fiber," by which he meant that the pacemaker, like a myocardial cell, would gradually reach a critical potential and fire to induce an escape beat in the absence of a normal beat emanating from the sinus node. In later years, he learned to present the point in simple and compelling terms. Noncompetitive pacing was a way of "electronically duplicating nature's escape mechanism." Setting the imitation of nature as a standard proved an effective way to popularize the new design because it implied an invidious distinction between noncompetitive, which was natural, and asynchronous, which was artificial.[73] The belief that patients would be better off if treatment emulated "normal physiology" was fundamental to the appeal of other new pacing modes introduced in the mid-1960s and later. But, inevitably, the effort added complexity to the pacemakers and made them more difficult for some physicians to comprehend.[74]

To test the noncompetitive pacemaker, first in animals and then clinically, Berkovits turned to Louis Lemberg and Agustin Castellanos Jr., cardiologists at Jackson Memorial Hospital in Miami.[75] For several months in 1963–64, the group used a prototype of the new pacer that was entirely external. Lemberg remembers that "it wasn't miniaturized. We wheeled it to the side of patients. . . . It was a big hospital — I remember wheeling it all over the place to get ECG records." Meanwhile, Berkovits developed a miniaturized version intended for long-term use, and this became American Optical's first implantable device.

After using the implantable model with animals in 1965–66, Lemberg felt that the time had come for a clinical trial. "We had the pacemaker ready to go — it was all set. . . . I was waiting for the classical patient, which was a patient that's in intermittent block. Not one that has complete block [because] you can't tell if it worked or not." In a patient with fixed complete heart block, the new pacemaker would emulate an asynchronous pacer, so he needed someone whose block came and went. To document the effect of noncompetitive pacing, the Miami group had acquired a Holter monitor, a wearable ECG recorder that would create a record of every heartbeat over 24 hours. In the summer of 1966, Lemberg finally found a patient with intermittent block. Just days before the scheduled implantation in Miami, surgeon Dwight Harken, in Boston, convinced Berkovits that he had a postsurgical patient who needed a noncompetitive pacer. The first clinical use of an implantable noncompetitive pacer thus took place in Boston — but the patient did not have intermittent heart block. It was Lemberg's case, a few days later, that demonstrated the clinical effectiveness of noncompetitive pacing.[76]

American Optical's implantable "demand" pacemaker went into production in 1968. Despite the broad patent awarded to Berkovits in 1967, several other manufacturers announced new models designed to avoid competition with the heart's own electrical impulses.[77] By 1969, noncompetitive pacemakers had almost completely superseded the earlier asynchronous models.[78]

## NEW TECHNOLOGY AND NEW THINKING IN THE 1960S

In cardiac pacing, new technology and new thinking about heart rhythm disorders have often encouraged each other in a reciprocal process. This is evident in the case of noncompetitive pacing. As the number of pacemaker-dependent patients began to rise into the hundreds during the 1960s, physicians encountered a few mysterious cases in which pacemaker patients had died suddenly. They hypothesized that some people had, not fixed complete heart block, but intermittent block with some normally conducted beats. Berkovits, an engineer, conceived of a way to solve the problem of pacemaker competition and then invited physicians' comments on the idea and their collaboration in clinical trials. Earlier choices in the design of implanted pacemakers, accumulated clinical experience, and intensified research into disturbances of heart rhythms all contributed to physicians' ideas about intermittent heart block and to the invention of the "demand" pacemaker. Eventually, physicians reached a consensus that asynchronous pacing had indeed caused VF in some patients; but according to Berkovits, it was only *after* noncompetitive pacing had come into widespread use that comparative mortality data provided compelling corroborative evidence to support this hypothesis. Before noncompetitive pacing, it had been more of a suspicion.[79]

During the 1960s, innovative physicians concentrated on identifying, describing, and then solving critical problems in pacing. At first these tended to be problems with hardware; but Furman, Parsonnet, Chardack, and other early leaders also worked on the implantation procedure and even on organizing pacing services and record keeping in the hospital. Throughout the decade, the pioneer physicians in pacing also published a stream of papers, contributed to surgery textbooks, consulted with the manufacturing firms, and subjected new pacemaker models to clinical testing. They did not draw a sharp distinction between their roles as clinicians, inventors, and proselytizers. They consistently sought to establish pacing as a safe and effective treatment and were prepared to take on whatever projects showed promise of advancing that purpose. The key features of the reliable pacemaker of 1970 were as follows:

— The pacemaker was a fully implanted device that managed the heartbeat in patients suffering from complete heart block.

— The pulse generator was normally implanted in the patient's chest or side above the diaphragm.

— A battery consisting of zinc-mercuric oxide cells supplied power.

— Electronic circuitry consisted of discrete, transistorized components mounted on a circuit board.

— The implanting physician normally introduced the pacing lead via the cephalic vein in the patient's chest. The tip of the lead was placed in the right ventricle of the heart.

— Via the ventricular lead, the pacemaker sensed as well as paced; a sensed ventricular contraction (indicated by the R wave of the ECG) inhibited the pacer from delivering a stimulus (at this time, the term *demand pacing* was widely used).

— Routine follow-up procedures were coming into use to check on the condition of each patient and assess the remaining service life of the battery.

## CHAPTER 6

# THE INDUSTRIALIZATION
# OF THE PACEMAKER

A handful of physicians contributed important innovations during the 1960s, but business firms were increasingly shaping the cardiac pacemaker and the field of pacing. The decade witnessed the rise of a pacemaker manufacturing industry in the United States and western Europe that consisted of seven or eight principal companies. As the years passed, these companies increasingly took the initiative in product improvements. Through their marketing efforts, they also introduced pacing to physicians in other parts of the world, thereby preparing the way for future market growth. As the Swedish observer Patrik Hidefjäll has written, by 1970 the pacemaker had become an industrial product.[1] What differentiated one firm from another was less the technical features of its products than doctors' perceptions of the reliability of those products and the degree of support that the firm could offer to the implanting doctor. The major unresolved technological issue in the eyes of engineers and physicians had to do with improving or finding a substitute for the mercury batteries that powered virtually all implanted pacemakers.[2]

So intently focused were physicians and engineers on the problems of developing the pacemaker into a consistently reliable medical device that they understandably spoke little about the broader implications of the shift from academic research labs to corporations.[3] Certainly the engineering and manufacturing of pacemakers in quantity required multifarious skills that only a business corporation could bring to bear. But as companies such as Medtronic and Cordis built their pacemaker businesses in the 1960s, they introduced a host of new actors: sales and marketing people, industry analysts from the financial world, and investors. The field developed feverish preoccupations with growth, innovation, and patent protection that surprised some of the doctors and engineers who had helped to create cardiac pacing in the first place.

## THE RISE OF THE AMERICAN PACEMAKER INDUSTRY

The implantable-pacemaker manufacturing industry had its start in the intense collaborative relationships that developed between physician-inventors and medical electronics companies in the late 1950s. The modest size of the field of cardiac pacing and of its manufacturing firms made it fairly easy for innovative men and women to bridge the normally separate worlds of medical research and corporate product development.[4] The informal and face-to-face character of the work in that era is remembered nostalgically, but it has proved difficult to perpetuate such intimacy in a large and highly regulated field of medical practice and industrial competition.

Until the mid-1970s, barriers to entry remained low in the pacemaker industry, and about two dozen companies worldwide introduced pacemakers to the market.[5] It was not difficult by 1963 to put together an implantable asynchronous pacemaker. In that era, national regulatory requirements for costly and time-consuming controlled clinical trials to assess the safety and efficacy of life-sustaining medical devices were minimal. Though manufacturers secured patent protection for some devices and components, many of the basic parts of early pacemakers were standard products purchased from suppliers; others, including the blocking-oscillator pacing circuit of the early implantables, were in the public domain. The small size of the market until the late 1960s gave manufacturers time to lower their failure rates and to scale up their production. At first, executives of the manufacturing companies did not recognize that market success would depend not only on the technological superiority of the products but on a company's ability to develop close ties to the growing number of physicians who implanted them.

## MEDTRONIC

Earl Bakken and his brother-in-law Palmer Hermundslie founded Medtronic in Minneapolis in 1949. They defined their business as one that built electrical appliances to order for the research lab and the hospital. Bakken also serviced electronic equipment in hospitals and distributed products regionally for Sanborn, a leading manufacturer of ECG machines. As he made the rounds of hospitals and medical schools, he "began to get acquainted with the researchers all around the [upper Midwest]."[6] Out of his experience on these travels, Bakken realized that the company able to provide reliable support to the researcher or the clinician was often the company that had the greatest success at selling.

The early phase of Medtronic's history might be likened to the mythical log-cabin childhood of many a U.S. presidential candidate: the business was located in a pair of garages behind the home of Hermundslie's parents in northeast Minneapolis. By spring 1958 Medtronic had six employees,[7] and by the end of the decade, thanks to its transistorized pulse generator and the Hunter-Roth myocardial pacing lead, the firm had gained a modest national reputation. In the fall of 1960, William Chardack and Wilson Greatbatch offered to license their implantable pulse generator to Medtronic. The agreement, signed in October 1960, set licensing fees, ensured that every Medtronic implantable pacemaker would bear on its shell the inventors' names, and gave them veto power over design changes. Production began less than a month later; by the end of December 1960, the company had taken orders for 50 implantables at a price of $375 apiece.[8]

Medtronic's association with Chardack and Greatbatch was crucial to the firm's subsequent success. Chardack proved an able champion of the implantable pacemaker and was probably the most influential pacing specialist in the world between 1960 and 1968. His announcement of the first clinically effective implantation in 1960, his invention of the coiled-spring myocardial lead in 1962, and his meticulous analyses of his successes and failures encouraged others to try cardiac pacing. Engineer Greatbatch joined the Medtronic board of directors. The pair from Buffalo were "key consultants" to the firm, overseeing its implantable pacemaker program and keeping in touch with clinicians around the United States. They visited Minneapolis every few weeks, critiqued the engineering staff's latest ideas, and reviewed every proposed change in pacemaker design.[9]

As late as 1963, Medtronic shipped only about a hundred pacemakers a month. Bakken and others seem to have had little expectation that the pacemaker business would soon be growing rapidly. Finances were always a problem until the mid-1960s; despite rising income from sales of its products, the company repeatedly incurred net operating losses. When the need for money exceeded the two partners' personal cash (Hermundslie had already put a mortgage on his house), they sold debentures (bonds convertible into stock) to business friends in their north Minneapolis neighborhood and raised $215,000. At one point, Bakken entertained the idea of trying to sell Medtronic to the Mallory Battery Company in order to realize some kind of return on his 12-year investment of time and money. But when the Arthur D. Little consulting firm advised Mallory that "the worldwide, all-time market for pacemakers would be about 10,000 units," the deal fell through.[10]

Medtronic's annual reports listed numerous causes for the growing losses: spending on product development, the high cost of initial produc-

TABLE 4. SALES INCOME AND NET INCOME AT MEDTRONIC, 1960–1972

| Fiscal Year* | Sales Income All Products | Net Income (Loss) |
|---|---|---|
| 1960 | $180,984 | ($16,093) |
| 1961 | 298,015 | (53,063) |
| 1962 | 518,462 | (144,135) |
| 1963 | 984,828 | 72,923 |
| 1964 | 1,587,098 | 151,108 |
| 1965 | 2,471,803 | 199,268 |
| 1966 | 3,410,370 | 307,784 |
| 1967 | 5,012,848 | 469,923 |
| 1968 | 9,946,085 | 1,073,258 |
| 1969 | 15,322,593 | 1,314,777 |
| 1970 | 22,507,113 | 2,137,613 |
| 1971 | 30,565,532 | 2,455,456 |
| 1972 | 38,484,083 | 3,973,077 |

Source: Medtronic annual reports.
*The Medtronic fiscal year ran through April 30 of the year indicated.

tion runs, and increased expenses of marketing the products, including the cost of sending employees to medical conventions. Then, in early 1962, a Twin Cities bank provided $100,000 in long-term financing, and a local venture-capital firm invested $200,000 more. By 1964, net income had risen to $151,000 and working capital to $405,500 (see table 4).[11]

The cash shortage led Medtronic's leaders to focus the company more clearly — an exercise that some competitor firms did not force themselves to undergo. Medtronic manufactured 21 products in 1962, but it soon reduced its commitment in monitoring-device technology, its area of greatest strength before 1958, and emphasized "prosthetic electronic equipment" including some devices outside the cardiovascular area, such as a carotid sinus nerve stimulator and a bladder stimulator.[12]

Despite its reputation among doctors as a company at the technological forefront, Medtronic actually did not emphasize product innovation in cardiac pacing until after 1968. Instead, it focused on quality control in manufacturing. The priority given to *reliably manufactured products* over *technologically advanced products* reflected a distinctive Medtronic understanding of the pacemaker business. The businessmen on the Medtronic board of directors believed that most physicians were cautious about new technology and often too busy to focus on the intricacies of implantable electronic devices. This implied that new technological features would not sell pacemakers unless they helped make the devices safer and easier for the physician to comprehend and use.

In the late 1960s, it was still the case that most implanters of pacemakers were surgeons. One manager at Medtronic recalls that "they were *very* interested in being on the frontier. Just doing a pacemaker *was* being on the frontier. There weren't that many people doing pacing back at that point in time, so you were sort of a cutting-edge guy if you were doing pacing." But "the subtleties of pacing, most of the time they didn't get all that involved in. . . . There was a small number of surgeons who really did want to understand what was going on"; most implanters, however, were oriented toward action, toward performing a definitive procedure and walking away. For people with this outlook, product simplicity and reliability were tokens of quality.[13]

By 1970, this analysis of the customer had become firmly established at Medtronic. In working with engineers and product planners, Chardack and Greatbatch consistently emphasized simplicity and reliability rather than recommending a string of new features for the product (see appendix A). Earl Bakken shared this outlook; engineer Bob Wingrove remembers that whenever he proposed a new design feature, Bakken would ask, "How long will that last in the body?" This cautious approach was frustrating to some at Medtronic, but for the 1960s it proved a sound instinct. Companies that introduced radical innovations sometimes had difficulty producing their devices in numbers for the market and found that many physicians resisted novelty.[14] Despite its technological caution — or because of it — Medtronic cemented its overall position as market leader in the late 1960s and gained a particularly strong position among doctors in community hospitals who might implant only 10 or 12 pacemakers per year.

Greatbatch's caution was founded on an engineer's awareness that the more features an engineer tried to design into a pacemaker circuit, the more ways the pacemaker had to malfunction; thus his aphorism that is still remembered at Medtronic: "The most reliable pacemaker is the one with no components." Chardack, the surgeon, was acutely aware of how much he and the company did not understand about the failure modes of implanted pacemakers. In addition to reporting in print on complications and early deaths among his own pacemaker patients, he persuaded Medtronic in 1964 to publish a detailed performance report. The report showed doctors that Medtronic pulse generators functioned on average for only 12 to 24 months before needing to be replaced. Some people in marketing opposed Chardack out of concern that the disparities between the company's earlier claims and the actual performance of its pacemakers could hurt sales. Instead, physicians apparently appreciated the candor of the report, and it probably helped to boost sales.[15]

Bakken *was* receptive to product innovations that would help make implantable pacemakers safer and easier for the physician to implant. Working in their labs in Buffalo, Chardack and Greatbatch developed the two most important new design features for Medtronic pacemakers during these years—the coiled-spring myocardial and endocardial leads (chapter 5) and an implantable noncompetitive pacemaker (discussed below). Other companies also introduced significantly improved pacemakers; they were not, however, able to grow their U.S. market share. This tended to bear out the belief that having the most advanced technology did not, in itself, sell pacemakers.

The preference of physicians for simplicity and reliability also suggested to Medtronic's management that technical support for the physician could become an important marketing tool and might prove indispensable as the number of implanters grew. But here the company had a problem, for in the early 1960s, rather than creating its own sales force, Medtronic had contracted with distributors who would buy the pacemakers from the company and then resell them. In this way the cash-poor firm got its money quickly, whereas the distributors had to carry the accounts receivable. The company used a similar approach in western Europe, where it arranged for Picker International, a distributor of medical devices, to handle Medtronic pacemakers.[16] The use of distributors made financial sense in the early years, but it isolated the firm from its customers. The practice flew in the face of Bakken's belief that the company should stay in close contact with the doctors. In his experience, ideas for product improvement and for new medical needs that technology could address came from the doctors. Moreover, a salesman who understood the Medtronic pacemakers, who knew about heart-rhythm disorders in some depth, and who had no other products to push was far more likely to persuade a surgeon to try pacemaker implantation than was a salesman who represented several companies and handled many products.

At the urging of his board of directors, Bakken hired a director of marketing and sales in early 1967. Charles Cuddihy had "never heard of a pacemaker. . . . The only thing I knew about medicine was aspirin." Cuddihy, who had spent 10 years with RCA in California selling semiconductors to the aerospace industry, was astonished at the Midwestern culture he found at Medtronic. To Cuddihy, the company seemed "conservative, almost like a Baptist congregation." In an interview he said: "I thought, we are not aggressive enough. We have to go out and get to the doctors, educate the doctors who were not pursuing [cardiac pacing]. There were not many doing it. . . . So what we decided to do was have our own sales force and train them, train them, train them." With the blessing of Bakken and the board of directors, Cuddihy immediately began to terminate agreements with U.S. distributors who carried Medtronic products, acquiring the best of them and severing the relationship with the others. For both the domestic and the European market, Medtronic would have its own sales force.[17]

The company also developed other ways to maintain the close relationships that Bakken had nurtured with academic medical researchers and to reach out to the growing number of nonacademics who were implanting pacemakers. Medtronic began to offer short courses in cardiac electrophysiology for physicians and other medical personnel, maintained an aggressive program of attending medical meetings in the United States and Europe, and encouraged company engineers to develop ties to researchers in U.S. medical schools. But perhaps most important, Medtronic sales reps often observed pacemaker implantations and offered advice to the surgeon when invited to do so. This practice, too, emulated what Bakken had done in the 1950s. Some executives were uneasy about the company's exposure to possible legal liability, but being present in the operating room seemed an excellent way to demonstrate company support for the surgeon, to build long-term relationships between the sales rep and "his" physicians, and to encourage more physicians to get into cardiac pacing. Other companies soon adopted the practice with their own sales reps, and by the late 1970s most doctors had come to take it for granted. With little discussion, hundreds and then thousands of pacemaker implanters permitted company reps to join them in the operating room or the cath lab and the follow-up clinic.[18]

After a time, Medtronic's Cuddihy began to realize that Bakken, in addition to being "a technical guy," was "very much a teacher." Bakken's teaching centered on a corporate mission statement that he had prepared in 1961 when the board pressed him for a written outline of the Medtronic business focus. Three decades later, Bakken recalled that despite the financial struggles of the early 1960s, he and others had had "a very exciting, fulfilling feeling" because their products were saving lives. "It certainly was an extension of my thoughts as a child with wonder about electricity and its role in life." A shy man, Bakken had imagined as a boy that he would "get a job at Honeywell . . . and hide in the corner and do my thing there." The pacemaker had changed his life, and he wanted to let it change the lives of other Medtronic employees, too.[19]

According to the mission statement, Medtronic's first goal was "to contribute to human welfare by application of biomedical engineering in the research, design, manufacture, and sale of instruments that alleviate pain, restore health, and extend life." Point 2 stated that the company would "strive without reserve for the greatest possible reliability and quality in our products; . . . and to be recognized as a company of dedication, honesty, integrity, and service." The other points were equally idealistic. It is an old joke at the company — and apparently true — that some board members had to remind him to add a statement about making "a fair profit" as a means of achieving the other goals.[20]

Bakken not only wrote the Medtronic mission statement, he insisted that directors and executives hold retreats to discuss its implications, made sure that it was printed in company literature, and personally introduced it to every new employee. In the mid-1960s, Bakken began the practice of meeting individually or in small groups with all new Medtronic hires to explain the mission statement; eventually these meetings would take him on trips to company sites on every continent. He alone did this practice until the early 1990s, when other executives began to lead some of the meetings.[21] Discussing the Medtronic mission with new employees proved an effective way to assimilate them to the company culture as Medtronic grew. The mission statement can also be regarded as a marketing tool, for by its resemblance to professional standards in medicine, it impressed doctors that Medtronic was not just another supplier but a partner in their work.

As the years passed and Medtronic continued to grow, the stories of the 1950s — the blue babies at the University of Minnesota, Dr. Lillehei and the myocardial pacing wire, the external pulse generator with its blinking red light — took on a mythic character; and as a long-time Medtronic board member remarked, "companies live on myths." A business writer who studied Medtronic in the 1980s found that "everyone knows the story. The Garage . . . symbolized an unfettered state where technical genius and creativity could be applied for the betterment of mankind." Stories of the early days also reminded employees that the company's success depended on close collaboration between engineers and daring, innovative physicians. The mission statement and the person of Earl Bakken himself linked the garage-based business of the 1950s with the much larger and necessarily more formally structured company that emerged in the 1970s.[22]

## CORDIS

Cordis Corporation of Miami introduced an implantable pacemaker slightly later than Elema-Schönander and Medtronic but under similar circumstances. Cordis was founded in 1959 by William P. Murphy Jr., a physician with some engineering training who had worked in hematology and kidney dialysis, invented several pieces of medical equipment, and in the early 1950s cofounded the American Society for Artificial Internal Organs. "Cordis was very much a reflection of individuals rather than a corporate philosophy or corporate mission," Peter Tarjan (later chief scientist with the company) has commented. "The initial definition of the company was pretty much Bill Murphy's concept." Murphy had a knack for identifying needs that could be filled with new technologies. One emerging market that he identified quite early was intensive cardiac care in the hospital. "We named [the company] Cordis [because] it seemed to me that there was an enormous need for all kinds of gadgetry in the world of cardiology." Murphy designed several pieces of equipment, including a cardiac programmer, a machine that enabled the physician to program events such as injection of dyes into the bloodstream to occur at precise moments in the heart cycle. When he needed help with the engineering, he would retain consulting design engineers. But all of this early equipment used vacuum tubes. Late in 1959, Murphy hired engineer J. Walter Keller, who had helped design transistor control systems for U.S. guided missiles and for hydrofoils. When Keller joined the company, Cordis had twelve employees; its offices and labs took up several houses on a Miami street and its machine shop was located in a garage.[23]

"People frequently came to Cordis, and came to Bill Murphy, to do things for them," according to Keller. One of the people who came through the door was Philip Samet, a cardiologist at Mount Sinai Hospital in Miami Beach, who requested help in building an electronic device that would assist in his laboratory research by simulating a dangerous heart conduction disorder known as Wolff-Parkinson-White syndrome. This project was Walt Keller's introduction to the idea of emulating the conduction system. He told Murphy that he thought it would be possible to design an implantable pacemaker that linked the atrium and the ventricle rather than simply driving the ventricle while ignoring the atrium, but Murphy felt that Cordis could only take on such a project in association with physicians.[24]

Late in 1960, "two doctors walked in one day, a surgeon and a cardiologist, who wanted to bridge the AV node." David Nathan, the cardiologist and former chief of medicine at Mount Sinai, had come up with the idea and enlisted surgeon Sol Center. "I was always of the opinion, when Bill Chardack would give a lecture about his single-chambered stimulation of the heart in complete heart block, that what he was doing was just eliminating the consequences of complete heart block; but he was not returning the heart to a physiological state," Nathan said. "And I felt that if the heart was going to be physiological, it had to have atrioventricular synchronization." Nathan was coming back to the concept that had so vexed the participants at the Rockefeller Conference in September 1958 (chapter 3): rather than delivering the ventricular stimuli at a fixed rate, he wanted to put a sensing electrode on the right atrium and let a sensed P wave trigger the pacemaker signal to the ventricle. This arrangement would coordinate the contractions of atria and ventricles — atria first, pumping blood into the ventricles — and would permit the ventricles to vary their pumping rate.[25]

Murphy agreed to build the "AV synchronous" pacemaker that Nathan and Center needed, but he apparently believed that this would be a laboratory research project like Samet's and would not yield a commercial product. He assigned the task to Walt Keller, who designed not just a laboratory device but a fully implantable pacemaker. The new pacer could stimulate the ventricle off sensed P waves but could revert to asynchronous pacing if the atria went into fibrillation. This required a more complex circuit design than anything in a standard fixed-rate pacemaker: Keller used 12 transistors. After months of trials with dogs, Center implanted one of the new pacemakers in a human being in June 1962, just two years after the first Chardack-Greatbatch implantation.[26]

Like Medtronic, Cordis had defined itself as a medical electronics company; then an idea from the physician community propelled it into the design and manufacture of implantable pacemakers. According to Keller, Bill Murphy "didn't think that pacing would be anything other than an image maker for us." Yet even at this early stage in the formation of the industry, the Cordis Atricor pacer stood out for its sleek physical appearance and its distinctive features. When introduced to the market in 1963, Atricor was the first implantable pacemaker able to sense intrinsic electrical activity within the heart and vary its behavior accordingly.[27]

Murphy and Keller looked for further opportunities that would position Cordis as the leader in pacing innovation. Cordis added an asynchronous pacer and a model that would withhold its stimulus if it sensed a ventricular contraction (Cordis called it a "standby" pacemaker). In 1967, Keller initiated a development project that would lead several years later to the first pacemaker that could be programmed from outside the patient's body to adjust its pacing rate and stimulus amplitude. In the 1970s, many small firms would seek to carve out a position for themselves in the industry by adopting the Cordis strategy of inventing attractive and patentable new features for the pacemaker.[28]

This strategy was not without its problems. In the pre-Medicare era, when many patients paid for their own pacemakers, Atricor was more costly than the other models ($850 compared with about $400 to $600 for asynchronous pacers in 1963). AV synchronous pacing was a highly sophisticated concept for the time, but it proved difficult to embody the idea in a fully satisfactory device. The complexity of the circuitry necessitated a bulky pulse generator and reduced the life of the battery. The implantation was more complex and drawn out: the surgeon had to attach a second lead on the outside of the right atrium and, for good measure, a third lead on the ventricle for use if the main ventricular lead should fail. There were reports of problems with erratic sensing of the P wave and some unhappiness with the occasional abrupt drops in pacing rate when the device responded to changes in the atrial rate. Atricor turned out to appeal to the more venturesome physicians, usually those who were most heavily committed to cardiac pacing and who performed the highest number of implantations; Murphy found that, as a commercial product, "it was a dud." The reason, he said, was that "mostly it wasn't understood, and we did not have the ability to teach the profession why [AV synchronous pacing] was a good thing to do." Cordis did not move beyond a small niche position until, in the mid-1960s, it added an asynchronous pacemaker—the kind that "could be implanted by almost anyone who knew how to make an incision."[29]

From 1963 to 1975, Cordis developed close relationships with several innovative physicians in pacing, including surgeons Victor Parsonnet and Dryden Morse and cardiologists Warren Harthorne and David MacGregor. But the Cordis set of businesses was less sharply focused than Medtronic's. Although the firm's key products were cardiac pacemakers and angiographic catheters, Murphy's interests ranged beyond these items. For example, one in-house business was a line of valves for the treatment of hydrocephalus; from this, the company got into implantable neural stimulators in the 1970s. Cordis Laboratories, a subsidiary, developed and sold blood complements and enzyme-tagged immunological test systems—a business that put constant stress on corporate finances, as did a joint venture with Dow Chemical to manufacture and market blood dialysis equipment that Dow had developed. Murphy also acquired several small companies that owned technologies of interest to him. Overall, Cordis offered a somewhat eclectic collection of products sold in different markets. Management resisted narrowing down to implantable prosthetics or cardiovascular technologies. For its 1970 fiscal year, Cordis reported net sales of $9.7 million from all its businesses, compared with Medtronic's $22.5 million.[30]

Known for pacemakers that extended the boundaries of pacing, Cordis had problems with quality-control and production. This is conceded by two of its leading engineers, Keller and Tarjan. Top management was often distracted from the pacing business as they attended to Cordis's other efforts. Some of the early people in charge of production probably lacked the skills to oversee the manufacture of implantable devices with transistor circuitry. The company hired production engineers who were not able to institute the kinds of controls that Medtronic was putting into place. Keller attributed the weak production engineering to Cordis's location in Miami, a site he compared unfavorably to Minneapolis for high-tech talent. Scaling up for larger-volume production also proved difficult and the company was poorly prepared for the surge of demand that came its way in 1967, when it beat Medtronic and other companies to the market with its "standby" pacemaker. Unable to keep up with physicians' requests for this device, Cordis lost an opportunity to establish itself as the peer of Medtronic. Despite Keller's sophisticated designs and Murphy's beautiful blue pacemaker shells, the industry's judgment was, "Cordis can't produce."[31]

## ELECTRODYNE AND GE

Two American manufacturers played modest roles in the early and mid-1960s but proved unable to challenge Medtronic's dominant position: the Electrodyne Research Corporation and General Electric (GE). Electrodyne, of Norwood, Massachusetts, had been closely associated with Paul M. Zoll for a decade. Electrodyne's owner and chief engineer was Alan Belgard, who had developed a commercial model of Zoll's original external pacemaker and later had worked with the cardiologist on a heart monitor and a closed-chest AC defibrillator. In the Electrodyne implantable pulse generator, known as the TR-14, Belgard used the familiar blocking oscillator circuit and a mercury-cell battery.[32]

For a time around 1961–63, Electrodyne held about a 25 percent share of the small pacemaker market in the United States, but thereafter its position deteriorated. Belgard's firm was hampered by its limited capital resources and weak sales structure, and Zoll proved a technological conservative who adamantly opposed new ideas, including the two most important of the decade, noncompetitive pacing and the transvenous lead. Coming at a time of intense development in the field of pacing, this posture proved disastrous. In 1965, Belgard sold Electrodyne to Becton, Dickinson, a large and diversified medical-equipment manufacturer, in hopes that the parent company would commit resources to an ongoing effort in pacemakers. Belgard stayed with the new owners for eight more years, but he was disappointed. "With Becton, Dickinson," he later commented, "we got into too many areas. One that suffered was implantable pacemakers. The field was growing and changing very rapidly. Medtronic made a better decision; 90 percent of their work went into pacemakers." Becton, Dickinson closed its pacemaker business in 1971.[33]

GE had developed an implantable pacemaker in its electronics laboratory in cooperation with heart surgeon Adrian Kantrowitz of Maimonides Hospital in Brooklyn. This project began in 1960, apparently in response to the announcement of the Chardack-Greatbatch pacemaker. The initial model was implanted in May 1961 and, as was common with these early devices, the designers made improvements based on the experience of the early patients.[34]

The GE pacemakers had one remarkable technological feature — an external control unit that communicated with the implanted generator by magnetic induction. When taped to the skin on the patient's abdomen, the controller enabled the physician to set the pacing rate anywhere between 64 and 120 beats per minute. Kantrowitz viewed rate control as a means to safeguard the elderly patient. The heart would normally respond to major surgery by beating more rapidly for several days, so he would set the pacer at 80 or 90, then scale it back later to a more moderate rate. On the other hand, a patient who developed a fever could be given a higher rate in imitation of the normal response of the body. "Several of our younger patients," Kantrowitz added, "thoroughly familiar with the purpose and correct handling of the external control, increase their pulse rate as needed during periods of unusual physical or emotional stress." Generally, though, the intention with the GE pacemaker was not to vary an individual patient's heart rate from moment to moment, but rather to permit the physician to optimize the rate based on the patient's general condition.[35]

Because no division in GE had experience with the manufacture and marketing of implantable prosthetic devices, the task fell to the X-ray Department, which produced the pacemaker at its plant in Milwaukee. GE did not acknowledge the fundamental differences between equipment like X-ray machines and the implantable pacemaker. X-ray machines had been in use for 60 years, hospital personnel were thoroughly familiar with them, and most large hospitals had technicians who managed and repaired them. Pacemakers, by contrast, were something entirely new and were marketed not to the hospital radiology department but to the individual surgeon. Because the field of pacemaker implantation was so new, thoracic surgeons of the 1960s needed persuasion, education, and constant support from the manufacturer. GE had an attractive pacemaker, but its marketing strategy doomed the company to a small share of the market.

## AMERICAN OPTICAL

The American Optical Company (AO), of Southbridge, Massachusetts, had a history extending back to the nineteenth century as a maker of eyeglasses and optical instruments. The company had reorganized in 1956 to create a central R&D office and to disentangle its eyeglasses and instruments businesses. By 1960, AO was investigating medical, defense, and space applications in laser and fiber-optic technology. An industry analyst commented that "the management . . . has become highly oriented toward new product development in recent years." With Barouh Berkovits as its lead engineer in the medical-equipment field, AO introduced three important new medical appliances in the early 1960s: a heart monitor, the first closed-chest DC defibrillator, and the DC cardioverter for terminating atrial and ventricular tachycardias. These devices were targeted to one of AO's chosen new markets — the larger hospitals that were becoming centers of high-technology medical care.[36]

As we saw in chapter 5, Berkovits next designed a noncompetitive pacemaker. He and the company always described this design as "demand" pacing. AO in fact regarded the term as proprietary and objected when other manufacturers tried to use it. The Cardio-Care Demand Pacemaker, introduced in 1968, was American Optical's first implantable device. Berkovits and AO held an important patent, but Cardio-Care did not succeed in the market for reasons that are explored below.

## THE INDUSTRY IN REVIEW

Almost every firm in the pacemaker industry introduced some important technological innovation. These new features included the coiled-spring lead (Medtronic); sensing of the heart's own electrical activity and dual-chamber pacing (Cordis); the transvenous lead (Elema; and in the United States, Medtronic); rate adjustability (GE); noncompetitive pacing (AO and others); thin-film technology (Devices, Ltd., U.K.); and hermetic sealing (Telectronics, Australia). Collectively these innovations sketched out many of the main paths of future technological development in cardiac pacing.

Medtronic's dominant position did not rest on unquestioned technological superiority but on other factors. As noted, the company's early association with illuminati like Lillehei, Chardack, and Greatbatch was fundamental. Medtronic was also fortunate in its location. With the University of Minnesota Medical School close at hand and the Mayo Clinic just 75 miles away, the firm grew up in an area known for medical research and invention. The Twin Cities also boasted a sizable high-tech industrial sector centered on Minnesota Mining & Manufacturing, Honeywell, Sperry-Univac, and the supercomputer manufacturer Control Data. The area had thousands of technicians, production engineers, and workers with skills in electronics. As a result, Medtronic could turn for help to many small high-tech shops. If Medtronic engineers needed to conduct technical studies on, for example, biocompatible materials or the strength and electrical properties of a new lead design, it was often possible to contract the project out to a local business with special competence.[37]

In the early 1970s, Cordis employed 31 domestic salesmen for pacemakers, whereas Medtronic had more than 150. Behind this disparity lay quite different concepts of the pacemaker and of the physician who implanted it. After 1966, Earl Bakken and Palmer Hermundslie no longer tried to remain personally in charge of all aspects of company business, but gradually brought in business managers to guide Medtronic. These men tended to view the pacemaker as *an industrial product* that would be manufactured in quantity and sold to users for whom the device would in most cases remain a black box. They saw the pacemaker as a machine with a limited set of applications, all of them already known. They were, of course, interested in improving the pacemaker. Thus Medtronic was the first company to introduce transvenous pacing leads to the U.S. market and was quick to accept noncompetitive pacing.[38] At production and marketing, Medtronic excelled all other firms in the industry. Medtronic, not American Optical, produced trouble-free noncompetitive pacemakers and won physician acceptance for this new pacing mode at the end of the 1960s. In the next decade, however, the company would prove reluctant to adopt innovations that might render obsolete its own competencies or those of its clients.

Cordis's Bill Murphy and Walt Keller, on the other hand, implicitly defined the pacemaker as a device that afforded the doctor and the manufacturer an occasion for the display of technical virtuosity. Keller recently remarked, "The asynchronous pacer was nothing. Anybody can make a metronome."[39] It is true that as early as 1964, the asynchronous pacemaker was no longer a daring new technology—but as an industrial product, it was far from obsolete. Doctors continued to implant "metronomes," and Medtronic made millions of dollars on them during the latter years of the 1960s.

Murphy and Keller, especially the latter, saw many ways to improve the design of pacemakers so that the devices would do more things while drawing less current. No engineer in the industry surpassed Keller for economy and elegance in circuit designs. But it took a skilled and bold physician to put the Cordis pacers to best use, partly because defects were more of a problem than with Medtronic products and also because the functioning of the pacers themselves and the situations for which they were appropriate to implant were more difficult for many physicians to master. Murphy wanted Cordis to be known as a kind of Bell Labs for medical technology, a champion of innovation and a place earning the highest respect from physicians for its scientific and technical creativity. He seemed less concerned about market share and less interested in staying with any particular product and driving it to maturity. Cordis would repeatedly introduce an attractive pacemaker and then, instead of focusing on production and marketing so as to milk the product for a few years, push on to introduce yet another model.[40]

## EXPLOSIVE GROWTH IN THE IMPLANTABLE PACEMAKER INDUSTRY

The demand for cardiac pacemakers and leads began to take off once the Medicare program came into full operation in 1966. For the world as a whole, about 40,000 units were implanted in 1970, about 150,000 in 1975. A market analyst from 1975 estimated worldwide revenues from sales of pacing equipment at about $170 million — *worldwide* having a somewhat restricted meaning, for pacemakers were sold almost entirely in the United States, Canada, and the liberal democracies of western Europe. These countries had health-care systems committed to the use of modern medical technology.[41] Another way to gauge the growth of demand in cardiac pacing is to look at the largest supplier, Medtronic. The company had shipped about 1,200 pacemakers in 1962–63; this number rose to 7,400 in 1966–67 and to 25,500 in 1969–70. From 1967 through 1975, Medtronic sales revenue grew at nearly 50 percent per year, reaching $78 million in 1974 and $112 million in 1975. The company consistently held 60 percent or more of the U.S. market.[42]

Based on an assumption that 80 percent of Medtronic's 7,400 pacemakers went to the U.S. market in 1966–67 and that Medtronic supplied about 70 percent of the U.S. demand, we get an estimate of 8,450 U.S. pacers implanted in 1966–67; this includes both first-time and replacement procedures. For 1974–75, just eight years later, doctors performed an estimated 74,600 implantations, an increase of 800 percent.[43] The precise number is unimportant, but no one can doubt that an astonishing period of growth had begun. What, we may ask, accounts for this growth in the clinical use of pacemakers, a trend that few observers had foreseen?

Growth in cardiac pacing after the mid-1960s was the product of several factors: the demonstrated effectiveness of pacing in treating a set of disorders for which alternative treatments did not exist; the need to replace implanted pulse generators when their batteries failed; the coming of Medicare in the United States; and the framing of new uses for pacing. This surge of demand for pacemakers was behind the growth of Medtronic during the 1960s, and in the 1970s it would bring new firms into the industry.

Some of these same factors also fostered rapid growth in other cardiovascular fields besides pacemaker implantation. As W. Bruce Fye has recently shown in his history of American cardiology, facilities and programs in cardiovascular medicine and invasive procedures invented between 1955 and 1980 often showed explosive patterns of growth, whether or not their benefit to patients had been firmly established.[44] By considering in some detail just why cardiac pacing took off after about 1966, we may gain some insight into a much broader pattern in the treatment of heart disorders.

Improvements in the implantable pacemaker and the introduction of new techniques of implantation led to a broad acceptance of pacing as a standard treatment by the end of the 1960s. As pacemaker therapy found its place among the treatments available to specialists in heart disease, many more practitioners, including some nonsurgeons, were emboldened to enter the field. Many others did not implant the devices themselves but came to regard pacing as a practical treatment and began to refer their patients to specialists. The risk of complications and death from the procedure declined greatly once implanters could place the pulse generator in the patient's chest and introduce a transvenous lead. Noncompetitive pacemakers, which came into general use after 1967, reduced many of the unpleasant side effects that some patients had experienced with the earlier asynchronous pacemakers. Once these improvements were in place, physicians began to use pacing to treat *nonlethal arrhythmias*, such as lesser degrees of heart block or simple slowing down of the sinus rhythm of the heart. Pacing began to be understood as a way to improve an older person's quality of life even if the patient was not at death's door.

A rising number of first-time implants also implied a future need for replacements. Victor Parsonnet estimated in 1970 that about 46,000 patients carried pacemakers in the United States and Canada. At that time, the longevity of implanted pulse generators remained "pitifully poor," with more than half the units requiring replacement within 2 years of implantation. As physicians applied pacing therapy to more patients, many of whom were less severely ill than the patients of the early 1960s had been, the average life expectancy of patients with pacemakers rose to 10 years from the time of implantation. The very different life expectancies of pulse generators and patients implied that, on average, a new patient might require four or five pulse generators over the course of his or her remaining life. For the industry, ten thousand new patients in a given year meant several times ten thousand pulse generators over the next decade.[45]

The invention of pacing in the 1950s and its spread in the 1960s coincided with the postwar spread of prepaid hospital insurance in the United States. A response to the growing use of expensive technology in hospitals, insurance tended to reduce cost constraints on doctors and hospitals by creating a situation in which none of the three direct parties to the medical transaction — physician, patient, and hospital — had a pressing interest in economizing. As Rosemary Stevens remarks, "hospital expenditures and reimbursement mechanisms drove each other, in an expansionary spiral." By 1960, about two-thirds of the American public enjoyed coverage under some type of private hospital insurance; but the remaining one-third, including the elderly, lacked insurance and often found that the cost of hospital care was outdistancing their ability to pay out of their own pockets.[46]

In the aftermath of the Democratic electoral landslide of November 1964, a broad coalition of interest groups — organized labor, various industrial associations, Blue Cross and the private health insurance industry, hospitals, and the American Association of Retired Persons — was finally able to persuade Congress to create a federal program that would cover most costs of hospitalization and doctors' fees for Americans over the age of 65. The 89th Congress passed the Medicare bill and President Lyndon Johnson signed it into law on July 30, 1965. One year later, the federal government through the Medicare program began to pay costs associated with pacemaker implantation and aftercare in patients aged 65 and older, or about four-fifths of the population of pacemaker patients.[47] Medicare part A (hospital insurance) paid for the pacemaker itself and for hospital services and procedures, including workup and the primary or replacement implantation procedure. Medicare part B covered 80 percent of physicians' fees, outpatient follow-up care, and subsequent office visits to check on the pacer's performance.

By guaranteeing payment of "reasonable and customary" charges, Medicare greatly reduced the cost of cardiac pacing to the elderly patient, provided no incentive for the hospital or the doctor to elect not to implant a pacer in marginal cases, and signaled that care providers need not be greatly concerned about economizing in the choice of hardware.[48] This is not to imply that cardiac pacing was a tremendously costly treatment: improvements in pacemakers and the advent of new implantation techniques in fact substantially reduced the cost per patient between 1965 and 1975.[49] The point is, rather, that the existence of Medicare encouraged the spread of pacing as a treatment. Between 1967 and 1972, the number of first-time implants tripled, and overall expenditures on cardiac pacing soared. One can reasonably conclude that policymakers and the public had intended this kind of result since Medicare so clearly encouraged the acceptance and use of new medical devices and procedures.[50]

Physicians usually omit mention of Medicare in writing about the history of pacing, perhaps because they ask a different question: not why the field grew, but how it progressed in scientific knowledge and therapeutic practice. But "progress" and growth were intertwined. The creation of new knowledge about heart arrhythmias helped expand the universe of potential patients because many of the newly defined rhythm problems could be treated with pacing. The most important example came in the late 1960s, when academic cardiologists began to speak of a major disorder of the heartbeat that they called sinus node disease, or the sick sinus syndrome (chapter 7).[51] Clinicians reacted promptly: by 1975, at least one-third of the primary pacemaker implantations in the United States were being carried out to manage rhythm disorders arising from sinus node disease. The decision to use pacemakers for these new disorders accounted for half or more of the growth of pacing in these years. An industry analyst commented in 1975, "Each new indication represents a new application area resulting in another increase in the potential market."[52]

Finally, the rapid growth of cardiac pacing was rooted in the deep structure of the U.S. health-care system. Much medical care in the 1960s and 1970s was highly procedure-oriented: the culture of medicine defined the specialist's role in terms of diagnosing discrete diseases or syndromes and carrying out some defined procedure, such as an operation, that would definitively solve the problem. Those who performed procedures, including surgeons and invasive cardiologists, commanded higher incomes than nonspecialists who diagnosed and treated everyday illnesses and injuries.

Like other procedures of heart medicine, pacemaker implantation was readily popularized (see table 5). It began in a handful of academic-research centers, then was adopted in regional medical centers throughout the United States and later made its way into the smaller community hospitals. Once implantation had been redefined as a routine procedure that could safely be carried out in a hospital cath lab, almost any surgeon or internist who wished to implant pacemakers in the United States could do so, regardless of subspecialty or training. In contrast to prevailing practices in Eu-

TABLE 5. SOME INDICATORS OF GROWTH IN AMERICAN
CARDIOVASCULAR MEDICINE

| Innovation | Number at First | Number Later |
| --- | --- | --- |
| Coronary care units in hospitals | zero in 1960 | 2,300 in 1972 |
| Training programs in cardiology | 72 in 1961 | 253 in 1976 |
| Bypass procedures | zero in 1967 | 120,000 in 1979 <br> 350,000 in 1991 |
| Angioplasty procedures | zero in 1979 | 320,000 in 1991 |

Source: W. Bruce Fye, American Cardiology: The History of a Specialty and Its College
(Baltimore: Johns Hopkins University Press, 1996), chapter 8 and appendix A6.

rope, no national accrediting boards and few hospitals defined and enforced
implantation privileges in the United States. Under the fee-for-service regime that prevailed in the 1960s and 1970s, doctors who accepted more patients and performed more procedures raised their incomes. The prevailing
arrangements for third-party payment through insurance or Medicare dis-
guised both to the doctor and to the patient the true cost of the procedure.[53]

Despite the robust growth of the pacemaker industry, third-party payment arrangements, and the absence of price competition, the manufacturers' profit margins were generally lower than in similar industries, such
as scientific instruments. Thus for its 1970 fiscal year, Medtronic had after-
tax earnings of $2.138 million (9%) on sales of $22.51 million. This represented a sharp percentage rise in earnings since the early 1960s, but after
1970 margins stabilized at about 8 to 10 percent annually. Medtronic and its
competitors incurred significant R&D costs, but considerably higher marketing costs. Sales commissions and other marketing expenses averaged 25
to 30 percent of the list price of a pacemaker during the 1970s, while R&D
expenses ran around 5 to 8 percent. Companies unable to field a large and
aggressive sales force did not flourish, no matter the novelty or reliability
of their pacemakers. The patent struggle over "demand" pacing between
1967 and 1974 clearly demonstrated this principle.[54]

## THE STRUGGLE OVER DEMAND PACING
A pivotal event in 1967 jeopardized the competitive position of Medtronic
and Cordis and eventually led all the manufacturers to shift their strategies.
On October 10, the U.S. Patent Office issued a patent to Barouh Berkovits,
the director of research in the medical division at American Optical, for the

design of the "demand" pacemaker. The patent covered not just a particular circuit but the very concept of allowing a sensed R wave to inhibit the pacer from firing. The Berkovits patent thus created an enormously disruptive situation in the pacemaker industry, for the company that now held this trump card had never introduced a single implantable pacemaker to the market and would not make its implantable "demand" pacemaker generally available for another year.[55]

Greatbatch had taken note of the early papers that described Berkovits's experimental "demand" pacemakers and completed work on an implantable noncompetitive pacer ahead of Berkovits. Sometime in 1966 or late 1965, Chardack implanted the device in Buffalo, and Greatbatch later licensed it to Medtronic. This pacemaker went into production in May 1967 — five months ahead of the Berkovits patent — and even used the word "demand" on the pulse-generator jacket. Around the same time, Cordis also introduced a noncompetitive pulse generator that Keller had designed; the company called it a "standby" pacemaker.[56] The entire community of implanter physicians was in the process of shifting rapidly from asynchronous to noncompetitive pacing, but physicians usually chose pacemakers bearing the more familiar Medtronic or Cordis name over the units from American Optical. AO proved unable to capitalize on the Berkovits patent.

With the awarding of a broad patent to Berkovits and American Optical, one of the two industry leaders shifted its strategy. Cordis downplayed its line of "standby" pacemakers and tried to encourage physicians to switch to a (patented) alternative design called "triggered" pacing, wherein a normally conducted heartbeat did not inhibit the pacemaker from firing but rather caused it to fire immediately into the absolute refractory period of the heart cycle. The clinical effect was the same because this pacing mode avoided competition just as "demand" pacing did.[57] Medtronic, for its part, had no fallback technology like the Cordis triggered pacemaker. The Medtronic "demand" pacemaker sold briskly from the start (8,000 units in the first 16 months) and Medtronic's overall sales jumped sharply in the 1968 fiscal year. The firm was becoming heavily dependent on a technology to which it lacked a secure right.

Thomas E. Holloran, a lawyer by training, was executive vice president at Medtronic and had been a board member since 1960. He remembers that "the shot across the bow was a letter [from American Optical] telling Medtronic it was infringing and threatening injunctive action." Holloran believed that Medtronic should try to negotiate a licensing arrangement with AO. He flew to Boston to talk to the lawyers for the other firm, but "it was not a very pleasant or satisfactory conversation." Months passed; the talks "were very hard-nosed and were going nowhere except into litigation." Eventually, AO's parent, Warner-Lambert, stepped in to broker a licensing agreement. "But it was very much in your face, we're going to put you out of business at the start of these conversations." Under the agreement signed in 1968, Medtronic was able to go on manufacturing its own line of non-competitive pacemakers but agreed to pay annual fees to American Optical. The fees ran into the hundreds of thousands of dollars (more than $700,000 in 1971–72) and significantly cut into Medtronic's earnings.[58]

Holloran, Bakken, and other executives at Medtronic were astonished that the Patent Office had awarded Berkovits such broad patent protection. They noted that engineers had invented and patented apparently similar concepts before Berkovits had submitted his application.[59] To add further confusion, the Patent Office in November 1969 awarded a patent to Great-batch for *his* noncompetitive pacemaker.[60] This unexpected victory for Greatbatch suggested that perhaps the Berkovits patent was not as broad as it appeared to be; it strengthened the hand of people at Medtronic and Cordis who believed that the companies should challenge the Berkovits patent in court. In 1970, while continuing to pay royalties, Medtronic (joined by Cordis) filed suit against American Optical in order to determine the validity of the Berkovits patent. AO countersued in 1971.

In Holloran's words, "it looked to me like it would be a typical patent suit in which they would sue to enjoin and for damages, and we would argue that the patent was not valid, [that] it was anticipated, that somehow we were designing around it." But as the process of discovery went forward, with the legal teams on each side compiling thousands of pages of documentation and depositions, Medtronic management continued to harbor misgivings. Medtronic was the world and U.S. market leader in pacemakers, and by 1970 noncompetitive pacers were the core of its business — virtually its only product, in fact. The company was doubling in size every year or two and had an intense need for outside capital. "I think we went back to the [financial] market five or six times," Holloran said. "It would have been difficult [to raise money] were we to be enjoined [from selling noncompetitive pacemakers] or were it to appear we were going to face the consequences of a willful violation with a trebling of damages."[61]

The executives began to question their lawyers very closely about their chances of winning in court. Norman Dann, a vice president at Medtronic, recalls that the legal advisers hedged: Medtronic had a better than even chance of winning. That was not good enough for the company management. If Medtronic lost, AO would truly have the company over a barrel. There was always the risk that AO would limit Medtronic's production of noncompetitive pacemakers or even refuse to license Medtronic to produce them at all.[62]

Early in 1974, Medtronic settled with AO out of court. Medtronic conceded the validity of the Berkovits patent; AO gave permission for Medtronic to manufacture and sell noncompetitive pacemakers under license through the life of the patent. Medtronic paid $1.78 million in licensing fees it had been withholding for 1973 and 1974. The parties also worked out a payment schedule through 1985, when the patent would expire: Medtronic would pay fees that rose every three years from $550,000 in 1974–76 to $850,000 in 1983–85.[63] Subsequently, American Optical licensed the technology to almost all manufacturers in the United States and abroad. AO itself never gained a significant market share in pacemakers.

To Greatbatch, the decision to settle was "a monumental mistake"; he believed that "if the case had gone to trial, my patent would have held up." Yet in retrospect, the settlement proved highly favorable to Medtronic. Not only did it bring to an end doubts and concerns about the long-term future of the company's pacemaker line, it achieved this at remarkably low cost. The total on paper was about $10 million, or 1.5 times the company's net income for the 1974 fiscal year, but the more distant payments should be discounted steeply because years of inflation intervened; by 1983, $850,000 was worth about half of what it had represented 10 years earlier. Medtronic also succeeded in severing the link between the volume of its own pacemaker sales and the size of its payments to AO. Cardiac pacing continued to expand in the 1970s and, as it did so, Medtronic's revenues continued to rise so that the negotiated annual fees proved but a minor burden to the company. As Greatbatch conceded, "maybe Tom Holloran wasn't so dumb after all!"[64]

The success of American Optical in defending its patent and winning a stream of income from other companies alerted all pacemaker manufacturers to the strategic uses of patents. It was obvious, first of all, that companies should not allow themselves to become overly reliant on one technology but should diversify and constantly innovate. To accumulate their own portfolios of patented devices and components, the manufacturers began to redefine their relationships with the medical-research groups whose ideas had ignited the industry. Until the late 1960s, physician-researchers had often brought to the manufacturers prototypes, or at least the concepts for new products. As a Medtronic annual report observed, the company had developed new pacing products largely by "looking at products developed by others in hospitals and research institutes that might well fit our manufacturing and marketing talents." In the 1970s, manufacturers increasingly followed the lead of Cordis and AO by developing their own patentable technology in line with their own strategic planning. The AO patent dispute helped to bring product planning and development firmly within corporations' walls.[65]

Paradoxically, the struggle over noncompetitive pacing also demonstrated that market success did not depend wholly on being technologically innovative. American Optical held a master patent for demand pacing, but other companies beat AO to the market and offered noncompetitive pacers that were at least as good as its model.[66] AO was a respected manufacturer of medical instruments for the research lab and the hospital, but it proved less effective at responding to the needs of physicians. One AO distributor from the mid-1960s later commented that AO had never figured out the medical-device business once its market began to change from research labs to clinicians. The AO medical division was created only in 1963; many of its engineers and managers had simply been reassigned from the corporate R&D office or from other divisions and had little knowledge of the evolving medical-device market. Despite their technical success with the demand pacemaker, they did not develop strong and lasting contacts with physicians. Critically, they did not understand how important it would be to offer service support to physicians.[67] Through Berkovits, American Optical played a brief but important part by helping to define and achieve reliable pacemakers in the late 1960s. But the company lacked many of the attributes required for sustained success in the pacemaker industry.

## END AND BEGINNING

Pacers that incorporated all the major improvements of the 1960s were available from several companies by 1970. Manufacturers of implantable pacemakers had responded effectively to the entrepreneurial opportunities opening up in the 1960s — a decade that saw the expansion of health-insurance coverage and the creation of Medicare in the United States and higher state spending on health care in western Europe. The typical pacemaker of 1970 was a representative artifact of the postwar industrial democracies: a mass-produced electronic device aimed at a growing mass market, like the transistor radio or the color TV.

Medtronic had committed itself most clearly to the idea of the pacemaker as a mass-produced industrial product. The principal Medtronic product innovation of the early 1970s was the TeleTrace transmitter, a portable device that enabled the pacemaker patient to send an ECG by telephone to a physician or follow-up service. Advertising for TeleTrace emphasized that it would help to simplify and make routine the follow-up of pacemaker patients, while reducing the probability of emergency or premature replacement of pulse generators. TeleTrace was highly compatible with the company strategy for continued growth: transform cardiac pacing from a rather daring treatment into a safe and transparent therapy.[68]

With the achievement of reasonable reliability by 1970, the pacemaker did not settle down and cease changing. New entrepreneurs were entering the field, and their interests were better served by transforming the pacemaker than by trying to compete under the existing rules. The dominant companies also had reason to promote further technological change, because the end of innovation would bring commodification of the pacemaker — that is to say, technical knowledge available to all; pulse generators and leads all more or less identical in performance and hence interchangeable. Such a situation, had it developed, would have encouraged still more entrants and led quickly to price competition, declining profit margins, volatile market shares, and, for the physician, a decline in the prestige that came from association with a leading-edge treatment. Manufacturers and physicians had a common interest not only in the expansion of pacing, but in its ongoing technological transformation.

Events in the next few years took the pacemaker industry and its associated medical community into a troubled period, but one that did not lead to commodification. Manufacturers and doctors demonstrated once again the remarkable malleability of cardiac pacing; between 1972 and 1983, they redefined and reconstituted pacing once again. By the end of the 1970s, the newer pacemakers had become more flexible and more complex. Without exception, the important innovations came from firms other than Medtronic.

## CHAPTER 7

# THE PACEMAKER BECOMES
# A FLEXIBLE MACHINE

A succession of radical innovations in cardiac pacing broke up the brief era of stability that had begun in the late 1960s. New themes emerged: doctors' uneasiness about the growing complexity of cardiac pacing, the public's anxieties about the possible dangers of a life-sustaining technology, and broadly shared concerns about the growth dynamic of the field of pacing. These matters will occupy the next three chapters. The fact that these issues had moved to center stage by the middle and late 1970s indicated that the postwar era of unquestioning enthusiasm for new medical technology was coming to an end.

In the 1970s, the device manufacturers drove technological innovation with some of the most important new features coming from small start-ups. Companies innovated partly to adapt the pacemaker to a wider range of heart-rhythm disorders, but also because innovation brought prestige to a company and helped it build its portfolio of patents; and because, through innovation, the industry avoided the commodification of the pacemaker. Manufacturers sought to balance several goals: meeting the expressed wants of their customers, displaying their virtuosity in design and manufacture, and continually reassuring the public that implantable pacemakers were safe and "friendly" — all this in the context of evolving public attitudes about modern medicine and shifting government policies toward the device industry.

Once a manufacturer had introduced a new pacemaker that included some significant novelty, doctors had choices to make. They could get acquainted with the new model and try to put its capabilities to work for their patients; they could avoid the new device; or they could treat it as if it were an older model by ignoring or underutilizing the new features. In this complex set of interactions between manufacturers and physicians, the uses of the pacemaker repeatedly came up for redefinition.[1]

## A NEW DISORDER OF HEART RHYTHM
By about 1970, manufacturers and pacemaker implanters had grown accustomed to working with noncompetitive pacemakers and transvenous leads.

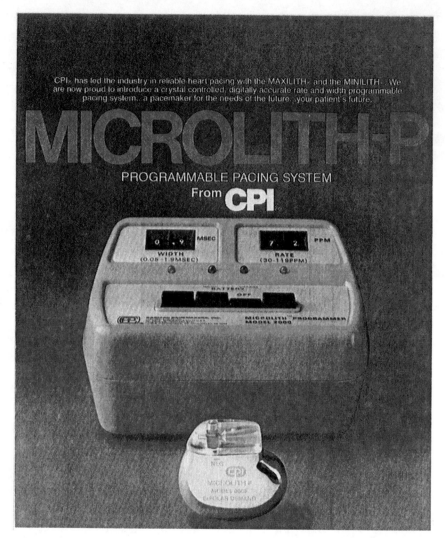

CPI's Microlith-P Pacemaker and Model 2000 Programmer, 1978. The Microlith-P embodied the major innovations of the 1970s in cardiac pacing: hybrid circuitry, a long-lived lithium battery, a hermetically sealed titanium shell, and noninvasive programmability for the rate and amplitude of the pacing stimulus. CPI offered a six-year product warranty on this pacer and emphasized that its Model 2000 programmer was easy for the busy doctor, nurse, or technician to set up and use. (Courtesy of Guidant Corp.)

Sudden device failures were becoming uncommon. After a decade of development, the field of implantable cardiac pacing had attained a measure of stability. At a pacing conference in November 1968, virtually every paper assumed that heart block remained the sole indication for permanent pacing.

That same year, academic cardiologists began to discuss a new disorder of heart rhythm that they called sinus node disease, or the *sick sinus syndrome* (SSS). These names lumped together several known disturbances of the heartbeat involving default of the sinus node, the source of the electrical impulses that trigger atrial and then ventricular contraction — the heart's natural pacemaker. If heart block is a failure to pass the heartbeat on from the atria to the ventricles, SSS is a set of disorders in the formation of the heartbeat.[2]

The sick sinus syndrome covered a diverse list of symptoms: persistent and noticeable slowdown of the firing rate of the sinus node; an inadequate pickup in the heart rate when a person became more active; sinus slowdown alternating with a rapid atrial rate. All could begin episodically but later become fixed. In more severe forms, the impulse might fail to spread beyond the sinus node. In the absence of their normal signal, the atria might fibrillate transiently or continuously and the ventricles might adopt a slow rate of contraction dissociated from the atria.

Many elderly people have a bit of sinus bradycardia, and their hearts may not speed up as promptly or as energetically as the hearts of the young. These are mild symptoms of SSS. But investigators reported that some people experienced more troubling symptoms, such as dizziness, fatigue, transient blackouts, and organ failure. Most of the symptoms, however, were not unique to SSS and could vary greatly from one patient to another. The same patient could manifest a range of symptoms from one office visit to the next, with some patients showing no clear symptoms at all except for slight irregularities in the ECG tracing.[3] Because of the erratic course of the "disease," diagnosing a failing sinus node could be difficult, especially in its early stages.

Physicians had detected and described some of these abnormalities decades earlier, but the sinus node had come in for renewed attention in the early 1960s, in tandem with the growth of pacing for heart block. The pacing community began to pay attention to the syndrome at the end of the decade. From papers in the journals and presentations at medical meetings, they learned that sinus node disorders, though often difficult to diagnose, were not rare. There also seemed general agreement in the early 1970s that most such disorders did not present the same danger of sudden death as did complete heart block.[4]

A consensus quickly took shape that pacing was the best treatment; and precisely because "the exact progress and timing of the complications of [SSS] are still unknown in great detail," it seemed prudent to "consider installing a pacemaker, for safety's sake," even before the patient had experienced the more serious symptoms. According to the leading expert on sinus node disease, the clinician "need not wait" for symptoms "to be intolerable"; as soon as "symptoms of any note" appeared, "a pacemaker had best be installed." Indeed, she added, "periodic or sustained SB [sinus bradycardia] can no longer go unchallenged, *even if asymptomatic.*" These statements opened the way for a substantial expansion of cardiac pacing by adding a large new class of rhythm disorders and by redefining the pacemaker as a prophylactic device, insurance against possible (but unpredictable) future deterioration in a patient's condition.[5]

Pacing for SSS came on with a rush in the early 1970s. In the first of his repeated surveys of cardiac-pacing practice in the United States, published in 1971, Victor Parsonnet did not inquire about SSS. By mid-decade, at least one-third of the new pacemaker implantations in the U.S. were being carried out to manage the condition. This new indication for pacing coincided with a doubling of the number of first-time implantations between 1972 and 1975.[6] It is not difficult to account for the growing use of pacing for sinus node disorders: the standard pacemakers of that era seemed as appropriate for SSS as for heart block. Doctors could simply carry out more procedures of the sort they already knew how to do. The sick sinus syndrome did not, at first, seem to require new kinds of pacemakers.

**NEW IMPLANTERS ENTER THE FIELD**

At the beginning of 1960, no physician had implanted a pacemaker in the United States, but five years later, approximately two hundred physicians or physician groups were doing so. As pacemaker reliability improved and the implantation procedure came to be better understood, more physicians entered the field. Medicare payments to hospitals and physicians got under way in July 1966; this also gave a powerful boost to physician acceptance of the new procedure. The number of implanters almost tripled between 1965 and 1970 and nearly doubled over the next five years. By 1975, about twelve hundred physicians or groups were actively implanting pacers in the United States.[7]

Over the decade from 1965 to 1975, the average number of implantation procedures that a doctor or a group performed in a year also rose substantially, from about 25 per implanter or group in 1965 to about 60 in 1975. These averages disguise the diversity in cardiac pacing practices. A few groups implanted 100 or more pacemakers in 1975; Parsonnet cites a team of four physicians in New Jersey who implanted 350 pacemakers in 10 different hospitals. On the other hand, the manufacturers reported to Parsonnet in 1972 that between 25 percent and 61 percent of implanting physicians (estimates varied) treated fewer than five new patients a year. As the number of implanting physicians rose, the proportion doing only a few procedures a year rose as well.[8]

Surgeons continued to implant most pacemakers in the United States. Although the transvenous pacemaker lead in theory had opened the field to cardiologists, experience had shown that in some proportion of patients — perhaps 30 percent — the cephalic vein was too narrow to accept a lead. In that case, the implanter would use the external jugular or the subclavian vein. But opening these veins and introducing the lead was a more complex and a riskier procedure, one usually left in the hands of a surgeon. A cardiologist would typically diagnose a treatable disorder of heart rhythm and refer the patient for a pacemaker; he or she might participate in the implantation by helping to position the lead in the patient's right ventricle and supervising tests of the electrical properties of the lead.

At the end of the 1970s, manufacturers brought out introducer kits that greatly eased the procedure of introducing a transvenous pacing lead. The kit includes a hollow needle through which a temporary guide-wire is introduced to the vein. Over the guide-wire, a sheath-dilator is inserted. The dilator and guide-wire are then removed and the lead is inserted through the sheath. Finally, the physician removes the sheath by tearing it down its length, leaving the lead in place. By making it possible for nonsurgeons to use the subclavian vein, the introducer kit was of utmost importance in bringing cardiologists into the practice of implanting pacemakers and, later, defibrillators.[9]

In the 1970s, many surgeons who implanted pacemakers were only tenuously connected to the community of pacemaker physicians. Some worked with pacemakers primarily because it was a new procedure that had gained considerable attention in the national press and in the medical world; many had a rather superficial knowledge of microelectronics, battery chemistry, tissue reactions to implanted electrodes, and the subtleties of heart arrhythmias. Pacemaker implantation enhanced their reputations and added to their incomes without being as stressful for the surgeon or as risky for the patient as some kinds of heart surgery.

In western Europe, the growth of physician interest in pacing took a different course. Contrary to the conventional image of U.S. health care as being uniquely receptive to new procedures and devices, some countries in western Europe by 1972 had implantation rates per million population that matched or exceeded rates in the United States. For the region as a whole, rates varied greatly by country (appendix B). However, the most striking difference between the two sides of the Atlantic lay not in national implantation rates but in the distribution of pacemaker procedures within each hospital system. In Sweden, for example, all 2,172 implantations took place in 23 medical centers; in West Germany, the estimated 9,600 procedures were confined to 129 hospitals. But in the United States, as Parsonnet commented, "the concept of the pacemaker center as a place to which all patients with pacemakers are referred is unknown." Instead, "pacemakers are implanted in hospitals of all sizes and affiliations and by many independent types of specialists, usually surgeons." Almost no U.S. hospital had a pacemaker center, defined as an appropriately staffed service devoted exclusively to pacemaker work.[10]

The emerging national pattern in the United States — a rising number of implanters, many of them performing only a few procedures each year — dovetailed with the standard pacemaker design of that period. Occasional implanters, defined as doctors who did fewer than 12 pacemakers per year, needed reliable and comprehensible pacemakers requiring few choices and little post-implantation management.

Occasional implanters tended to think of the pacemaker as a finished product carrying with it a settled body of medical doctrine. They were cautious about using new and complex equipment unless the innovation addressed their insistent demand for smaller, more reliable pacemakers or until a substantial body of clinical experience had been accumulated and reported in the medical journals. As early as 1973, some expressed dismay "over the profusion of new pacemakers and designs and wondered how they could get unbiased opinions regarding manufacturers' claims, and details on the potential risks of new units."[11]

Senior pacing experts such as Victor Parsonnet, Michael Bilitch, and Seymour Furman often found themselves in the role of intermediaries between these practitioner implanters and the manufacturers. They bridged the worlds of the manufacturer and the clinician, helping industry people to assimilate and use new clinical information and speculating on future directions in clinical electrostimulation, while informing their physician colleagues about the possibilities and limitations of new pacemaker designs and what to expect in the future.[12]

Beyond their oft-repeated requests for long-lived batteries and small pulse generators, the occasional implanters did not actively seek to influence the field of pacing. They willingly accepted new hardware and new techniques promising greater simplicity and reliability, but as a group often proved rather slow to adopt other innovations. Academic specialists in cardiac pacing of course shared the occasional implanters' desire for simplicity and reliability. But the occasional implanters seem to have had a larger vision in mind: the words *simplicity* and *reliability* implied for them something like an industrial process, a notion that most patients in heart block could be treated in much the same way, using the same techniques of implantation and pacemakers that all had more or less the same capabilities. This assertion is difficult to prove because the occasional implanters did not publish many papers, but as a hypothesis it helps make sense of their resistance to some major innovations of the 1970s and later.

An early sign that academic physicians and the more numerous practitioners had significant differences in outlook and practice showed up in a brief paper that Bilitch, an academic cardiologist at the University of Southern California, presented at an international conference on pacing held in the Netherlands in 1973.[13] Bilitch had interviewed 104 practitioner pacemaker implanters in southern California. He discovered that about 3 physicians in 5 had "some sort of overall follow-up plan," but that many of the plans were "unrealistic." Bilitch judged a plan to be realistic "if the physician demonstrated that he performed regular pacemaker evaluation, took into account the functional characteristics of the pacemaker . . . and performed one or more tests which could be expected to assess pacemaker system performance and integrity." Out of Bilitch's 104 practitioners, only about 3 in 8 used "realistic" follow-up procedures. But this proportion actually varied quite a bit by specialization: 18 of the 30 cardiologists followed realistic plans, but just 1 of the 8 general surgeons did so. Performance also varied by number of patients followed: 65 of the practitioners followed fewer than 25 patients each, but it was those who followed 25 or more who were more likely to have realistic arrangements for follow-up.

Overall, Bilitch's report was disheartening, especially because he argued that his 104 respondents were similar to practitioner implanters all across the United States. He found that many doctors knew little or nothing about then-standard techniques of assessing the condition of the implanted pacemaker. The widely used noncompetitive pacemakers, which fired only "on demand," could readily be switched to the asynchronous mode by placing a magnet on the patient's chest over the implanted generator; then the doctor could count the number of pacemaker stimuli per minute. But some doctors in the Bilitch study did not know about use of the magnet, and others used it only sporadically. Bilitch advised that it would be pointless to add more sophisticated technology for patient follow-up until pacemaker physicians had accepted the quite simple techniques and equipment then available. "More complex operational systems," he concluded, "will often not be used in the intended manner and may inadvertently place the patient at risk because of the lack of expertise of the physician who is following the patient."

## NEW MANUFACTURERS

According to one analysis of the pacemaker industry, world sales of pacemakers and related equipment totaled about $200 million in 1975 and $600 million in 1980; for the United States, sales were $105 million in 1975, $345 million in 1980, for an average annual growth rate of 27 percent. But the high inflation of the late 1970s complicates these data, so the growth in the number of pacemakers sold—about 11 percent per year from 1975 to 1980—gives a more realistic picture of the U.S. market.[14]

In 1973, Medtronic held about 58 percent of the U.S. market (in pacemakers sold) and Cordis about 23 percent.[15] The two companies had somewhat different profiles: Cordis was known for products that were slightly out of the mainstream and appealed to specialists in large pacemaker clinics, whereas Medtronic was firmly established among physicians who implanted relatively few pacemakers. The dominance of Medtronic and Cordis led Becton, Dickinson, General Electric, and American Optical to pull out of pacing entirely rather than hang on with less than 5 percent of the market while exposing themselves to the risk of lawsuits and adverse publicity in the event of product failures. Becton, Dickinson closed its Electrodyne pacing subsidiary in 1971 and GE sold its pacing business to Telectronics, an Australian firm, in 1976, giving Telectronics a stronger foothold in the U.S. market. Warner-Lambert, having acquired American Optical in 1967, spun off AO's pacemaker business as a new firm called American Pacemaker in 1975. American Pacemaker never again played a significant role in the industry and was eventually absorbed by another company.

Industry observers no longer defined pacemakers merely as "medical equipment" or as "electronic devices." To compete in the industry, a firm required distinctive capabilities in design and manufacture because the pacemaker was implanted within the human body. Most established medical manufacturers kept their distance from the pacemaker industry, but the lure of strong and steady market growth attracted eight entrepreneurial start-ups between 1971 and 1974.[16] Five of the newcomers eventually fell by the wayside, but the other three managed collectively to take a huge share of the domestic pacemaker market away from Cordis and Medtronic. They accomplished this by combining innovative technology with aggressive selling practices and by capitalizing on the product woes that the two established firms encountered in the mid-1970s.

## NEW TECHNOLOGY

Tiny companies can be more dangerous competitors than large ones. The newcomers to cardiac pacing sought to differentiate themselves quickly and decisively from Medtronic by introducing many novel (and patentable) features and by frequently announcing new and improved models. As a result, when cardiologist Doris Escher and her colleagues at Montefiore surveyed the field in 1978 they observed that 16 pacemaker brands were then available in the American market— 14 from U.S. companies and 2 from abroad— and that manufacturers were offering about 130 pulse generator models. This was five to six times as many models as had been available eight years earlier, and the newer models collectively had a greater array of features. Doctors also had to be able to troubleshoot earlier models no longer available but still implanted in thousands of patients. At the Montefiore pacemaker center, staff were following 1,404 patients who carried 87 pacemaker models from 13 manufacturers.[17]

Technological innovations between 1973 and 1980 rendered obsolete the reliable pacemaker of 1970. First, manufacturers resolved the problem of pacemaker longevity by accepting the lithium battery, a long-lived power source free of the problems associated with the nuclear generator. Soon afterward, the device companies introduced noninvasively programmable pacemakers, followed within a few years by multiprogrammable units. The lithium battery greatly simplified the job of the pacemaker doctor. But at the same time, some manufacturers began to modernize their pacemaker circuitry by employing integrated circuits. Instead of operating according to the manufacturer's settings, the pacemaker was becoming a highly flexible tool that the physician could repeatedly reconfigure. For economy's sake, we can refer to the result of these technological innovations as *the flexible pacemaker.* Implanting physicians accepted the lithium battery immediately but proved considerably more reluctant to explore the uses of programmable pacemakers. For many of them, the new dominant design might have been called *the complex pacemaker.*

## WILSON GREATBATCH, CPI, AND THE LITHIUM BATTERY

The invention of a long-lived pacemaker battery based on lithium chemistry precipitated a cascade of radical changes in cardiac pacing. As the 1970s began, the mercury battery was the principal roadblock to acceptable pacemaker reliability. American and European battery makers had extended its worklife beyond two years, but physicians still found it unsatisfactory, primarily because hydrogen gas, a byproduct of the chemical reaction in the cell, had to be vented to the outside of the pulse generator, and this meant that the generator shell had to be gas-permeable. Because it was not hermetically sealed, the generator was open to the invasion of body fluid, which could then come into contact with the circuitry and cause it to short out. Virtually everyone viewed the mercury battery as the most important impediment to reliable cardiac pacing. Nuclear pacemakers were much in the news, but as of 1972 doctors had had almost no chance to evaluate this alternative to the mercury battery.[18]

The device companies had made improvements to subsystems of the pacemaker so as to use the battery more efficiently. Noncompetitive pacing saved energy because the pacer would fire only when needed and the sensing function required little power. Led by General Electric and Medtronic, manufacturers also introduced a function called *rate hysteresis* to their noncompetitive pacemakers: they configured the pulse generator to prolong its escape interval between the last natural R wave and the first pacemaker impulse. This meant that the device would not begin to pace during minor variations in the natural heart rate such as occur during sleep; it would go to work only during a true episode of bradycardia or block. This too saved on the battery.[19]

Still another way to get around the limitations of the mercury battery was to redesign the pacemaker electrode. Manufacturers understood that a smaller electrode would require less energy to stimulate the heart. At Cordis, there was considerable work on the design of electrodes that led, for example, to a model with a 2-millimeter tip in 1969. However, almost all pulse generators of that period had fixed pulse width, pulse amplitude, and pacing rate; consequently, manufacturers could not take full advantage of more efficient electrodes without substantially redesigning their circuitry.[20]

Wilson Greatbatch had understood for years that the mercury-zinc battery significantly limited the longevity and reliability of his invention, the implanted pulse generator. In 1964, he had developed a method for assessing the condition and estimating the remaining life of a mercury battery by an X-ray procedure. As a board member and technical consultant at Medtronic, Greatbatch had tried to push the company toward exploring entirely new battery chemistries, but Earl Bakken and other company executives resisted the idea. According to Greatbatch, Bakken was "very reluctant to antagonize Mallory. . . . He said, that's my only source [for batteries]. . . . I don't want to get them mad at me." Finally, in 1968, Greatbatch terminated his formal ties to Medtronic, in part out of impatience over its technological caution on pacemaker batteries. Temperamentally unsuited for the role of an organization man, he sold his patents to Medtronic and resigned from the board of directors but remained on cordial terms with Bakken and others at the company.[21]

During the late 1960s, working out of his small R&D company near Buffalo, Greatbatch instituted a wide-ranging search for alternatives to the mercury-zinc cell that led him through biological power sources, rechargeable nickel-cadmium batteries, and nuclear generators. He also seriously explored cells based on lithium as the anode with various cathodes, but results were poor: lithium-nickel sulfide cells could not be sealed; lithium-sulfur dioxide required pressurization, and several test cells burst.[22]

By coincidence, engineers at a small battery manufacturer in Baltimore — the Catalyst Research Corporation (CRC) — had invented a cell with a lithium anode and a cathode of iodine in 1967. CRC did not find a commercial use for this product until word of the invention reached Greatbatch in 1970, when Fred Tepper from CRC called him. Despite his earlier difficulties with lithium, Greatbatch realized on examining prototypes that the CRC battery would be suitable for implanted pacemakers. He soon signed an agreement with Catalyst to develop and produce the battery under license.[23] By 1973, his own firm of Wilson Greatbatch, Ltd. was manufacturing lithium pacemaker batteries based on the CRC design.

Because of his prestige in the pacing community and his effectiveness as a champion of technology he believed in, Greatbatch was able almost single-handedly to turn the industry to lithium; in fact by 1978, a survey of pacing practices indicated that only 5 percent of newly implanted pulse generators still used mercury-zinc batteries.[24] But this success did not come easily at first. In 1971, Greatbatch took CRC's lithium battery around to several pacing firms including Medtronic, the Swedish manufacturer Siemens-Elema, a French company, Ela, and a new Italian firm, LEM Biomedica. Some of these manufacturers were willing to bench-test prototypes of the battery but none was prepared to abandon the mercury battery on Greatbatch's urging. As one Medtronic executive later described his company's position:

> We, of course, were very interested in extending [battery] longevity; and we had worked with the Mallory Company . . . for years and years in an effort to take this [mercury] battery, which was not designed for pacemakers, to

make it reliable, and to extend the life. We were really committed to that battery and we felt comfortable with it, even though it provided a limited lifetime.

Medtronic, being a market leader, was perhaps conservative. It was our feeling that physicians would not be quick to accept a new technology, a new battery, even though they were not necessarily happy with the life of a mercury [battery]. . . . And so we were not quick to move into the lithium power source in our units. Even though we had a market share of 50 to 60 percent, we were not willing to subject patients, whose lives were dependent on cardiac stimulation, to a new, as yet unproven technology.[25]

Tom Holloran, executive vice president at Medtronic in 1972, remembers that "Bill Greatbatch sent us a test block of batteries. They were encapsulated in epoxy, and as an indication of his belief in their longevity, he encapsulated a cicada in the block to indicate that these were 17-year-long batteries."[26] Medtronic began to test successive versions of the CRC-Greatbatch battery—but neither Medtronic, Elema, nor any other established company was willing to abandon mercury. Lithium was not only an unproven technology, it would entail the redesign of pacemaker circuitry.[27]

LEM Biomedica, a start-up company in Florence, Italy, obtained prototypes of the new battery, developed appropriate circuitry, and produced a lithium-powered asynchronous pacemaker in 1972. After a rather brief period of bench-testing during which it functioned well, surgeons used the LEM model in two clinical cases in 1972 without Greatbatch's knowledge. When the pacing rate in the second unit dropped gradually from 74 to 63 beats per minute over 16 months, surgeon and patient agreed to remove it. Subsequently, LEM engineers redesigned the circuitry to compensate better for changes in internal cell impedance. This case, though probably unknown in the United States, illustrates why established companies were reluctant to pioneer a new pacemaker battery.[28]

One entrepreneur on the periphery of the U.S. pacemaker industry, however, had enormous faith in Greatbatch. This was Manuel A. Villafaña, an exuberant man who had joined Medtronic in 1967 and had opened Medtronic's South American sales and technical support center in Buenos Aires two years later. He left Medtronic in 1970 but learned about the lithium battery through a chance encounter with Greatbatch in 1971. Villafaña had known Greatbatch for years; his attitude was, "If Bill Greatbatch says it works, it works." Villafaña decided to found a company that would manufacture lithium pacemakers. He raised $50,000 from private sources, lined up more financing from a venture-capital firm in Minneapolis, and with three colleagues, two of them former managers at Medtronic, founded Cardiac Pacemakers, Inc. (CPI), in 1972. The new company opened its doors at a suburban site within five miles of Medtronic headquarters.[29]

According to Greatbatch, "I warned Manny that lithium was too new to risk his company on. I told him to go to mercury and then transcend into lithium. He said, 'No, if I'm going to buck the big boys, it's got to be with something completely new, that solves all the problems, but is too risky for them to undertake.'" Within a few months, engineers at the new company led by Rich Kramp had put together an implantable noncompetitive pulse generator. Clinical trials began in November 1972, and CPI released the first clinically reliable lithium pacemaker as its initial product in 1973. As a Medtronic manager commented years later, Villafaña and his associates were "willing to roll the dice on lithium. . . . And they were right—lithium turned out to be a good chemistry." Villafaña recalled, "we were open 24 hours a day; we were working three shifts. We couldn't make them fast enough."[30]

The battery that CPI pioneered had a lithium anode, an iodine cathode, and a solid-state electrolyte of lithium iodide. Some manufacturers experimented with other cathodes such as silver chromate, copper sulfide, and bromide. These batteries had somewhat different properties and varied actuarial survival performance, but all enjoyed significant advantages over mercury. Pacemaker physicians soon came to recognize two key virtues. Although lithium batteries stored less energy than mercury batteries of the same size, the voltage decline with lithium was extremely gradual until the very end of battery worklife. This meant that the pacemaker was able to utilize a far greater proportion of the energy stored in the battery before the voltage dropped to the point where replacement was called for. For this reason, manufacturers were able to promise seven years of worklife with a lithium battery. Second, the chemical reaction in the battery generated no gas byproduct, so the pulse generator could at last be hermetically sealed in a metal can. As manufacturers adopted hermetically sealed shells, made usually of titanium, failure of pulse generators from incursion of body fluid became a thing of the past. Greatbatch was correct when he wrote that "the pacemaker battery is no longer a significant clinical problem."[31]

For the physician and the patient, the lithium battery made life easier because it meant longer pacemaker life with fewer unexpected failures, yet entailed no changes in physicians' implantation procedures. But for the established manufacturers, the success of the new battery inaugurated a period of lurching change in the industry. Lithium satisfied an important and long-standing demand, and CPI carved out a significant market share.

By 1977, the company had gained 11 percent of the growing U.S. market. This share would probably have been even larger, but CPI had problems keeping up with the growth of demand. Other manufacturers gained time to gear up their own production of lithium pacemakers; CPI's market share topped out in 1979, then began to fall.[32]

Medtronic and Cordis delayed their adoption of the lithium battery, with unfortunate results for both companies. Medtronic announced its first lithium pacer (Xyrel) in the spring of 1976 but did not reach full production until February 1977. Medtronic had committed itself to a mercury model called Xytron as its flagship product in 1974, taking the position that mercury pacers could routinely attain worklives of five years or more.[33] Then in 1975 some Xytrons began to fail prematurely. This disaster accelerated physicians' shift to lithium pacemakers and away from Medtronic. Cordis developed a proprietary battery of its own based on lithium and copper sulfide. Introduced in 1976, the battery led the company into a product failure that seriously damaged its reputation. These difficulties at Medtronic and Cordis opened a huge opportunity for smaller firms in the industry. They were not slow to take advantage of it.

## CORDIS, HYBRID CIRCUITS, AND "PROGRAMMABILITY"

Integrated circuits, along with the lithium battery, undergirded the revolution in pacemaker functionality after 1975. The two innovations might appear unrelated; certainly the key innovators worked entirely separately. But in retrospect it is clear that the lithium battery and integrated circuitry were intertwined innovations, because long-lived pacemakers virtually required some adjustable features so that pacers could be repeatedly retuned as patients aged and their rhythm problems changed. The new circuitry made this adjustability possible.

Implantable pacemakers of the 1960s had used discrete circuitry in an epoxy encapsulant. To interconnect circuit components, a manufacturer would mount them on a printed circuit board that had an electrical conductor pattern etched on its surface; several boards could be stacked "cordwood-style."[34] Discrete circuitry placed a high demand on the battery and took up space within the pulse generator, so designers faced strict limits on circuit complexity. Pulse generators were confined to fairly simple behaviors, as in a noncompetitive pacer from around 1970 that could recycle its timing circuit if it sensed an R wave or deliver a ventricular stimulus at a preset rate if it did not sense one. The constraints on circuitry contributed to the sense, pervasive among implanting physicians, that the pacemaker had attained a stable design. But the larger field of semiconductor electronics had moved on: by the last years of the 1960s, the "transistorized M.D." was already an archaic figure.[35]

As the pacemaker companies adopted lithium batteries, they also began a transition from discrete to hybrid circuits in an effort to satisfy physicians' demands for smaller and more reliable pulse generators. The hybrid circuits used in the pacemaker industry (and manufactured by outside suppliers such as Motorola and RCA) represented partial integration, with some circuit elements printed on a ceramic substrate and others individually mounted. These assemblies were certainly more compact and were intended to be more reliable because the manufacturing process was partly automated and the entire circuit package could be hermetically sealed. A hybrid circuit also drew less power from the battery because it used power only when performing an action such as opening or closing a switch. Overall, hybrid circuitry enabled manufacturers to downsize their generators while also providing more reliability at no sacrifice of pacemaker longevity.[36]

Hybrid circuitry also permitted more flexible pacemaker behavior. The pacemaker that broke decisively with the dominant design circa 1970 was the Cordis Omnicor model. Omnicor included an integrated sensing amplifier and two integrated digital logic circuits in an overall hybrid design. It behaved like a standard noncompetitive pacemaker, but it was adjustable: the physician could *noninvasively* choose from six pacing rates and four stimulus amplitudes. The new circuitry made this flexibility possible — but equally important, Cordis engineers, pursuing a suggestion that president Bill Murphy had made, designed a miniature magnetic reed switch. With a handheld programmer, the physician transmitted a series of magnetic pulses that opened and closed the switch. A counter noted the number of changes in the position of the switch and associated this number with a corresponding value for rate or output. Despite its new capabilities, Omnicor made less demand on the battery than a conventional noncompetitive pacer.[37]

At the time, most pacemakers delivered their stimuli at preset amplitudes and durations, usually 10 milliamperes and 1.5 or 2.0 milliseconds, but these settings far exceeded what was required to trigger a heartbeat. Engineers at Cordis designed Omnicor to be tuned down to a lower stimulus amplitude after the chronic threshold of stimulation had established itself, usually a week or two after the implantation. Reducing the output would save on the battery and further extend the life of the pacer.[38]

Medtronic engineer Bob Wingrove had breadboarded a prototype programmable pacer and demonstrated it to Bakken and other Medtronic executives in 1969, but the company did not take the idea further at that time. In contrast, management at Cordis was receptive to daring and innovative projects. Walt Keller, the senior pacemaker engineer at Cordis until 1968, had been intrigued by CMOS (complementary metal oxide semiconductor) transistors for some years and believed that they represented the future for pacemaker engineering. He inaugurated the Omnicor project in 1967 with Murphy's strong support.[39]

First implanted late in 1971, Omnicor was introduced to the market in 1973. Leading pacemaker physicians, including Parsonnet, Furman, and Morse, had been involved in clinical testing and fully supported rate and output programmability. These men were interested in pushing the field of pacing toward more sophisticated technology; they looked forward to a time when they would be able to control additional parameters such as stimulus duration and the pacemaker's sensitivity to intrinsic heart activity. They presented the new pacer to the larger community of implanters through published reports, explaining in detail how the physician could use rate and output adjustability to optimize the pacemaker for each patient's individual needs.[40]

Omnicor and other pacers with adjustable features were called "programmable," but this term is somewhat misleading. For years they were hard-wired, not software-driven, and could not be reprogrammed in the sense of being given new instruction sets. But Omnicor, though simple by later standards, portended a coming era in which engineers would design complex, fully integrated circuits comprising thousands of elements. There would be no more discrete components; everything, including resistors and capacitors, would be on the chip. The new circuitry promised smaller and longer-lived pulse generators that would be capable of more flexible action than any pacer of the 1960s.[41] All of this undermined the dominant design in pacing, which emphasized reliability, simplicity of design, and ease of use.

Hybrid circuitry and programmable features represented a radical shift in direction for the field of cardiac pacing, but the consequences did not play themselves out overnight. The Cordis patents on the "digital counter driven pacer" kept other firms from offering programmable models of their own for nearly five years.[42] Meanwhile, Cordis introduced a follow-up model that added a third programmable parameter, stimulus duration. Practitioner implanters remained cautious about programmability and about the reliability of Cordis products: the company actually lost market share between 1972 and 1977. Early Omnicor units had had a high failure rate from moisture entrapped within the hermetically sealed circuit. Even though the company corrected that problem, the episode reinforced an impression among physicians that Cordis pacers were less reliable than those from Medtronic.[43] More broadly, many implanters seemed uncertain how to make use of programmable features. Adjusting the output downward to prolong battery life seemed less pressing once lithium batteries had come into use. Adjusting the rate shortly after implantation helped to make pacing a more comfortable experience for the patient, but many doctors never changed the factory settings.

In retrospect, programmability appears to have been an innovation coming from the Cordis engineers and the more advanced physicians in the field; they could see possibilities for future uses that had not yet occurred to most implanting physicians. Pacemaker specialists recognized that were the typical pacemaker to function for five or more years, a doctor or technician should alter its behavior from time to time as the patient aged or the rhythm disorder changed. By reprogramming the device, they could repeatedly adjust its settings for the individual patient — but this was not yet a vision that the community of implanting physicians generally shared.[44] Omnicor was a time bomb, a token of things to come in cardiac pacing.

## PACESETTER'S RECHARGEABLE BATTERY

A new company, Pacesetter Systems, of Sylmar, California, put into play a series of pacers with rechargeable batteries. The first implanted pacemaker (1958) had been powered by a nickel-cadmium rechargeable cell, and several investigators had announced experimental rechargeable systems during the 1960s.[45] These earlier Ni-Cad versions had proved unsatisfactory because the batteries ran down quickly when kept at the temperature of the human body. Even with frequent recharging, the batteries had a shorter life expectancy than mercury batteries. In 1968, Robert Fischell, of the Applied Physics Laboratory at Johns Hopkins University, and cardiologist Kenneth B. Lewis had begun a collaboration that led in 1973 to a new kind of Ni-Cad battery able to function more effectively at body temperature and hermetically sealable. Alfred E. Mann, a California entrepreneur with a background in the aerospace industry, had provided some financial support to the Hopkins group. Eventually Mann founded a small company to develop a pacemaker for the rechargeable battery; this was the origin of Pacesetter Systems. The rechargeable pacemaker reached the market in the summer of 1973, just as CPI introduced its lithium pacer.[46]

The Johns Hopkins battery clearly had some great advantages over mercury and nuclear power sources. The Pacesetter pulse generator was small, making it easier to implant. Many patients could manage the recharging unit at home rather than returning to a physician's office every few days. The patient donned a vest that held the recharger in place just above the implanted pulse generator, then plugged the power cord into a wall socket and waited 90 minutes for a week's charge. A blue light flashed when the device had been fully charged, but a yellow light flashed and a buzzer sounded if the recharger was not properly positioned. Fischell claimed that the battery could take repeated recharging for up to 20 years, though some people were unpersuaded. But despite the vest, any rechargeable pacemaker had the usual disadvantage in actual use that it placed a weekly discipline and a considerable responsibility upon the elderly patient. Fischell, however, expected that further development would yield batteries that could last for several weeks between rechargings.[47]

The rechargeable battery project had been a technical success, but Pacesetter was never able to get its sales beyond three hundred units per month. The arrival of the lithium battery at almost the same moment, and Wilson Greatbatch's effectiveness as a champion of lithium, made the rechargeable pacemaker seem unnecessary. Lithium batteries with life expectancies of seven or eight years would last as long as most cardiac pacemaker recipients and spared both patients and physicians the rituals and possible problems of recharging. Certainly, many physicians resisted the idea of allowing patients to recharge their own pacemakers, despite survey evidence that many patients enjoyed this task and felt competent to do it. Concerns about the company's possible liability if a patient failed to recharge his pacemaker properly or the vest failed also contributed to Mann's decision to go to lithium batteries in 1975.

Mann recognized that the company now would have to invest heavily to become a real competitor in the industry. He estimated the cost of developing a programmable model at $10 million or more. "No rational man should do this," Mann thought to himself—but he anted up $10 million of his own to go forward with the project that led to the Programmalith pacemaker. Amazingly, Pacesetter Systems not only survived but by the mid-1980s had established itself as fourth- or fifth-largest manufacturer (in dollar sales) of pacemakers in the world. At a banquet in San Francisco in 1981, Mann accepted the Phoenix Award from a trade magazine for bringing the company "from the ashes of its earlier commercial failure . . . into the top ranks of the cardiac pacemaker industry."[48]

The success of Pacesetter Systems testified not only to its excellence in technology but to the competitive turmoil in the industry, the product of low barriers to entry and of physicians' uneasiness about Medtronic and Cordis. In the 1970s, new companies could find engineers with the know-how to design and manufacture pacemakers and could hire away salesmen from the older companies. The existing firms had not yet erected the wall of patents that would later effectively bar entry to the industry. Pacesetter was a small operation with few assets beyond the know-how of its engineering and sales staff, but the company survived the battery wars of the 1970s. Acquired successively by Searle, Sterling Software, Siemens, and St. Jude Medical, Pacesetter was in the pacemaker business to stay.

## INTERMEDICS INTRODUCES A
## MULTIPROGRAMMABLE PACEMAKER

By 1978, all manufacturers had essentially completed the transition to lithium batteries. The small new firms in the industry collectively had gained about a 28 percent share of the U.S. pacemaker market, but Medtronic and Cordis could be expected to try to regain the 10 percent that they had lost.[49] To expand their market share further, the newcomers would have to stay ahead of Medtronic and Cordis. They chose to differentiate themselves from the two larger firms by challenging the Cordis monopoly on programmable pacing. The manufacturers had reason to think that doctors would accept programmability more readily at the end of the decade than they had a few years earlier, primarily because the number of conditions warranting a pacemaker had continued to grow (a point to be discussed below). Rather than dealing with numerous pacemaker models, each dedicated to a particular kind of heart-rhythm problem, physicians would prefer to work with flexible pacemakers that they could adjust to meet the needs of the individual patient—or so the manufacturers hoped. To be sure, this approach called for more sophistication on the part of the doctor, but manufacturers seem to have concluded that their own representatives in the field could help implanters understand and use sophisticated pacers effectively.

In 1977, Cardiac Pacemakers, the small Twin Cities firm that had pioneered the lithium battery, began clinical studies on a new lithium pacemaker that weighed 75 grams (2.6 oz.) and could be programmed on rate and pulse width by means of an external programmer that was more user-friendly than the original Cordis programmer.[50] To the surprise of few in the pacemaker industry, Cordis sued, claiming that CPI had infringed its patent on the programmer, and CPI countersued. In a complicated series of rulings and maneuvers, a federal court found that the CPI programmer had not infringed on Cordis's design and another court held a key Cordis patent invalid, and Cordis then dropped its remaining claims.[51] CPI had broken the Cordis monopoly on programmable pacers, and by mid-1979 all other major firms in the industry had brought out units of their own. CPI enjoyed a high reputation for reliable products; its Microlith-P programmable pacer sold well and the company's market share reached 15 percent in 1979.

The rate of change in cardiac-pacing technology was accelerating, with major innovations coming every three or four years and manufacturers anxious not to be left behind. It now appeared that every company with a claim to being more than a niche manufacturer must have a lithium-powered pacer with rate and output programmability as its lead product. (Medtronic came out with its Xyrel lithium pacer in 1977 and added rate programmability for this model the following year.)[52] But before the industry and the physicians had had an opportunity to grow accustomed to this state of things, another of the aggressive new companies leapfrogged past the CPI Microlith-P and similar devices.

A former Medtronic sales representative, Albert Beutel, had founded a pacemaker company called Intermedics in 1973. Beutel borrowed $130,000, established an office in Freeport, Texas, and located an electronics engineer who was able to design and oversee the production of a small pacemaker with hybrid circuits and a mercury battery. Intermedics had its product on the market within less than a year, then shifted to lithium. The company's third pacer, InterLith (1976), was hermetically sealed, carried an eight-year warranty, and weighed just 65 grams. Since many physicians treated small size as a proxy for high-tech excellence, InterLith proved to be a breakthrough product for Intermedics.[53]

Then in 1978, four years after its first product, Intermedics introduced the most radical new pacemaker design in at least a decade. This device, CyberLith, could be reprogrammed on four parameters, the highest number yet and an indication that manufacturers would compete with one another by introducing ever more flexible pacemakers. It was easy enough to reprogram this and other pacers: The physician chose the desired settings using buttons or keys on the programmer. A magnet or wand positioned over the implanted pulse generator switched on the receiver in the generator, and the programmer transmitted the new data through the patient's skin to the generator. The challenge lay in choosing appropriate settings for a particular patient.[54]

CyberLith's truly distinctive feature was not its four programmable parameters but a novel, two-way telemetry system that enabled the doctor or a technician to download information about the behavior of the implanted device. Telemetry is defined as the transmission of signals or data from one instrument to another by radio waves or some other means. Pacemakers had always transmitted an RF signal with each stimulus that could be picked up on an AM radio, and for a decade many physicians had used that signal to determine the stimulus rate and check whether the battery was about to fail. Programming instruments for pacers like Omnicor and Microlith-P transmitted signals *to* the pacemaker to induce a change in its behavior. Cyber-Lith's telemetry system, however, greatly increased the amount of information about performance of the pacer that the device transmitted back to the external programmer. Among the items included were stimulation rate, battery voltage and impedance, lead impedance, and the integrity of the encapsulation. Intermedics sales revenues, just $2 million in the 1975 fiscal year, reached $68 million in 1979.[55]

Robert R. Brownlee, an engineer with Intermedics, and Frank Tyers, a surgeon at Pennsylvania State University, had invented the pacemaker telemetry system in 1974–75. It consisted of a low-power transmitter contained within the pulse generator that conveyed data in the form of transient changes in its high-frequency signal. Transmission distance was about two inches from the implanted generator to a receiver in the external programmer or to a transtelephone device that could send the data on to a distant pacemaker center. The implanted transmitter used energy only when activated by a magnet-controlled reed switch. Brownlee and Tyers saw bidirectional telemetry as an enhancement to conventional pacemaker follow-up procedures, a means of "noninvasively assessing the condition of the implanted power cell, circuitry, lead, and hermetic enclosure." Intermedics cofounder Charles Del Marco spelled out what this meant: "Not knowing the voltage and the impedance of the power source while following a pacemaker patient is comparable to flying a plane and not having a gauge on the fuel tank."[56]

But if it were true, as the inventors said, that "the number of specific physiologic and pacemaker parameters that can be monitored is limited only by the imagination of those implementing the system," then telemetry might be employed for other tasks besides the periodic check on a pacer's condition.[57] A physician, nurse, or technician could use telemetry to confirm that the pacer was behaving as intended after being reprogrammed. The pacer could store and yield up such information as its model number, date of implantation, and the rhythm disorders for which it had been implanted. It could transmit an intracardiac electrogram that the physician could read in combination with a standard surface ECG. It could even be configured to save in memory and later download a record of unusual events in the heart, such as episodes of tachycardia. Several years (and more advanced circuitry) were required for these and other uses of telemetry to become available, but by 1977 pacemaker engineers at more than one company were beginning to conceptualize the possibilities.

The radical innovations from Cordis, CPI, and Intermedics challenged all the manufacturers to compete on the basis of technological innovation. In the year 1978, 12 firms (some based outside the United States) applied for U.S. patents relating to programmability. A firm that was too deliberate would find that one or another of its rivals had locked up a new feature with patent protection or had reached the doctors first with an attractive new model. With so many aggressive companies in the industry, a firm could gain or lose as much as 10 percent in market share in a single year. Looking back on that period many years later, Earl Bakken remembered that his company, Medtronic, had become "a pretty big company . . . in the pacing market. So we didn't change things in the pacemaker easily. We wanted to study each change and be sure that it was at least better than what we were doing." Events by 1978 had overtaken this deliberate approach and threatened Medtronic's position as market leader. Despite the company's significant loss in market share, Bakken rather liked the new conditions: they reminded him of the early days at the company, when he had made big decisions in a hurry. "Like that first pacemaker I made [in 1958], doing it in four weeks. We didn't study it a lot, we just did it. Even starting Medtronic was just an evening conversation, [and] we did it." On a single day, November 6, 1978, Medtronic filed 15 patent applications on programmability. Medtronic would use patents, and hence its research capability, to protect its market share and beat back its challengers.[58]

## PHYSICIANS AND THE NEW TECHNOLOGY

Some improvements in pacemaker technology clearly addressed the needs of the surgeons and cardiologists who implanted pacemakers. With lithium batteries and hermetically sealed pacemaker cans, the average worklife of implanted pacemakers improved from two years or less to six or eight years, longer than the life expectancy of the typical elderly pacemaker recipient. Patients were subjected to fewer replacement procedures, and the ratio of replacements to all implantations dropped sharply. Hybrid circuitry also enabled the manufacturers to downsize their pulse generators so that they were easier to implant. Whereas new models of the early 1970s weighed 140 grams (5 oz.) or more, those of the early 1980s often weighed less than half as much.

It is less clear whether most practitioner implanters felt that they needed multiprogrammability. The device companies viewed this innovation as an important step toward more reliable, higher-quality pacemakers.[59] The manufacturers also hoped that the very complexity of these pacemakers would help to distinguish the top companies from low-end "garage producers" that were entering the industry during the 1970s. They quickly recognized, too, that complex devices tended to cement the relationship between the physician and the manufacturer. This was because each company's line of pacemakers possessed some features—a unique programmable parameter, perhaps a distinctive kind of behavior as the battery ran low—that set them slightly apart from other brands. Manufacturers reinforced the distinctiveness of their products by designing programmers that would work only with their own pacemakers, then offering them gratis to the hospitals and clinics where implantations and follow-up activities took place. Once familiar with the CPI or the Intermedics programmer and the pacemakers it could manage, the typical physician was likely to remain loyal year after year to the pacing products of that company. Only at large centers such as Parsonnet's hospital in Newark did implanters make a continuing effort to use all the major pacemaker brands.[60]

Physicians who specialized in pacemaker implantation and who performed a high number of procedures welcomed the new flexibility of the pacemaker, but for reasons somewhat different from those of the manufacturers. Specialists tended to see in programmability a means to optimize the pacemaker for the needs of the individual patient. This included, of course, the goal of extending pacemaker life and thus sparing the patient replacement procedures. But pacing specialists also believed that programmability promised to make pacing a useful treatment for a wider range of heart-rhythm problems. Parsonnet listed dozens of specific but uncommon clinical situations in which the physician might find it helpful to invoke one or another programmable feature.[61] Programmability was useful even in the more familiar conditions of complete heart block and SSS: many patients showed changes in their conduction diseases over time, but now the physician could fine-tune the pacemaker repeatedly to a patient's changing condition. In light of the longer life expectancy of pacemakers powered by lithium batteries, this flexibility seemed particularly important.

The practitioner implanters in community hospitals were a different story. By decade's end, they numbered in the thousands (a survey in 1981 yielded an estimate of 5,600 implanters working at 3,670 hospitals in the United States).[62] How did they view programmability? Beginning in 1969, Parsonnet had surveyed pacemaker physicians and hospitals in the United States and Canada. In the second half of the 1970s, the Parsonnet surveys began to pick up some patterns that leaders in pacing found worrisome. First, many doctors did not reprogram their patients' pacemakers. In the 1981 survey, most doctors reported that they usually implanted multi-programmable pacemakers and declared programmability to be "clinically important" — yet the same survey indicated that some 47 percent of these devices were not reprogrammed within the first three months after implantation, and that 30 percent were *never* reprogrammed. Instead, doctors would leave them on their factory settings even when these were contraindicated by the patient's condition. This was not a transitory finding: subsequent surveys yielded similar percentages. A plausible interpretation of these results is that many physicians did not view programmability as a way to optimize the pacemaker for the needs of the individual patient, but saw it rather as a tool for troubleshooting pacemaker problems if they developed later on. The majority of implanting physicians apparently believed that, for most cases, noncompetitive pacing at the factory settings did an adequate job.

To the experts, it seemed that practitioner implanters were underutilizing the technology. Practitioners tended to do pacemaker work at community hospitals, where they often lacked the technical support staff likely to be found at large teaching hospitals. While they would implant multiprogrammable units, they lacked an up-to-date understanding of cardiac arrhythmias and often did not have a clear concept of what to do with the programmable features of the newer pacemakers. Therefore, they simply left the pacers on the default settings for rate, amplitude, and other parameters. With dozens of pacemaker models available in 1978, occasional implanters also found it easiest to stick with one company as the supplier of all their pacemakers. Parsonnet, Michael Bilitch, and Warren Harthorne were concerned that many implanters thereby allowed themselves to become dependent on that company's products and its technical support staff.[63]

Specialists felt particular chagrin over the fact that many implanters allowed manufacturers' sales representatives into the operating room or cath lab where the procedure took place. The reps often handled threshold testing and programmed the pacemaker for the physician. Furman, who was perhaps the most influential doctor in the field, commented that "the dependence of an implanting physician on a salesman for routine procedures is unwholesome"; he cited the rapid technological changes in cardiac pacing as the root of the problem and chided hospitals that "apparently remain unwilling to provide pacer technical assistance as they do for a host of other medical efforts."[64] Senior specialists in pacing also believed that these failures to deploy the available technology effectively occurred because too many doctors were inadequately trained and were implanting pacemakers on a part-time basis. About one-quarter of implanters were self-taught in pacemaker work; many in this category reported that they had learned solely by observation, apparently without even having read a textbook. Performing fewer than 30 procedures a year as one element of a practice in general surgery, cardiology, or internal medicine simply did not enable a doctor to keep his or her skills honed and to gain familiarity with the new technology.[65]

Manufacturers recognized the technological caution of most pacemaker physicians. One manager commented, "I don't expect our customers to innovate voluntarily or autonomously. If we leave it up to our customers, things will go really slow. With all due respect to physicians, most of them have to be conservative about they way they treat people, and most of them aren't technology-oriented. Most of them don't see the possibilities."[66] According to this formulation, many pacemaker implanters remained technological conservatives partly because they did not entirely understand the tremendous power that the new microcircuitry could place in their hands. In the ongoing development of cardiac pacing as a complex system of treatment for heart-rhythm disorders, some observers believed that practitioner implanters had become a bottleneck to progress.[67] Underlying this concern was a notion of "progress" eminently compatible with the goals of the device manufacturers: unending innovation in pacemaker technology and a broadening range of uses for pacing as a treatment for ever-more-subtle rhythm disorders of the heart.

## CHAPTER 8

# SLOWING THE PACE:
# THE INDUSTRY'S TIME OF TROUBLES

Ever since the early 1970s, physicians and manufacturers in the field of heart-rhythm management have had to operate in a highly unsettled political environment in the United States. The second decade of implanted cardiac pacemakers proved a watershed because the federal government shifted course to try to ensure the safety and effectiveness of medical devices that would be implanted within the body. The Medical Device Amendments of 1976 required, for the first time, that device manufacturers carry out extensive laboratory and clinical tests of new life-sustaining devices and secure the approval of the Food and Drug Administration (FDA) before they would be allowed to introduce these products to the market. A consumer protection law above all, the Device Amendments embodied a rising public anxiety about the potential dangers of medical technology. Device manufacturers and many physicians feared that the new legislation would slow the pace of innovation in the industry, but these concerns turned out to be exaggerated.

The 1970s also brought other changes to pacemaker manufacturers and doctors. Product recalls at Medtronic and Cordis enabled new firms such as Intermedics to gain sizable market shares. As competition intensified, manufacturers raced to introduce new components and features that would burnish their reputations and steal a march on rivals. New circuitry and other improvements enabled the manufacturers to introduce smaller, more reliable pacemakers — but these new devices were also more complex and spelled new challenges for doctors. Despite the improvements in pacing's capacity to treat heart-rhythm problems, the general public was given a picture of a field that was troubled and possibly corrupt. Charges that physicians were implanting "too many pacemakers" made national headlines in the early 1980s and portended a slowing in the growth rate of cardiac pacing.

## PACEMAKER RECALLS DISRUPT THE INDUSTRY

Since the inception of implantable pacing, every manufacturer had produced units that subsequently malfunctioned. In a few cases, especially in the early 1960s, these failures had led to patient deaths. The early failures can generally be ascribed to the firms' inexperience at designing and manufacturing implantable devices and doctors' inexperience at managing them. By the 1970s, the basics of pacemaker manufacture and treatment were well enough understood that new companies such as CPI or Intermedics could quickly enter the industry and introduce competitive products. But partly because pacing had gained acceptance as a "normal" procedure and the basic engineering knowledge was widely shared, product failures received a much higher level of unsympathetic public attention than had been true a decade earlier.

A series of product failures during the 1970s contributed to the disruption of the pacemaker industry: they hastened General Electric's withdrawal from the business entirely, set Cordis on a downhill slide from which it never recovered, altered physicians' perceptions of Medtronic, and gave aggressive new companies an opportunity to capture significant slices of the market. These widely publicized problems quickened the companies' acceptance of the lithium battery, which permitted hermetic sealing of the entire pulse generator and not the circuitry alone. Several deaths were attributed to pacemaker failures, and though some of these could not be confirmed, the recalls raised doubts in the public mind about implantable pacemakers, leaving the industry in a particularly weak position during the shaping of the legislation that empowered the FDA to regulate medical devices.

The first major recall of a sizeable number of pacemakers occurred in January 1972, when General Electric notified physicians that 125 of its model A2073 implantable "standby" pacemakers had experienced a circuitry malfunction that caused the device to drive the heart at abnormally rapid rates. GE advised doctors to replace the units on an accelerated schedule even if they appeared to be functioning normally. As more reports came in, GE issued a second letter covering all its implantable standby pacers manufactured in the summer of 1971, a total of 574 units. Early reports suggested that these defective pacemakers had caused seven deaths and two injuries, but a later review of the data blamed only four of the deaths on device failures. A third GE advisory covered 257 more of the same model. Finally, at the request of the FDA, GE on February 1, 1975, extended its advisory to all units of model A2073 still in use, a total of 1,241 pacers. One or two more deaths, beyond the original four, were later ascribed to failures of this model.[1]

Pacemaker manufacturers did not use the FDA's term *recall*, but spoke of "physician notifications" or "advisory letters." An advisory told physicians that some units in a batch manufactured within a specified time frame might be defective, but it was obviously not possible for anyone to know how many units or which specific units would fail. The companies recommended that physicians monitor suspect pacemakers closely. Most units under suspicion functioned normally and were not explanted (i.e., removed from the patient) until the end of their expected worklife. Thus the great majority of the GE standby pacers listed in product advisories worked properly, but the handful that malfunctioned apparently caused several deaths. General Electric exited the pacemaker industry in 1976, selling its business to the Australian firm Telectronics.[2]

Cordis got into even more serious trouble with malfunctions and product advisories. Between July 1973 and early 1975, the company issued advisories on more than one-third of an estimated 60,000 Cordis pacers implanted during that period; most of the advisories concerned its most advanced models, including Omnicor, the first programmable pacer. Cordis later estimated that 394 deaths had occurred among 18,460 of these patients. As in the case of the GE pacemakers, the Cordis failures involved moisture, but here it was moisture trapped inside the hermetic seal rather than moisture that had entered later. An outside firm, Fairchild Semiconductor, had supplied Cordis with its hybrid circuits in sealed packages. Inspections at Cordis did not reveal the problem.[3]

Unlike GE, Cordis held off sending advisories to physicians for several months. "We had an alert within our organization and with the transistor manufacturer," one executive said. "We didn't alert the doctors because we didn't know what to tell them—we were working night and day trying to find out what the problem was." When the advisory went out, Cordis sales reps also visited doctors to explain the problem to them. Some doctors, however, later claimed that the reps downplayed the seriousness of the problem. William Murphy, president of the company, defended his salespeople, "who are aggressively instructed to tell exactly what they know about the product and never exaggerate its capability." But Murphy also implicitly conceded that a company might have reasons to downplay a problem: "If we go to a doctor and tell it like it is, we're liable to be dragged through the courts. It's a serious impediment to communication."[4]

In 1975, the FDA sought an injunction to close the Cordis implantable pacemaker plant on the ground that the company had a poor quality-assurance program that could not spot component defects. Although a U.S. district court turned down the FDA's request and allowed Cordis to remain in the industry, these events undermined the reputation of the company. In a paper presented at the American College of Cardiology meeting in spring 1976, cardiologist Michael Bilitch compared the 18-month survival rates of various pacemaker models and reported that only 81.2 percent of the Cordis units functioned for 18 months — a far lower figure than for pacers from other companies.[5]

The reports of recalls and FDA intervention exposed the pacemaker companies to political attacks and raised the question of how a manufacturer should appropriately handle the discovery of serious flaws in a batch of pacemakers. Federal law before 1976 provided limited guidance. It was not even clear whether a manufacturer was obligated to inform the FDA of an impending advisory letter. As the scale of the industry grew, manufacturers found it more difficult to track every pacer to a particular hospital, physician, and patient. In many cases, the implanting physician left it to the patient's personal doctor to manage the pacemaker after implantation; thus an advisory might not reach the doctor who was most likely to be in touch with the patient. Advisory letters were often ambiguously phrased, partly because manufacturers often were not certain of the cause and scope of a problem at the time they began to inform physicians. Advisories were certain to come to the attention of the news media, with the result that doctors, even before they had received the letter, might find themselves besieged by patients seeking information and reassurance.[6]

All of these uncertainties were manifest during a product crisis at Medtronic. At the time that CPI introduced its lithium-powered pacemaker and Cordis its programmable model, Medtronic was readying for the market a new model of its own: Xytron, the first Medtronic product to bear a distinctive model name. Priced at $1,195 when introduced in January 1975, Xytron had several appealing features that included a hybrid circuit and a narrow pulse width to prolong battery life. It was smaller and lighter than any earlier Medtronic generator, and the circuitry was hermetically sealed in a blue can, shaped like the number *1*, that the physician could see through the clear epoxy of the outer generator shell. What most appealed to physicians, though, was the three-year warranty that Xytron carried.[7] Despite these attractive features, a pacemaker carrying a mercury-zinc battery and pacing in the noncompetitive mode with no programmable parameters could not be considered a state-of-the-art product. Marketed aggressively, Xytron was in fact a model intended to give the company some breathing room while it prepared a lithium-powered pacer for introduction in 1976 or 1977.

Even before launching Xytron, Medtronic had discovered that the hermetic seal on the can that contained the circuitry would occasionally fail and allow body fluids to leak in. "Epoxy is very permeable to moisture, and then because you had a voltage, and you had body fluids, and there was a little bit of solder flux, you would get something called dendrites [metallic crystals]. And it would short out."[8] At first, Medtronic management thought that the problems were limited to an early batch of generators manufactured before October 1974; the first product advisory letter (December 1975) covered only 2,377 units. In choosing to issue a narrow advisory, management unwittingly prepared the way for a public-relations disaster later on. The company had to issue three more advisories in August and November 1976 and October 1977 that covered 47,000 additional Xytrons manufactured up to August 31, 1975.[9]

Despite the advisories, Medtronic stuck with Xytron, and the device had good sales. But behind the public facade, the company was in turmoil because it could not determine the extent of the problem. One executive at the company recalls, "We used to joke that [the Xytrons] were going out the front door only slightly faster than they were coming in the back door." A debate raged over whether to issue a blanket advisory covering all units manufactured up to August 1975 or to warn doctors only about production lots known to have some failed units. Because explanted Xytrons often were not returned to the company to be opened up and analyzed, management was working with incomplete information.[10]

From start to finish, Medtronic manufactured at least 100,000 Xytrons and sold most of them. The four advisories covered some 50,000 units, and doctors probably explanted and replaced one-quarter to one-third of these. A few hundred implanted Xytrons actually failed, but without causing any deaths. By 1977, Medtronic had abandoned Xytron in favor of its new lithium pacer, Xyrel. But the firm had suffered serious blows both to its reputation for superior product engineering and to employee morale. As Charles Cuddihy, who was executive vice president, saw it, "We didn't handle it well at all. There was so little experience that one could tap into. . . . A lot of time was spent with people taking numbers and massaging them . . . in a way trying to kid ourselves." Some loyal customers turned to other companies. Bakken later recalled that the failure and the huge loss of market share that ensued were "devastating."[11]

## FDA REGULATION OF MEDICAL DEVICES

Publicity about pacemaker failures in the early and mid-1970s may have been sensationalized, but it was certainly true that cardiac pacing, like every medical treatment, carried potential risks as well as benefits for the patient. At the time of the Medtronic advisory letters, the U.S. public was nearing the climax of an extended national discussion over how best to reduce the risks of life-sustaining medical devices while continuing to enjoy their benefits. The movement to regulate medical devices had gathered momentum in the late 1960s as part of a broader push by consumer and environmental groups to project the national government into safeguarding Americans in the workplace, the consumer marketplace, and the shared natural environment. The Medical Device Amendments of 1976 greatly expanded the regulatory power of the FDA over medical devices. It continued the thrust of earlier legislation (1969–72) that had created the Environmental Protection Agency, the Consumer Product Safety Commission, and the Occupational Safety and Health Administration. These new regulatory enactments had enjoyed bipartisan support in Congress, and Republican presidents signed them into law.[12]

Since the New Deal era, the FDA had regulated medicines and drugs under the Food, Drug, and Cosmetic Act (1938); but if it considered a medical device unsafe, the agency had to sue the manufacturer and prove its case in court. As the FDA's battle with Cordis showed, this was an expensive and cumbersome process in which the FDA was unable to take action until after a device had been put on the market, and the burden of proof was always on the agency rather than the manufacturer. Until the end of the 1960s, the FDA focused its attention on seizing quack devices, especially those using dangerous substances such as uranium and those making false claims about curing common afflictions.[13] But by the late 1960s, the mainline medical-device industry had burgeoned to almost 1,500 manufacturers, and shipments were valued at more than $1.5 billion. FDA policy began to shift in response to the more intense use of what health policy analyst Susan Bartlett Foote calls "legitimate, therapeutically desirable medical devices," such as intrauterine devices (IUDs) and heart valves. Perhaps as a way of testing the boundaries of its authority, the agency in 1967–68 stepped in to recall suction catheters that were found to be unsterile, justifying the action with the claim that the statutory definition of a drug was broad enough to include medical devices and procedures. When device companies challenged this reading of the law, the U.S. Court of Appeals for the Second Circuit and the Supreme Court, in separate cases, upheld the FDA.[14]

Both the device industry and the agency now saw the need for revisions to the Food, Drug, and Cosmetic Act that would clearly delineate the scope of FDA authority over different kinds of medical devices. In a message to Congress on October 30, 1969, President Richard Nixon urged that "certain minimum standards should be established for medical devices" and that "government should be given additional authority requiring premarketing clearance in certain cases." He instructed the Department of

Health, Education, and Welfare to bring in recommendations for new device legislation. The director of the NIH National Heart and Lung Institute, Dr. Theodore Cooper, chaired a committee that surveyed the problem and submitted a report in April 1970. The Cooper Committee brought to light an FDA survey of the medical literature for 1963–69 that had tallied 10,000 injuries, including 731 deaths associated with medical devices. The manufacturers were not wholly responsible, since they could not prevent misuse of a device in the hospital; but "many of the hazards seemed related to problems of device design and manufacture." Widely disseminated and discussed, these numbers added to the political momentum that led to the Medical Device Amendments six years later. Congressional hearings in 1973 documented additional injuries and deaths from medical devices.[15]

As for pacemakers specifically, the Cooper Committee reported that they had caused 89 deaths and 186 injuries during the 1960s.[16] Most deaths and injuries had occurred in the early (pre-1964) phase of development as a result of hardware failures such as broken wires; a few patients had also died when asynchronous pacemakers caused ventricular fibrillation. Most of these deaths probably could not have been avoided unless manufacturers had conducted several more years of animal tests before introducing pacemakers into clinical use. The early deaths were, in a sense, part of the price society paid for the growth of knowledge about artificially managing the heartbeat and implanting electronic devices within the human body. The public had apparently been willing to accept those deaths because the technology was understood to be at an early stage of development and no alternative treatment existed for people in heart block. Instead of attacking inventors and surgeons, the national news media had applauded them for putting primitive devices into clinical use despite the loss of some patients. Over the decade of the 1960s, cardiac pacing had certainly grown safer in the sense that the number of deaths declined as a proportion of those carrying implanted pacemakers. But the absolute number of deaths was higher because so many pacemakers were now in use. Pacemaker-related deaths also received more publicity in the 1970s than earlier, and public expectations of safety had risen. Public perceptions of risk are often unrelated to the actual incidence of injury and death. By 1970, Americans' uneasiness about novel technologies, including medical devices, had become a significant political factor.[17]

Every year from 1969 to 1975, members of Congress introduced bills expanding the reach of the FDA to include medical devices. The device industry at this time was a heterogeneous collection of large and small firms that manufactured a huge range of products; no overarching trade organization represented the industry as a whole until the creation of the Health Industry Manufacturers Association (HIMA) in 1976. By then, it was far too late to head off some form of regulation. The much-publicized pacemaker advisories of 1972–75 left pacemaker manufacturers in a particularly poor position to influence the public debates.[18]

Signed into law by President Gerald R. Ford in May 1976, the Medical Device Amendments created a model for the regulation of medical devices that departed in important respects from the model for drug regulation. The amendments did not treat all devices alike but classified them into three categories on the basis of presumed risk. Class 3, the most highly regulated, included life-sustaining devices implanted within the human body. All device manufacturers were required to meet general regulatory requirements concerning record keeping, labeling, good manufacturing standards, and the reporting of product failures. But class 3 devices were also subject to "premarket approval" by FDA: before a class 3 device could be put on the market, the manufacturer was required to establish that it would be safe, reliable, and effective. In practice, this meant that the manufacturer must first apply to FDA to have the device designated "investigational," then carry out studies involving human subjects to establish that the device, properly used, would benefit and not harm patients.

FDA oversight did not require that each device be proved cost-effective or more effective than other treatments, only that it be safe, reliable, and effective. Companies did not have to prove a device *absolutely* safe, but they did have to provide "reasonable assurance of safety and effectiveness," meaning in practice that the anticipated benefits of a new device should outweigh its risks. The amendments also recognized that it was not realistic to demand controlled clinical trials — the gold standard for pharmaceuticals — for all medical devices. It broadened the definition of valid evidence to encompass partially controlled trials and even left the FDA free to accept case studies in some cases.[19]

In the months just before Congress approved the Medical Device Amendments, pacemaker manufacturers argued that premarket approval would slow the rate of technological innovation in pacing nearly to a standstill. But the final bill included a large loophole, section 510(k), exempting devices that were "substantially equivalent" to devices already on the market in 1976. The FDA interpreted substantial equivalency quite broadly: between 1976 and 1983, out of 19,431 new devices tentatively assigned to class 3, only 365 were required to undergo full premarket testing.[20]

The new rules as a whole tended to favor the established companies in the pacemaker industry. Established firms had already begun to keep detailed records of testing, manufacturing practices, and product complaints; hence, complying with the new FDA rules that applied across the board proved less onerous for them. Because their existing products were generating income, they were better able to incur the higher development costs and the sometimes lengthy delays in bringing new devices to market, even though Congress had included an "investigational device exemption" that enabled companies to bill hospitals and clinics for new devices during their clinical testing phase.

A study commissioned by the FDA in 1981 concluded that the amendments had not raised insuperable barriers to market entry, slowed the process of innovation, or adversely affected the device industry in other ways. This was not the opinion of most industry observers: according to one, "There is little question that device legislation will slow the rate of innovation in the industry and increase costs of doing business. . . . From the initial clinical test, it now may take a device company two to three years to commercialize a new [class 3] product, versus one year prior to device legislation." Yet Congress had certainly not intended to undermine innovation. Arguably, the worst excesses of the Device Amendments flowed not from the wording of the 1976 law itself but from the deliberate pace at which FDA issued regulations under the law and carried out its reviews of new devices. As enacted, the amendments were intended not to punish but to help protect and stabilize important new industries like the pacemaker industry by saving them from future strings of product failures.[21]

## COMPETITION, QUALITY, AND MARKETING

Earl Bakken, of Medtronic, and Bill Murphy, of Cordis, among others, were unhappy about the prospect of federal regulation because they feared that the FDA's rules would slow down the pace of innovation in the industry and thereby deprive patients, at least for a time, of the benefits of technological progress. Leading physicians shared many of the concerns of corporate leaders. The "unfettered introduction of new circuitry and technology" had "produced a group of devices which have been a distinct service to the patient and the public health," Seymour Furman wrote. "Unimpeded progress in pacemaker technology has been far more rapid and effective than that of the more regulated drug area. . . . It remains to be ascertained whether more regulation will improve patient care . . . or will cause a slowing of progress at the expense of patient welfare."[22]

Coming from leaders in cardiac pacing, such remarks were to be expected. But how important *was* product innovation in pushing a company toward the forefront of the industry? A number of industry observers asked

TABLE 6. SELLING AND R&D EXPENSES AT MEDTRONIC, CORDIS, AND INTERMEDICS, 1980

| FY 1980* | Sales | Selling Expenses | R&D Expenses |
|---|---|---|---|
| Medtronic | $270.4M | $69.2M (23%) | $19.1M (7.1%) |
| Cordis | $109M | $23M (21.1%) | $13.4M (12.3%) |
| Intermedics | $105M | $30M (28.6%) | $4.5M (4.3%) |

Sources: U.S. Senate, Staff of the Special Committee on Aging, *Fraud, Waste, and Abuse in the Medicare Pacemaker Industry: An Information Paper*, 97th Cong., 2d Sess. (Washington, D.C.: GPO, 1982), 31–32. Medtronic's selling expenses are estimated based on David H. Gobeli, "The Management of Innovation in the Pacing Industry," Ph.D. diss., University of Minnesota, 1982, table A-2.

*In 1980 dollars, each company's 1980 fiscal year. Percentages indicate expenses as a percentage of sales.

this question during the 1970s, perhaps because they had noticed that the company with the largest share of the U.S. and world markets, Medtronic, was not introducing the most significant product innovations. What capabilities really made for success in the pacing industry?

Pacemaker manufacturers competed intensely throughout the 1970s, but the competition did not take the form of reducing prices on pacers. The industry was labor intensive, not capital intensive: research, engineering, quality control, and marketing were more important than physical plant, equipment, and materials (see table 6). If there had been a sharp downturn in their business, companies would have resorted to layoffs and accepted lower sales rather than cut prices. But in the 1970s, the industry prospered. Demand for pacemakers and leads climbed year after year; no alternative product existed for treating slow and erratic heart rhythms, and the cost of the technology was of little concern to the doctor or to the end user, the patient, because an insurance company or Medicare typically covered most of that cost. The implanting physicians, who ordinarily made the decision about what kind of pacemaker to use in a particular case, had been trained to think of themselves as their patients' advocates and to put a patient's needs foremost in their thinking. They were certainly not committed to any notion of choosing treatment or technology based on cost or other nonclinical considerations. Doctors thought in terms of the quality of the product. Price was, if anything, a token of quality. "A lot of doctors in this country relate the cheap product to the product of questionable quality and reliability," according to one marketing executive.[23]

Quality mattered greatly to doctors and to manufacturers—but what was quality? With several new companies and numerous new models, how could a doctor distinguish a top-quality pacer from one of lower quality? A 1980 survey of implanting physicians suggested that the doctors viewed the pacemaker manufacturers as innovative in general, but that they especially valued those product innovations that tended to make their job more predictable. The 94 respondents rated most highly the lithium battery; the hermetically sealed pacemaker can; shielding from outside electromagnetic interference; rate programmability; and circuitry that prevented the pacer from accelerating unexpectedly. Innovations that these doctors considered relatively unimportant included pacemaker telemetry and rechargeable and nuclear power sources. The author of the questionnaire, former Medtronic engineer David Gobeli, noted that the typical respondent viewed the company whose equipment he used as highly innovative. Gobeli's observation prompts the question, Did Medtronic and Intermedics enjoy large market shares because they were innovative, or did they have a reputation for being innovative because doctors were most familiar with their pacemakers? Recognizing that doctors valued reliability above all, manufacturers began to offer product warranties on pacemakers. CPI announced a six-year, full-replacement warranty on its lithium pacemaker (1974), and the other companies followed one by one as they switched to lithium.[24]

Another characteristic that defined quality was size. Many doctors viewed small size of a pulse generator, particularly thinness, as highly desirable. Doctors had wanted small generators because they were easier to implant and less likely to erode through the patient's skin. Now lithium batteries and hybrid technology permitted manufacturers to downsize, and downsize they did. Microlith-P from CPI, a top product introduced in 1978, took up only 36 cc in the body and weighed 75 grams (2.6 oz.). CyberLith, from Intermedics, was heavier but even thinner and occupied about as much space as two books of matches, stacked. Medtronic's last mercury-zinc pacemaker weighed in at 135 grams (4.7 oz.), but the firm's first lithium device, Xyrel (1977), came down in weight to 90 grams (3.1 oz.).[25] As they realized that small, thin pulse generators appealed to physicians, the manufacturers also began to emphasize the physical sleekness of the generators in their shiny titanium cans. Companies often abandoned circular and rectangular profiles in favor of graceful ovoid shapes. In their advertisements in the medical journals, they dispensed with illustrations of circuit boards and X-ray images of the interiors of the generators in favor of full-color photographs that presented the generators as things of beauty. Sometimes a gloved hand or a masked face would appear as well to convey a sense of scale and as if to remind the viewer that this handsome object was a piece of high-tech medical equipment. As pacemaker technology became more opaque to the doctors, the pacer was presenting itself to them as an object of consumer desire.

Observers in the 1970s often said that Medtronic held the largest market share more because of its effectiveness at marketing than because it had the most advanced pacers. In an industry where the key consumers — the doctors — were technologically cautious and often inexperienced, the company sales representative served as the most important point of contact between the worlds of medicine and industry. Medtronic had developed its own sales force earlier than its rivals and by 1975 had more than 170 reps stationed throughout the United States. Louis T. Rader's immortal remark about trying to sell UNIVAC computers in competition with IBM applies to the pacemaker industry vis-à-vis Medtronic: "It doesn't do much good to build a better mousetrap if the other guy selling mousetraps has five times as many salesmen."[26]

Company representatives throughout the industry generally had engineering, biology, or medical backgrounds and had been extensively trained. They did much more than sell pacemakers. The typical rep could analyze electrocardiograms, discuss recent clinical studies, interpret for the physician the accumulated performance data on a particular pacemaker model, and compare the distinctive features of his company's equipment with competing models. As a group, the reps helped to speed the acceptance of new technology and new indications for pacing. A majority of implanting doctors preferred to have the company rep in the operating room or the cath lab while they performed the implantation. A few physicians working in large hospitals banned company reps, but it was far more common for the doctor to ask the rep to be at hand to offer advice in case difficulties arose. Al Beutel, the founder of Intermedics, explained that "our salesmen are trained to go right into surgery to help the physician make sure the pacer is inserted properly. The pacemaker is an electrical gadget, and since electronic gadgetry is not often a doctor's forte, the salesman goes into the operating room to help. The salesman is a specialist, and his personal contacts with physicians are very important."[27]

Medtronic's arrangements for physician support involved more than the work of its sales force in the field. The company also maintained a staff of professional consultants. Any doctor or rep could call 24 hours a day for technical or medical consultation. When programmable pacemakers arrived on the scene, pacemaker companies developed the practice of donating the programmer to the physician. None of the companies did this enthusiastically — the programmer was an expensive piece of equipment — but the doctor with a particular company's programmer on hand and a thorough understanding of how to use it would probably use that company's pacers exclusively.

By mid-decade, Medtronic had identified as "a key target for sales activity" internists and cardiologists who had never implanted pacemakers but might be taught to do so. Medtronic opened its Bakken Learning Center in 1974 and began to bring physicians in for short training courses. These programs also familiarized doctors with the company's own pacers, leads, and programmers and enabled the aspiring pacemaker implanter to meet not only the regional field rep but some of the executives, engineers, and technical support people who stood behind the rep.[28] As all these marketing practices show, pacemaker companies were trying to develop long-term relationships with implanting physicians. Once familiarized with a company's devices and its sales and support people, a doctor was likely to stay with that company and implant its pacemakers—hundreds of them, perhaps—for many years. A company's market share embodied relationships with physicians built up over a decade or more.[29]

Product innovation was (and remains) an important factor in the competition for market share among the pacemaker companies, but the importance manifested itself in subtle ways. Most doctors were not terribly interested in what was *in* the pacer (for example, discrete or integrated circuits; a mercury or lithium battery). They cared about how the device *behaved*. But to maintain its credibility as a high-quality manufacturer, a company needed to establish and repeatedly reenact a reputation for innovation in technology. If it proved unable or unwilling to emulate a major innovation like the lithium battery or programmability in a timely fashion, a company could quickly see its market share erode. This had happened to Electrodyne in the 1960s with noncompetitive pacing and to Medtronic in the 1970s with lithium. In the eyes of doctors, it was not acceptable for a firm to have a reputation merely as a reliable manufacturer of solid, middle-of-the-road devices while its rivals always came up with the important new features. Intermedics impressed doctors with its multiprogrammable/telemetry innovations and gained market share even though many doctors apparently did not make heavy use of these features with their own patients.

Product innovation aided a company in other ways, too. A continuous stream of new features burnished the corporate image with stockholders and corporate analysts and boosted morale within the firm. Perhaps most important, an innovation kept the sales reps happy by giving them a reason to request meetings with doctors ("face time") and to persuade some implanters to switch from another product. More subtly, a pacemaker firm must maintain its capabilities in R&D because it might be called upon at any time to respond to a major innovation introduced by another firm. A patentable invention also gave a company an important trading card, something that it could offer in exchange for access to another company's patented features. More broadly still, product innovations helped maintain the special aura of the implantable heart-assist device — the public impression that these were remarkable inventions. To the extent that doctors, hospital administrators, the press, and the public continued to accept this image of the pacemaker, the devices were less likely to be degraded into mere commodities. For all these reasons, new features had significant value to the manufacturer.

## THE PACEMAKER SCANDALS AND THE WITHDRAWAL OF CORDIS

Al Beutel, the founder of Intermedics, recognized that he needed to build a sales force in a hurry. Himself a former Medtronic pacemaker salesman, he believed that the physician-salesman relationship was the key to success in the industry. Hundreds of new physicians were taking up cardiac pacing in the 1970s, and the typical implanting physician chose one company and relied heavily on its products.[30] Ideally, Beutel needed salesmen who had already developed relationships with many pacemaker implanters. During the period of the Xytron advisories (1975–77), he was able to hire away a number of top Medtronic sales reps, many of whom he knew personally. He told an interviewer in 1978:

> I got the best sales force by approaching the leading salesman at every good pacemaker company, who might have been making $30,000 per year, and asking him, 'Do you want to be an entrepreneur? Do you want to be your own businessman?' . . . I offered to let Intermedics loan him interest-free money to start a business, naming it whatever he wanted. I promised to show the salesman how to incorporate and how to get going. So what I essentially did was to start twenty-three separate companies in the U.S. And we pay the salesman a straight commission, $450 per pacer. If a guy sold ten pacers last week, then he gets a check for $4,500 this week.[31]

Beutel was "putting out NBA [National Basketball Association] offers to our guys," according to Medtronic's Cuddihy, "and we lost a number of our top people." Cuddihy had pioneered the aggressive approach to pacemaker sales; now his former employee Beutel was using it against Medtronic. "I spent most of my time on an airplane flying here and there, telling people, 'Don't go,'" Cuddihy recalled. It was not the money alone that lured sales reps from Medtronic to Intermedics, for Beutel also offered them a high degree of independence, and some reps switched out of concern that Medtronic was losing its competitive edge. When sales reps jumped from Medtronic to Intermedics, they often took their customers with them. By 1977, as a result of its well-designed and well-engineered products and its aggressive strategy to build up sales, Intermedics had gained 9 percent of the domestic pacemaker market; by 1980, its share had reached 16 percent.[32]

Everyone in the industry perceived Intermedics as aggressive, flamboyant, and market-oriented in the late 1970s. The head of market research was part of upper management; everything the company developed and introduced was based on what the sales reps and the market-research department reported about physicians' needs and desires. The Intermedics strategy was basically a more aggressive version of the approach that Medtronic had been developing for a decade, and it worked. By the end of 1980, Intermedics had gained second place in "the worldwide pacer sweepstakes"; meanwhile, the market shares of Medtronic, Cordis, and CPI had eroded. Intermedics' success tempted smaller companies such as Telectronics and Pacesetter to take aggressive marketing one step further: both companies began to offer kickbacks in an effort to induce doctors to try their pacemakers.[33]

The escalation in marketing tactics came to the notice of the U.S. Senate's Select Committee on Aging. After a newspaper article appeared in the *Philadelphia Inquirer* (April 19, 1981) describing shady practices in the pacemaker industry, Senator John Heinz (R-Penn.), chairman of the committee, ordered a staff investigation.[34] Published the following year, the staff report commented that "a primordial sales environment has evolved" in the pacemaker industry, and it quoted salesmen and physicians who described the industry's situation as "filthy" and "dog-eat-dog." "Because of their importance to the industry," the report continued, "successful salesmen are bought, pirated, and lured from company to company." The report described how unscrupulous salesmen would help doctors set up pacemaker practices in ways that would bring in the highest reimbursement fees from Medicare. It cited one former officer of a pacemaker company who claimed that "nearly 50 cents of every dollar of the pacemaker list price was dedicated to marketing activities," with half of that going to the sales rep as a commission.[35]

The report aired a number of specific charges about the industry that had been rumored and quietly discussed for several years: that some doctors, with the connivance of sales reps, were deliberately performing numerous unnecessary implantations and follow-up pacemaker checks in order to get large fees from Medicare; that the manufacturers' procedures for reviewing defective products were haphazard and slow; that they priced their products too high and made outrageously large profits; that hospitals put an added markup on pacemakers, often doubling the company price; that implantation fees were also excessive; and that doctors sometimes, rather than passing a refund on to the patient or to Medicare, pocketed manufacturers' cash payments for pacers that failed under warranty.

Perhaps most serious were the charges that at least two pacemaker companies had paid kickbacks in the form of cash, stock options, and other gifts to doctors who agreed to use their pacers. The select committee's staff report also noted with distaste the industry practices of giving away pacemaker programmers and wining and dining doctors at medical conventions. Competition in this realm had escalated grotesquely, with one firm renting the *Queen Mary* for a convention party, another hiring George Burns as host at a sit-down dinner for doctors, and a third bringing in the Dallas Cowboys cheerleaders. According to the report, Intermedics owned "a fleet of jets, helicopters, and a 55-foot Hatteras yacht that exists for the primary purpose of encouraging physicians to come 'visit the plant.' . . . Alternatively, if the physician is not a sailor, the suggestion is made that he might enjoy the use of the firm's hunting lodge." Overall, the Senate report gave a disturbing picture of doctors, manufacturers, and hospitals colluding with one another to get rich at the expense of Medicare and sick, elderly patients.[36]

As a result of these findings, the environment for pacemaker doctors and manufacturers changed rapidly. The FBI began an investigation of Pacesetter Systems and Telectronics that led to indictments for paying kickbacks. In 1983 and 1984, these firms and several of their officers pled guilty or nolo contendere and were fined and given suspended sentences. For his part, Senator Heinz added provisions to the 1984 Deficit Reduction Act to revise the guidelines on the frequency of transtelephone monitoring, force a review of Medicare reimbursements to doctors and hospitals for pacemaker work, and require that the FDA set up a national pacemaker registry so that defective products could be spotted independently of data made available by the manufacturers.[37]

A second hearing in 1985 concentrated on problems at Cordis, the company that was second in seniority in the pacemaker industry. In the mid-1970s, under William Murphy's leadership, Cordis had run afoul of the FDA when the agency tried to force the company to close its manufacturing plant until it was able to meet FDA's "Good Manufacturing Practice" standards.[38] The company resolved its technology problems and had no further recalls until 1983. But this experience caused an attitude of suspicion toward the agency to develop at Cordis. Murphy was "shocked" to discover that FDA inspectors were suspicious of Cordis. People at the company, he felt, were "obviously more knowledgeable than [agency staffers] were. . . . We had the everyday problems of making circuits work, of understanding batteries." FDA's people, "knowing nothing about it [were] coming around telling us how to run the business." To Murphy, "they were interested in dictating, not learning."[39]

Then, over a two-year period, Cordis had to announce major problems with its proprietary lithium-copper sulfide battery and misbehavior in some of its new dual-chamber Gemini pacers. These problems triggered a new FDA investigation that dragged on through the first half of 1984, slowed perhaps by the company's having three vacant positions in manufacturing quality control. The inspectors concluded that "the problems encountered combined with the decisions made to continue marketing . . . devices that later developed serious problems is indicative of severe deficiencies in the firm's operations and management"; they recommended legal action against Cordis. With FDA action pending, company president Norman R. Weldon discovered in mid-1985 that some managers had been altering internal memos that had raised warnings about technical problems with Cordis pacers. Weldon sacked those responsible and informed the FDA; in turn, the agency prohibited Cordis from testing or marketing new products for 18 months. With market share plummeting and lawsuits pending against the company, the Cordis board of directors sold the pacing business to Telectronics in 1987.[40]

## TOO MANY PACEMAKERS?
## PRESSURES TO CONTROL GROWTH IN PACING
Concentration on the technology of cardiac pacing gives an impression of breathtaking advances between 1973 and 1981 — but a different picture emerges if we look at the public's image of pacing during these years. After nuclear pacemakers, no technological innovation created even a ripple of attention outside the medical profession. No pacemaker physician or engineer held the spotlight as the pioneer heart surgeons and implanters had done 20 years earlier. From the early 1970s through the mid-1980s, the public was likely to encounter news coverage of four topics concerning cardiac pacing: observers' claims that Medtronic and the other firms enjoyed excessive profits; recalls of defective pacemakers and the new FDA regulation of the medical device industry; evidence of sleazy sales practices; and claims that doctors were implanting "too many pacemakers."

Throughout the 1970s, the use of cardiac pacemakers continued to grow in the United States. There is no universally accepted set of numbers, but it seems that implantations (new and replacement) rose from about 34,000 in

1971 to 83,000 in 1976 and 120,000 in 1981.[41] The rate of growth slowed somewhat after 1976 because average pacemaker longevity began to rise as manufacturers introduced lithium pacers, so that fewer replacement units were needed. Even so, the number of pacemaker implantations was rising at about 10 percent per year in the late 1970s. By 1980, the implantation rate per 100,000 population in the United States exceeded that in any other country, although the rates were certainly rising in West Germany, France, and elsewhere. Victor Parsonnet observed in 1976, when it appeared that the number of implantations was doubling roughly every three years, that "continued extrapolation at this rate of growth would seem to indicate that by the year 2,000 every one of us will be wearing a pacemaker."[42]

Parsonnet estimated that there were some 5,600 physicians who implanted pacemakers in 1981 and that they carried out procedures at 3,670 hospitals in the United States. It is clear that more than two-thirds of these implanters had taken up cardiac-pacemaker work since 1973. About two-thirds of the physicians in pacing were still surgeons, most of the others cardiologists — a distribution suggesting that the majority of pacemaker physicians still viewed the procedure of implanting the device as the most important aspect of pacemaker work and left postimplantation management in the hands of the patient's primary-care physician.[43] Doctors flocked to pacing because by now the standard hardware and implantation techniques seemed safe and straightforward. Medicare and private insurance reimbursed hospitals and physicians at the "usual and customary" rate in their area for various pacemaker-related procedures; this practice, in the view of the Senate Select Committee, powerfully encouraged the overutilization of pacemakers. Changes in the population also impelled cardiac pacing toward continued growth. The number of elderly people (those at highest risk for heart-rhythm disorders) was growing faster than the overall population growth rate in the United States.

Pacing had acquired a momentum that by 1980 had become a matter of concern. Thousands of people in hospitals, clinical practices, and business corporations had committed their careers to pacing. While they played varied roles and did not always agree about issues in the field, they shared a common interest in improving the performance of pacemakers and in discovering more uses for the devices.[44] Industry observers recognized that the principal cause of growth in the 1970s continued to be the framing of new uses for pacing. As new diagnostic techniques and technology spread, cardiologists had defined several heart-rhythm problems involving a slowdown of the normal heartbeat, a lack of coordination between the upper and lower chambers, or the onset of rapid rhythms. One observer commented that "bradycardias and tachycardias are becoming progressively easier to diagnose. The potential number of pacemaker patients is therefore not static but dynamic, and should rise as medical knowledge both theoretical and clinical increases."[45]

But the lengthening list of accepted indications for pacing did not result solely from the advance of medical knowledge. One senior pacing physician, Thomas Preston, of the University of Washington, observed that "prevailing attitudes about utilization are due in no small part to blandishments by manufacturers to pace all patients with these syndromes. Marketing strategies, beginning in the early 1970s, aimed at widening the pacemaker market through 'education' of physicians by direct exhortation by the sales force, support of symposia and speakers favorable to expanded indications for pacing, and commercially generated literature. . . . Medical indications were influenced and generated not entirely by scientific information about biological needs of patients."[46]

Everyone involved — physicians, support staff, hospital administrators, manufacturers, investors, and, not least, the patients — *believed* in cardiac pacing, not least because a person could make a lot of money by doing good.[47] Doctors implanted pacemakers because they understood themselves to be contributing to the well-being of patients and because they were generously paid for doing so. Medtronic adopted a company motto, "Toward Man's Full Life," that conveyed the technological utopianism of many in the field of pacing, on the manufacturing side and on the medical.[48] Federal reimbursement and regulatory policies had played a powerful enabling role in the expansion of pacing, as the Senate Select Committee pointed out.[49] Lest the point be overlooked, we should also remember that pacing was clinically effective: it kept alive patients in complete heart block whose prognosis without pacing would have been extremely poor, and beginning in 1967 with the coming of noncompetitive pacing, it enabled many of them to enjoy moderately active lives.

As concern built up in the 1970s and 1980s about the high cost of health care in the United States, many critics singled out advanced medical technologies as a principal cause of the financial crisis in U.S. health care. The same year as the Senate Select Committee's inquiry into "fraud, waste, and abuse," cardiac pacing came in for additional criticism when the Public Citizen Health Research Group, an organization under the Ralph Nader umbrella, announced that more than one-third of the permanent pacemaker implantations performed in Maryland hospitals in 1979 and 1980 had been "unnecessary" or "questionable." A blue-ribbon committee appointed by the Maryland Society of Cardiology reviewed the medical records of pacemaker patients and reached a different conclusion: only a few of the procedures had truly been "unnecessary."[50] But what did "necessary" and "unnecessary" mean? People outside the field often assumed that some stable, authoritative set of implantation criteria must exist, but in fact the medical definition of necessity in cardiac pacing had never remained fixed for long.[51]

## GETTING A HANDLE ON HEALTH-CARE COSTS: THE DRG SYSTEM

Until the 1980s, the Medicare program was structured in such a way that patients, doctors, and hospitals could make decisions about care without much serious consideration to cost. Expenditures under Medicare rose from $4.5 billion in 1967 to more than $50 billion in 1982, a considerably steeper rate than the general rate of inflation. Most observers in the early 1980s attributed the intense, unremitting cost pressures partly to the steady increase in the number of Americans over the age of 65 and hence eligible for the program, but they also pointed to the structure of Medicare itself. Before 1983, Medicare paid doctors and hospitals the "usual and customary" charge for a given procedure. Under this system, incentives to economize were weak. It was widely agreed that the arrangement encouraged hospitals to adopt and use advanced medical technologies because they could count on being reimbursed. Certainly the system contributed to the astonishing rise in the number of heart-related procedures and facilities in U.S. hospitals.[52]

Implanting and managing a pacemaker cost a mere fraction of kidney dialysis, an organ transplant, or several other procedures for which Medicare paid. The average expense of pacing a patient for a year, in constant dollars, probably dropped by one-half in the 1970s. The problem did not really lie in the expense per year for any one patient, but in the growing number of patients and their longer life expectancy. Much of the cost per patient occurred in the months and years after implantation.

At the Harvard School of Public Health, a painstaking study of 32 pacemaker cases in the late 1970s concluded that the cost of hospitalization, patient workup, the physician's fees, the pacemaker and leads, and pacemaker management over the lifetime of each patient (including transtelephone monitoring) averaged more than $17,000. Since an estimated 120,000 Americans received pacemakers for the first time in 1981, that multiplied out to a cost of more than $2 billion to pace one year's cohort of new patients for the rest of their lives. To put it another way, Victor Parsonnet estimated that about 500,000 Americans carried implanted pacemakers in 1981. If 80 percent of these were Medicare enrollees and the program covered 90 percent of the total cost for each patient, it would appear that

Medicare's obligation for these patients' pacing care (over several years) was more than $6 billion.[53]

Confronted with such numbers, technological optimists replied that the cost per patient was entirely justifiable. The pacemakers of 1980 were more sophisticated devices than those of a decade earlier; they functioned reliably for many years and required replacement less often. The cost of pacing over a given patient's lifetime had gone up because patients lived longer, but quality of life for those patients had unquestionably risen as well.[54] Defenders of high-tech medicine could also point out that rapidly rising costs of care for the Medicare population had numerous causes besides medical technology and third-party payment schemes. But even if these two factors did not fully account for the problem of cost, they were matters that public policy could address.[55]

In 1983, Congress moved to avert a financial crisis in Medicare by amending the Social Security Act to change Medicare to a prospective reimbursement system for hospital care. Phased in over the next three years, the new rules (the Social Security Amendments of 1983) grouped all hospital treatments into 467 diagnosis-related groups (DRGs), each of which was assigned a fixed rate of reimbursement regardless of the length of the patient's stay in the hospital or the resources used. The DRG system thus established a price ceiling for each reimbursable procedure and pressured hospitals to economize. If the hospital could treat the patient for less than the fixed amount, it came out ahead. If treating the patient cost more than the reimbursement amount, the hospital would have to make up the difference. In the 1990s, prospective payment was extended to physician services.[56]

Of the original 467 DRGs, four involved pacing, two covering initial implantations and two covering replacement or revision of a pacemaker. Doctors discovered that under the new system, Medicare would pay for pacemakers in some patients but not others. The program would cover implantations for life-threatening conditions such as chronic or intermittent heart block and some severe forms of the sick sinus syndrome. With some other rhythm disorders, the physician would have to make a case that the pacemaker would really benefit the patient. But Medicare would refuse to pay for pacemakers in cases of sinus bradycardia or heart block with no significant symptoms, a slow heart rate during sleep, transient ventricular pauses, or other borderline situations. No longer would Medicare entirely trust the physician's judgment as to what was "reasonable and necessary" treatment for the patient.

To try to ensure that pacing and other high-technology medical procedures were being used only when appropriate, the Social Security Amendments also established a system of Professional Review Organizations (PROs), contract organizations that monitored patterns of admission and discharge in hospitals and reviewed the quality of care that patients on Medicare received. Each hospital was required to contract with a PRO as a condition of receiving Medicare reimbursement. The law explicitly required PROs to monitor pacemaker implantations and the use of other medical devices and to guard against unnecessary procedures. To provide a basis for these reviews, professional associations such as the American College of Cardiology prepared written guidelines specifying acceptable indications for permanent pacing. The PROs did not discover much evidence of excessive implantation of pacemakers. For example, of 8,000 pacemaker implantations in New York City between September 1984 and mid-1986, the PRO judged only 15 (0.2%) to be unjustified.

At best, the DRG system was a crude way to hold down hospital costs. Critics pointed out that DRG rates were based on past medical practice. As newer technologies became available and treatments changed, a gap would open up between the DRG rates and the actual cost of state-of-the-art treatment. Still, the DRG system did contribute to a decline in pacemaker implantations. U.S. doctors performed an estimated 86,000 new implantations and 19,000 replacements in 1985, a decline of 27 percent from the 120,000 new implantations of 1981. Procedures per million population dropped from the all-time high of 518 in 1981 to 374 in 1985.[57]

The DRG system affected cardiac pacing in other ways too. Many physicians reported that they used single-chamber rather than the new dual-chamber pacemakers, discharged patients earlier, and reduced the frequency of follow-up checks on the pacemaker (however, about 30% of physicians responding to a survey in 1985 reported no effect on their pacemaking practice from the new rules). Some data from 1985 also suggested that as a result of the DRG system, large hospitals would gradually come to monopolize pacemaker work.

For the device manufacturers, the upheaval in Medicare reimbursement added to the uncertainties of the business environment. The DRG system and other pressures caused a sudden though temporary slowdown in the growth of pacemaker implantations and may have delayed the acceptance of new, more expensive pacemaker models in the 1980s. Slower growth intensified the struggle for market share, impelled the companies to seek new markets outside the United States, and drove them to diversify their product lines by acquiring other companies or by growing new businesses themselves.

Since Medicare was a leading payer of hospital bills in the United States, the new reimbursement system put pressure on hospitals to adopt cost-saving strategies. Consolidation of their purchases was one obvious strategy. From 1983 on, the pacemaker companies found that hospitals began to involve other people besides the physicians in purchasing decisions, to narrow the number of pacemaker brands in their inventories, and to pressure manufacturers for price reductions. A Medtronic executive noted that "in the past, the doctor would say, 'This is what I want,' and that was relatively unchallenged. Today [1984], because of the emphasis on cost, that's being challenged." The new arrangement tended to give a competitive advantage to the larger companies, if only because hospital purchasers could be more confident that these firms would still be on the scene to support their products years in the future.[58]

Industry executives were most concerned that the DRGs would remain static instead of being redefined as the technology of pacing developed. In 1983, several companies were about to introduce important product innovations into an environment that was becoming steadily more preoccupied with constraining the rising cost of U.S. health care. Would reimbursements under the DRG system adequately reward hospitals for selecting the new, more flexible and effective — but more expensive — pacemakers as they became available in the future? Would clinical researchers receive encouragement and financial support for defining new disorders of heart rhythm as they had done with the sick sinus syndrome? The regulatory changes between 1976 and 1983 indicated that physicians and device manufacturers must again rethink cardiac pacing. In the future, major innovations in the treatment of arrhythmias would come under scrutiny to prove themselves not only safe and effective but cost-effective as well.

## CHAPTER 9

COMPETITION THROUGH
INNOVATION: ACCELERATING THE
PACE OF CHANGE

By 1980, virtually all pacemakers depended on hybrid integrated circuitry and lithium batteries that would reliably manage the heartbeat for eight years or more. Using an external programmer, the physician could alter the pacer's behavior in various ways and, in a few models, also interrogate the implanted device through coded magnetic signals and download information about its performance, the condition of the battery, and the behavior of the patient's heart. But the life cycles of products were growing shorter and each generation of devices added new possibilities (or new layers of complexity) for the physician. By 1985, the six remaining U.S. pacemaker manufacturers had moved on from lithium and programmability to compete in a new technological arena, *dual-chamber pacing.* Now the top-of-the-line pacers boasted *two* leads threaded through a vein into the heart — one set up in the right atrium and the other in the right ventricle. Manufacturers and leading pacemaker physicians argued that dual-chamber pacing, by providing better coordination between the contractions of the heart's upper and lower chambers, more closely emulated nature and offered important physiological benefits to the patient. But to the consternation of the industry, many doctors avoided dual-chamber pacemakers almost entirely for several years; in the main, the profession stuck with the familiar, single-lead, noncompetitive pacers that had come into use more than a decade earlier.

Senior specialists in the field were troubled by what they viewed as a lack of technological sophistication on the part of many practitioner implanters. Between 1974 and 1979, a group of the leading specialists laid out guidelines of good medical practice in cardiac pacing, founded a journal, and organized a professional society; they hoped to improve standards of training, to raise the professional profile of cardiac pacemaker work, and to assert the independence of implanter physicians from the device manufacturing companies. These events and the lukewarm reception given to dual-chamber pacing once again suggest that device manufacturers, academic physicians heavily committed to pacing, and practitioner implanters saw the pacemaker itself and the medical field of cardiac pacing in rather different ways.

## THE ICHD REPORT AND THE PACEMAKER CODE

In the 1970s, the physician community grew rapidly and changed in composition. Pacing had acquired the reputation of being a straightforward, problem-free procedure; as we saw in chapter 7, this attracted many doctors affiliated with community hospitals and many who did pacemaker work as one part of their broader practices in general surgery or clinical cardiology.[1] Two distinct groups of doctors thus coexisted: a small group of specialists who worked in large medical centers, and a much larger group of occasional implanters. Yet despite its reputation, pacemaker work was far from simple. Diagnosing and treating the sick sinus syndrome was a more complex matter than managing complete heart block. SSS revealed itself more subtly in the patient's ECG and had a broader spectrum of symptoms. In addition, the long-term survival of many paced patients entailed new responsibilities for physicians, such as setting up routine procedures for periodic reexaminations. And from 1973 on, manufacturers had introduced numerous pacemaker models with enhanced capabilities.

At the beginning of the decade, an organization called the Inter-Society Commission for Heart Disease Resources (ICHD), a joint creation of the American Heart Association (AHA) and the American College of Cardiology (ACC), began to sponsor reports on the state of medical treatments for various diseases of the heart. The commission named a committee to report on the status of cardiac pacemaker implantation, but the eminent senior figures — Paul Zoll, William Chardack, and Dwight Harken, among others — were unable to reach consensus on most issues. Seymour Furman remembers a meeting at which the group debated at length the definition of a pacemaker but could agree only on "A pacemaker should pace the heart." Eventually, the ICHD disbanded the original committee and asked Victor Parsonnet to convene a new "pacemaker study group" of his own selection. Parsonnet chose Furman and Nicholas P. D. Smyth, of George Washington University in Washington, D.C. The three men were all surgeons with long experience in pacing.[2]

The report of the ICHD Pacemaker Study Group (1974) proposed a simple, three-letter identification code of Smyth's devising that implanters and manufacturers quickly adopted (see appendix C). Within a short time, everyone in pacing had learned to speak of VOO, instead of asynchronous pacing, and VVI instead of "demand." The pacemaker code provided a language that avoided manufacturers' proprietary terms (e.g., *demand, standby*).

It not only described but clarified the behavior of pacemakers to the many implanters of the devices, thereby helping to educate newer entrants to the field. And since patients, journalists, regulators, hospital administrators, and even nonimplanting physicians generally were unfamiliar with the code, it also helped differentiate insiders from outsiders.[3] From here on, I will use the ICHD code in referring to various modes of cardiac pacing. Most pacemakers of the 1970s operated in the VVI mode.

The ICHD report also provided physicians with a concise summary of how a pacemaker operated, clear explanations of a variety of technical matters outside the normal experience of many clinicians — pacemaker electronics, electromagnetic interference, battery chemistry — and detailed recommendations for how to conduct a practice of cardiac pacing. These guidelines covered such matters as estimating the remaining life of an implanted pacer, equipping the operating room or cath lab, and keeping records. The group presented an extended discussion of the staff, equipment, and facilities needed to implant pacemakers and track patients afterward. Much of this had been anticipated in the published work of Furman and Parsonnet, but the study group now updated earlier recommendations and issued firm advice not only to physicians but to manufacturers, regulatory agencies, and insurance carriers. In effect, they were presenting the first formal practice guidelines for the field of cardiac pacing.

The three surgeons had consulted with 30 physicians and several engineers and industry executives: the report thus represented the consensus of leaders in the field. It indirectly raised the question whether implanters were sufficiently invested in pacing by suggesting that an implanting physician and support team should, "to maintain an adequate level of skill and technical competence," perform at least 25 new implantation procedures per year and an equal number of replacements. This was an interesting recommendation to make at a time when at least one-third of implanters performed fewer than 25 procedures a year of any sort.

## DIFFERENT OUTLOOKS:
## PACEMAKER SPECIALISTS AND PRACTITIONER IMPLANTERS

From the 1950s on, physicians specializing in cardiac pacing had invented new hardware and new procedures. They also wrote chapters for medical textbooks, offered workshops at medical conventions, vetted the new models that the device companies introduced, and taught intensive short-term courses under the auspices of the ACC. In arguing that a hospital pacing program should rely on a handful of expert pacemaker physicians supported by trained nurses and technicians while denying privileges to others, the ICHD report reflected the outlook of these specialists and the philosophy of the DeBakey Commission, a presidential commission on heart disease, cancer, and stroke that surgeon Michael DeBakey had chaired in the 1960s. DeBakey's group had proposed a number of federally supported regional "centers of excellence" to which patients with the most complex and intractable diseases would be referred for the latest treatments.[4]

The practitioner implanters who worked in community hospitals readily accepted many proffered innovations in pacing such as the VVI pacing mode and the lithium battery, but proved indifferent to the experts' guidance on training, long-term pacemaker follow-up, and other matters.[5] Most of these practitioner implanters in the 1970s were surgeons. They defined their role as implanting the pacemaker and, if any arose, dealing with immediate postimplantation complications. After that, responsibility for managing the patient and the pacemaker over the long term devolved to the patient's personal physician, who was more likely a general practitioner than a heart specialist.

For these doctors and their pacemaker patients, a number of commercial services sprang up in the 1970s to provide transtelephone pacemaker monitoring and record keeping. The patient would be given a transmitter about the size of a portable typewriter. On a regular schedule, the monitoring service would call the patient, who would transmit a heart-rhythm strip by telephone. The service stored the patient's record and contacted the doctor when a check of the rhythm strip indicated that the pacer needed replacement or was not functioning properly.[6]

Numerous pacemaker implanters, working with patients who had heart block or other rhythm disorders, apparently understood the procedure as a definitive treatment; specialists, on the other hand, viewed the pacemaker as a "halfway technology" that cured nothing and required continuing management. Commercial monitoring services helped reduce the immediate problem of adequate long-term oversight of patients carrying pacemakers, but they also perversely enabled many physicians to remain uninvolved in a huge part of the overall medical practice of cardiac pacing. Moreover, surveys from the 1970s indicated that many patients received no pacemaker follow-up at all beyond occasional visits to the doctor for general checkups. As we saw in chapter 7, evidence also mounted during the decade that many physicians underutilized the capabilities of the new programmable pacemakers by simply leaving the devices on their factory settings. In other cases, implanted pacemakers were misprogrammed, sometimes because doctors misinterpreted ECGs. In surveys, low-volume implanters also reported "high" rates of electrode dislodgment or perforation (in 6%, or more, of their cases), a clear indicator that they lacked essential skills. Furman, Parsonnet, Warren Harthorne, and other specialists grew apprehensive that the future of cardiac pacing might fall into the hands of lightly trained implanters who relied too heavily on the manufacturers during the implantation procedure and neglected long-term management of their patients.[7]

## SENIOR SPECIALISTS AND THE FOUNDING OF NASPE

The Pacemaker Study Group had urged in 1974 that every hospital's credentials committee "should require documentation of a physician's training and capability to implant pacemakers and . . . clearly and specifically delineate each practitioner's clinical privileges." Implementing this proposal would require action from organizations not centrally interested in what was best for cardiac pacing. Hospital administrators and specialist societies had their own agendas, and restricting physicians' access to a prestigious and well-compensated activity such as pacemaker implantation did not rank near the top. Senior specialists, increasingly convinced that doctors with a primary interest in pacemaking should create their own professional organization without waiting for the attention of the major specialty groups, soon turned their energies to institution building. A new organization would recognize the distinctive character of cardiac pacemaker work — its inherently interdisciplinary character, its involvement with advanced computer-based technology, and the rapid evolution of its knowledge base and hardware.[8]

At an international symposium on pacing held in March 1976 in Tokyo, Furman mobilized support for a specialty journal that would publish the best new research on cardiac arrhythmias and pacing. The journal *Pacing and Clinical Electrophysiology (PACE)* began publication in January 1978 with Furman as editor-in-chief and an editorial board dominated by academic cardiologists and surgeons. *PACE* disseminated advanced knowledge, aired controversies, and helped to build consensus on good medical practice; to read the journal regularly identified a clinician or engineer as part of the professional community in pacing. Every issue of *PACE* also opened with an editor's column that gave Furman or some other leading figure an opportunity to comment on a significant issue in the field. A repeated theme of his editorials in the early years was the danger of industrial control of cardiac pacing.[9]

The ideas and personal contacts that led to the founding of the North American Society of Pacing and Electrophysiology (NASPE) also began to take shape around 1976. In the Boston area, cardiologists and surgeons interested in pacing formed a pacemaker club that met three or four times a year for lectures from experts and discussions of current controversies. Similar groups formed in a few other large cities. These local groups brought practitioners and academics together and fostered a sense of group identity, something difficult to accomplish at the huge annual scientific sessions of the ACC and the AHA.[10]

In the 1970s, most device manufacturers set up "consultant groups" consisting of leading surgeons and cardiologists who advised the companies about new treatment concepts and doctors' technology needs. Through these groups, leaders in pacing became better acquainted. Surgeons Victor Parsonnet, of Newark, Dryden Morse, from the Deborah Heart-Lung Center near Philadelphia, and Seymour Furman, from Montefiore in the Bronx, and cardiologists Warren Harthorne, from Massachusetts General Hospital in Boston, and Michael Bilitch, from the University of Southern California, began to talk about "forming an organization to cater to the needs of physicians involved in pacing."[11] Furman, Parsonnet, and Morse had been three of the earliest surgeons to enter the field of cardiac pacing, and Harthorne and Bilitch had been active in the field for more than a decade. All believed that cardiac pacing had attained a kind of identity distinct from both heart surgery and cardiology and were interested in seeing it attain a place in the roster of formal medical subspecialties. They thought that eventually the field should have its own certification process under the auspices of the Advisory Board of Medical Specialties. But neither the ACC nor the American College of Surgeons seemed interested in designating cardiac pacing as a new subspecialty.[12]

Beyond this quest for a distinct professional identity, the five doctors in one degree or another were concerned about the flow of ever more physicians into the field of pacing and the mounting, if indirect, evidence that some implanters seemed inadequately prepared for working with pacemakers. They were dismayed over many doctors' allowing company sales reps to attend and advise during implantation procedures and the rumors of sleazy arrangements between some doctors and some reps. All believed that the field needed authoritative voices that could delineate standards of good practice, lay out training guidelines for practitioner implanters, and represent the pacemaker specialists in dealings with the manufacturers, the national medical societies, and the North American public.[13]

Four of the five leaders met in Parsonnet's office in Newark to work out details for a new professional organization. They had fun thinking of possible names for the new group: the Great American Society of Pacing (GASP) was suggested, but this became North American (NASP) and, later, in a gesture to the new field of clinical electrophysiology, NASPE.

While Harthorne was out of the room, the others elected him the first president. They invited two Canadians, surgeon David C. MacGregor and cardiologist Bernard S. Goldman, to join the ad hoc committee of founders. Parsonnet had a database of the names of thousands of physicians who had implanted pacemakers. A mailing brought in membership dues from about a hundred of these, and the society sprang to life in a corner of Harthorne's office in Boston.[14]

The founders planned to adopt policy positions that would influence hospitals, medical schools, other professional societies, and the FDA. They expressed "growing concern over the increasing complexity of pacemaker systems . . . and the proper surveillance of an ever-expanding recipient population."[15] By laying out and updating practice guidelines, they hoped gradually to affect the way ordinary implanters handled their clinical work. Between 1983 and 1990, NASPE sponsored 19 policy conferences on subjects as varied as hospital resources for pacemaker implantation, guidelines for clinical investigations of new pacemaker models, the use of computers in pacing, the appropriate use of dual-chamber pacemakers, and the content of fellowship programs in clinical electrophysiology.[16] The organization grew steadily but slowly. It never attracted all pacemaker implanters. In mid-1998, about fifteen hundred American physicians, or fewer than 20 percent of all implanters in the United States, held memberships. With an overall membership of about twenty-five hundred in 1998 (including nurses, engineers, industry people, and non-Americans), NASPE was about one-tenth the size of the AHA and the ACC.[17]

NASPE's leaders were interested in raising the standards for allowing doctors to implant pacemakers.[18] In an era when pacemakers had become programmable and specialists used them to treat more complex arrhythmias, many fellowships in cardiology and surgery did not offer young doctors sufficient experience in pacing. Thus, in updating its guidelines in 1983, the Pacemaker Study Group noted the special knowledge required of pacemaker implanters and argued that "board certification in cardiology or thoracic surgery should not be used as an automatic credential for permanent-implantation privileges." The guidelines condemned the practice of allowing manufacturers' reps in the cath lab or operating room during implantation; they hinted that doctors not doing at least 25 new implantations and 5 replacements a year (fewer procedures than they had recommended in 1974) would find it difficult to keep their skills sharp and should exit the field.[19] But in the absence of an authoritative credentialing body, such suggestions — and they were offered repeatedly — made little headway.[20]

In 1985, in an effort to bootstrap NASPE into this position of authority, Furman and the executive committee of NASPE created a wholly owned subsidiary called NASPExAM that developed and biennially administered a competence examination in cardiac pacing for physicians and a second exam for nurses and other nonphysicians.[21] For his part, Parsonnet focused less on the exam than on broader questions of implanter and hospital competence to do pacemaker work. He and Harthorne believed that pacing ought to be confined to "centers of excellence," which they defined as institutions in which at least one hundred permanent pacemakers were implanted annually "by a limited number of physicians"; such centers should have "a full, modern cardiac service, catheterization laboratories, and open-heart surgery with training programs in medicine, surgery, and cardiology." Clearly, this would limit pacing to major hospitals in large cities.[22]

It was evident by 1980 that the well-prepared physician in cardiac pacing needed some surgical and catheterization skills, the ability to diagnose complex rhythm problems from ECGs and electrograms, and more than a hobbyist's knowledge of electronics. But even the specialists in pacing found it a challenge to keep up, for pacing was not only a medical field but a highly profitable segment of the medical-device industry that attracted new companies with new ideas. Tied to the advancing "revolution in miniature" in semiconductor electronics, the knowledge base and hardware of pacing continued to change.

### THE PACEMAKER INDUSTRY IN THE EARLY 1980S

New firms and intensifying competition had reshaped the pacemaker industry during the 1970s. Several companies dropped out, and Cordis and Medtronic lost market share (see table 7). Cordis by 1984 was sliding toward oblivion in pacing; Medtronic had seen its domestic share decline by well over one-third. Four smaller firms had established themselves partly through their aggressive selling practices, but also by offering advanced and reliable products at a time when physicians' loyalty to Medtronic and Cordis had weakened. The U.S. pacemaker industry in the early 1980s consisted essentially of six firms: Medtronic, Cordis, Telectronics, Pacesetter, CPI, and Intermedics.

Telectronics, founded in Sydney, Australia, in 1962, had done well in the British and French markets and in Japan, then gained a foothold in the United States by acquiring GE's pacemaker business in 1976. The company had introduced a number of respected features, notably a very small pulse generator (55 grams) in 1975. Later, Telectronics introduced a universal connector in 1978 that enabled the physician to use Telectronics genera-

TABLE 7. ESTIMATED MARKET SHARES OF PACEMAKER
MANUFACTURERS, 1975 AND 1980

|  | U.S. 1975 | U.S. 1980 | World 1980 |
|---|---|---|---|
| Medtronic | 57% | 36% | 35% |
| Intermedics | 1 | 16 | 10 |
| CPI | 5 | 14 | 11 |
| Cordis | 19 | 12 | 9 |
| Pacesetter | 3 | 5 | 3 |
| Telectronics | · 2 | 3 | 7 |
| All others* | 13 | 14 | 25 |

Source: Patrik Hidefjäll, "The Pace of Innovation: Patterns of Innovation in the Cardiac Pacemaker Industry," Ph.D. diss., Linköping Univ., Sweden, 1997, appendix A.
*Others included American Optical (1975), General Electric (1975), Biotronik, Coratomic, Simens-Elema, Vitatron, and so on.

tors with Medtronic and Cordis leads, and pulse generators and a programmer compatible with Cordis equipment. This was clearly a small manufacturer's strategy to secure a place for its products in a market dominated by others. In the late 1970s, Telectronics established its corporate headquarters in a suburb of Denver.[23]

Pacesetter Systems, with headquarters in Sylmar, California, expanded its niche in the market by introducing, in April 1979, its multiprogrammable pacemaker with telemetry called Programmalith. This pacer had six programmable functions and could be interrogated about several parameters. It offered direct competition to the Intermedics CyberLith and was more advanced than the competing Medtronic product, Spectrax (1980), a small multiprogrammable pacer that lacked telemetry. Programmalith included a portable printer that saved the doctor the trouble of copying down telemetered readings from a screen. The company offered four models with varying degrees of programmability and telemetry so as to address the preferences of doctors with different levels of sophistication. With Programmalith, Pacesetter was able to hire good sales reps and win acclaim from the medical community. The device saved Pacesetter from slipping behind and perhaps closing its doors as American Pacemaker, Starr-Edwards, and other tiny producers had been forced to do.[24]

Cardiac Pacemakers, the company that had introduced lithium-powered pacemakers, had entered on a difficult period by the early 1980s. The pharmaceutical giant Eli Lilly acquired CPI in 1978. CPI excelled in production: its electronics and battery modules were separately built and tested and it had developed an efficient and inexpensive procedure for assembling and hermetically sealing its pacers. By contrast, Xyrel, a Medtronic lithium pacer of the same period, had the electronic components spread around the battery and required complex welding. Cordis, too, had troubles with welding. CPI thus had gained a reputation, once held by its neighbor Medtronic, for the highest manufacturing quality in the industry. But CPI's market share and profits stagnated, the result apparently of Lilly's decision to do away with CPI's s own sales organization and distribute pacemakers through the Lilly sales network. The Lilly strategy was a poor one because selling pacemakers was a highly specialized activity requiring ongoing technical support for the physician. CPI's problems were compounded when the company bet on a new type of circuit technology that failed to meet reliability standards.[25]

In technology, Medtronic began to make important adjustments to its product strategy. Medtronic pacers of the late 1970s and early 1980s did well in the marketplace without pioneering major new functions. The newer companies shared the role of daring innovator: thus Intermedics introduced multiprogrammable pacemakers with telemetry in 1978, whereas Medtronic followed suit only in 1981. But Medtronic's product lineup suggests that the company understood very well that the overall pacemaker market consisted of several distinct segments and that much of its revenue came not from its most advanced products but from products that no longer qualified as state-of-the-art.[26] It was important, for example, that Medtronic have a small, attractive pacer with a lithium battery but *no* advanced programmable features — because many doctors were cautious about programmable pacers and preferred to stay with simpler devices that promised long life and paced in VVI mode. A product like Mirel (1978), which met these criteria, gained scant attention from the top implanters but brought in a stream of income for Medtronic.

Yet demand for Mirel and other nonprogrammable VVI pacers would presumably decline eventually. A second market segment existed by 1980 for VVI pacers that also had one or two programmable features. Created in 1978 when CPI introduced its Microlith-P pacemaker, this segment was also expected to shrink at some future time as doctors advanced to more complex pacers. The Medtronic product that competed directly with Microlith-P was Xyrel-VP (1978), but instead of including output as well as rate programmability, Medtronic offered only rate programmability and advertised several separate models, each with a different output setting. In the eyes of engineers and leading doctors, the Microlith-P was the more advanced product — but Xyrel in the late 1970s was the most widely used lithium pacemaker in the world.[27]

A third distinct market segment took shape around 1979 for the newer and more complex multiprogrammable pacers like CyberLith, Programmalith, and Spectrax. These products carried higher prices and yielded higher returns (and prestige) for their manufacturers, who obviously hoped to move the physicians on from the older to the newer technology.

A market segment had a natural history: over a period of years it was born, grew, shrank, and disappeared. The longevity of a market segment depended on how swiftly the manufacturers were able to persuade doctors to make the transition to more advanced technology with higher profit margins. A manufacturer sought to shift its emphasis in a timely fashion to each new segment as it opened up. Perhaps the best way to gain future market share was to be first into an entirely new segment, leaving it to the other firms to figure out how to master the new technology, if they could. This is what CPI had done with its lithium pacemaker in 1973 and Intermedics with its multiprogrammable model in 1978. Two new market segments opened up in the 1980s, the first anticipated and the second a shocking surprise to the industry.

## TWO LEADS IN THE HEART:
## THE "BIFOCAL" PACEMAKER—A TRANSITIONAL DESIGN

"Once programmability has pervaded the field, it's doubtful that there will be further major product upgradings in the next four to five years." That was the opinion of an industry observer in 1979 who apparently did not know that several of the manufacturers were scrambling to prepare dual-chamber pacemakers for market release.[28] A second pacemaker lead to sense or pace in the atrium was a physicians' concept that had been around since the 1950s. If pacers of the 1970s had gained flexibility, they still lacked this one capability for which specialists had long yearned. As we saw in chapter 6, the first Cordis pacemaker (Atricor, 1964) stimulated the ventricle after it sensed the P wave indicating atrial depolarization. The Cordis electrodes had been fixed on the outside of the heart rather than introduced through a vein to the inside. AV synchronous pacing (VAT, in the nomenclature of the ICHD report) was a sophisticated concept for the mid-1960s, but it had not led to a commercially successful product.[29]

Still, the idea of dual-chamber pacing continued to intrigue many physicians because stimulating the ventricles to pump after the atria had filled them would improve the heart's output of blood to the body. Bridging the atrium and the ventricle also seemed more "physiological" because the pacemaker would raise or lower its ventricular pacing rate in accord with changes in the atrial rate. In the late 1960s, American Optical's Barouh Berkovits invented a dual-chamber pacer that moved beyond Atricor by introducing two transvenous leads through two different veins.[30] With one lead in the atrium and the other in the ventricle, Berkovits's "bifocal" pacemaker could deliver stimuli to both chambers. It avoided the problem of reliably sensing the P wave by sensing only in the ventricle (AV sequential pacing or, in the ICHD terminology, DVI). An atrial pacemaker stimulus would cause the atria to contract. The ventricles would then contract normally, or if heart block occurred the pacer would deliver a ventricular stimulus after a preset AV delay. If the patient had a spontaneous ventricular rate that was faster than the pacemaker's atrial rate, then both the atrial and the ventricular stimuli were inhibited. The pacemaker thus used ventricular activity to program the atrial and ventricular pacemaker stimuli.

The AO "bifocal" pacemaker embodied an original approach but, like Atricor, failed to gain wide acceptance. The two leads were troublesome to many physicians; some probably found the device difficult to understand. Published reports indicated that the pacer would self-inhibit: if the atrial lead had been incorrectly placed, the ventricular lead sometimes sensed the atrial stimulus as a normal ventricular beat and recycled instead of pacing the ventricle. Finally, the bifocal pacemaker was quite large. One doctor who implanted a dozen of the pacers later recalled that several of the bulky generators had eroded through the chest wall.[31]

## THE RACE TO INTRODUCE AV UNIVERSAL (DDD) PACERS

In an era when most doctors were accustomed to implanting a VVI pacemaker for every form of slow heartbeat they encountered, dual-chamber pacing needed a compelling clinical rationale. Some specialists believed by 1980 that dual-chamber pacing would almost always be preferable to the familiar single-chamber stimulation. They realized that VVI pacing was not really appropriate for many patients who had disorders of sinus rhythm. The choice of the VVI mode implied that the physician was giving up on the patient's sinus node and atria and would simply strive to maintain a ventricular beat. But sinus node dysfunctions seemed to call for some form of atrial pacing to compensate for the failure of the sinus node, and this implied a pacemaker with leads in both chambers.[32]

With patients who had sinus dysfunction, simple ventricular pacing in VVI mode could sometimes cause a disorder known as the pacemaker syndrome. The ventricular pacemaker stimulus was sometimes conducted in a retrograde direction back to the atria, causing the upper chambers to contract against closed atrioventricular valves. This forced blood upstream from the atria into the pulmonary veins on the left side of the heart and into the systemic veins on the right side. Symptoms ranged from lethargy and breathlessness to severe drops in blood pressure to fainting. As Kalman Ausubel and Seymour Furman put it, "the pacemaker syndrome represents a clinical spectrum of intolerance to ventricular pacing."[33] Standards were rising — it was no longer enough to keep patients alive and give them the daily existence of semi-invalids. Furman and others believed that it ought to be possible for cardiac pacing to bring many patients back to something approximating good health.

Technological roadblocks had impeded dual-chamber pacing, but these were now being dismantled. Experience up to the early 1970s had clearly shown that effective dual-chamber devices, which required more complex circuitry and atrial leads, must await the downsizing of pulse generators and new lead designs. A decade later, lithium batteries and hybrid circuits had made it easier for engineers to design small pulse generators. Stable atrial leads proved more difficult. The P wave indicating atrial depolarization is a tiny signal, so the atrial lead must be properly and securely positioned to sense it reliably. But it proved frustratingly difficult to maintain a lead stably in the atrium, which is a smooth-walled chamber. In 1969, surgeon Nicholas Smyth had described an experimental electrode with a fixed J-shaped curve for atrial pacing; this proved to be a breakthrough, especially when Smyth later added tines that helped the tip lodge securely against the atrial wall. The device manufacturers eventually came out with a number of variants such as an atrial lead with a tiny screw-in tip.[34]

Medtronic was particularly well positioned to pioneer dual-chamber pacers. When American Optical withdrew from the pacing industry in 1975, Barouh Berkovits moved to Medtronic on condition that the company develop and aggressively market a dual-chamber product like the AO "bifocal" pacer. In 1979, Medtronic introduced Byrel, a smaller and more advanced embodiment of the concept. Although Byrel was not part of the main line of Medtronic pacemaker development, it embodied an important transfer of technological knowledge from the defunct AO pacemaker business.[35]

Medtronic had also invested heavily in research on transvenous pacemaker leads. The company in 1975 had hired engineer Paul Citron, a former member of Smyth's research group, to oversee development of atrial leads; the company thus created a connection with Smyth and arranged to license his key patents on tined leads. Another engineer at Medtronic, Ken Stokes, in 1978 announced a new lead that had an insulating sleeve of polyurethane. Implanters found that they could introduce two of the thinner, slipperier polyurethane leads down a single vein. If not wholly satisfactory, leads had come to seem less of an impediment to dual-chamber pacing by the end of the 1970s. These advances led Parsonnet, speaking for the specialist implanters, to write that routine implantation of a second lead was "a technique whose time has come."[36]

As early as 1976, Medtronic and Cordis had singled out dual-chamber pacing as a future growth area that they could use to regain their lost market share. By jumping into dual-chamber technology and proceeding rapidly to more sophisticated second- and third-generation devices, they hoped to ride the growth of this new market segment for many years. Experienced pacemaker doctors, assembled in advisory groups by Medtronic and Cordis, supported the push into dual-chamber devices and offered advice on specific design features. As a bridge to advanced "AV universal" pacemakers that could be configured to sense and pace in both the atrium and the ventricle, Intermedics and the German company Biotronik both introduced DVI pacers that competed with Medtronic's Byrel. Cordis, however, skipped this stage and rendered DVI pacing obsolete by introducing an AV universal (or DDD) pacemaker to the U.S. market in November 1982; the device, named Sequicor, had been in clinical stages of testing since 1980. The following year, Cordis stayed ahead by bringing out Gemini, a DDD model with telemetry and a longer worklife. Medtronic followed suit by introducing its Versatrax DDD pacer that same year. These new products incorporated all the major design innovations of the previous decade: lithium batteries, circuitry that permitted multiprogrammability and telemetry, and dual transvenous leads (see table 8). Could pacemaker technology progress any further?[37]

A look inside the shells of Sequicor and Versatrax reveals that Cordis and Medtronic followed quite different paths to DDD pacing. The first microprocessor, a general-purpose programmable circuit on a single chip, had been introduced commercially in 1971. By the late 1970s, more advanced microprocessors were widely available at greatly reduced prices. DDD pacemakers would be expected to behave in very complex ways involving sensing and pacing in both chambers of the heart and would require new programmable features beyond anything required in a single-chamber pacer. Pacemaker engineers could choose a complex custom-designed hybrid circuit or take the more radical course of installing a general-purpose microprocessor that would be programmed by software.[38]

For Sequicor, Cordis's senior pacemaker engineer Peter Tarjan elected to base the design on a microprocessor that the company purchased from

TABLE 8. MARKET INTRODUCTION OF DDD PACEMAKERS

| Company & Product | FDA Approval & U.S. Market Release | Generation* |
|---|---|---|
| Cordis Sequicor | November 1982 | I |
| Medtronic Versatrax | Early 1983 | I |
| Cordis Sequicor II | February 1983 | II |
| Telectronics Autima | February 1983 | II |
| Medtronic Versatrax II | May 1983 | II |
| Biotronik Diplos-03 | October 1983 | II |
| Cordis Gemini | Late 1983 | III |
| Medtronic Symbios | 1984 | III |
| Intermedics Cosmos | 1984 | III |
| Pacesetter AFP 283 | 1984 | III |
| CPI Vista | 1989 | |

*Source:* Peter H. Belott, "Clinical Experience with over 250 DDD Pacemakers," in *Modern Cardiac Pacing,* ed. S. Serge Barold (Mount Kisco, N.Y.: Futura, 1985), 439–81, esp. table 1.

*First generation: bulky generator, limited programming functions, limited longevity, no telemetry. Second generation: reduced size, improved programmability and longevity. Third generation: small size, numerous programmable parameters, extensive telemetry functions including diagnostics, more user-friendly programmers.

RCA. In effect, Sequicor was a rudimentary implanted computer controlled by software; this meant that the device was minimally hard-wired and could be entirely reconfigured by giving it new software instructions. Patrik Hidefjäll suggests that Cordis chose the microprocessor because the company could cut development time by simultaneously designing and testing the pacemaker and doing the software programming. The decision generally reflected the culture at Cordis, a company that since the 1960s had placed a high valuation on leading-edge innovations that would impress the top physicians and mark out the path to the future.[39]

Medtronic's culture and circumstances favored a more cautious approach. The company had a subsidiary called Micro-Rel that manufactured hybrid circuits for Medtronic pacemakers but was not yet in a position to turn out microprocessors on a production scale in 1980. Going to a microprocessor-based design would have entailed the uncertainties and possible delays of dealing with an outside supplier. In addition, the key pacemaker engineer at Medtronic, Jerry Hartlaub, believed that a microprocessor-based DDD pacer would tend to drain the battery too quickly. Hartlaub was also concerned about what computer scientists call "the halting problem" — the risk (tiny but real) that a software flaw might cause a microprocessor-based pacer to stop suddenly. "Software reliability," he pointed out, "is undecidable" because a program written to check a second program might itself contain errors. Hartlaub believed that custom-designed integrated circuits could manage even multiprogrammable DDD pacemakers adequately, but he also recognized that microprocessor-based pacemakers would eventually prevail.[40]

Sequicor and Versatrax were highly flexible pacemakers for the time. Although it lacked telemetry, Versatrax could be programmed to pace in any of five modes and adjusted or fine-tuned on seven other parameters. Both devices had received premarket approval from the FDA, but both exhibited crippling problems in the real world. As Hartlaub had predicted, the Cordis Sequicor experienced high current drain and a short battery life. Cordis engineers also discovered that their device required the physician to interpret large amounts of telemetered data and make numerous programming decisions; clearly, the external programmer must become more user-friendly.[41]

Versatrax had a more embarrassing flaw. A DDD pacemaker can trigger a tachycardia if the atrial electrode incorrectly interprets a ventricular event (such as a premature ventricular contraction or the pacer's own ventricular stimulus) as a natural P wave. In this situation of retrograde conduction from the ventricle back to the atrium, the pacer will deliver another stimulus in the ventricle after the A-V delay period. The new ventricular stimulus in turn may be conducted back and misread as another P wave, causing the pacer to fire again in the ventricle. Furman and his colleague John D. Fisher first described this problem, calling it an endless-loop tachycardia or *pacemaker-mediated tachycardia* (PMT). A doctor or technician could readily terminate the tachycardia by placing a magnet or the head of the programmer on the patient's chest to cause the pacemaker to cease sensing. But it was obvious that a pacer that exposed some patients to a constant risk of tachycardia needed fixing.[42]

With the original Versatrax pacemaker, the physician could not program this problem away except by choosing some more primitive pacing mode and losing the benefits of DDD pacing for the patient. Alone among DDD pacers of the early 1980s, Versatrax did not permit the physician to use programming to lengthen the interval during which the atrial electrode was insensitive (refractory) to electrical signals. Medtronic quickly introduced a new Versatrax with a longer atrial refractory period and in 1984 provided a programmable atrial refractory period in its Symbios line of DDD pacers.[43]

According to Michael Toffoli, a Medtronic manager involved with the Versatrax project, the designers had considered a programmable atrial refractory feature but rejected it "because we thought that these devices are getting so complicated that we don't want to include a feature that has no obvious benefit or role. . . . We had done an extensive clinical [trial] on the devices. . . . You know what we never saw in all those units? Retrograde conduction. . . . Nobody in the whole clinical on our first DDD ever saw retrograde conduction. So then we went into the marketplace—and then all of a sudden, you know, people started to observe pacemaker-mediated tachycardias. Well, how do you fix that? With the programmable atrial refractory [period]. And . . . we had made a conscious decision not to include that."[44]

The Cordis DDD pacemakers briefly revived the company's fortunes, but problems with the Gemini model surfaced in 1984 and contributed to the decision by the board of directors to sell the Cordis pacing business in 1987 (see chapter 8). On the other hand, success in DDD pacing enhanced Medtronic's position in the industry. Despite the problems with its initial product, Medtronic benefited from being early into the DDD pacemaker market. Medtronic had not been at the forefront of the industry in product innovation for many years, but Versatrax and Symbios hinted that the firm's cautious approach was changing.[45]

In the pacemaker industry, manufacturers introduced radical new products like the DDD pacers not only to enhance their own reputations and carve out shares in an emerging market segment, but to put their competitors to the test. Just how this worked is evident from the case of Cardiac Pacemakers, for the coming of DDD pacing severely undermined CPI's position in the pacemaker industry. CPI twice had to abandon patent applications on multiprogrammable pacemaker designs. Finally it secured a patent in December 1981, later than most of the competition. CPI also had problems switching to a new type of circuit technology that turned out not to meet reliability standards. As a result of these difficulties, the company trailed its competitors by more than five years in the dual-chamber market; more broadly, its reputation for innovation and quality was sullied.[46]

Intermedics chose a technologically radical strategy for its first DDD pacemaker rather than the more conservative approach of Medtronic or reliance on an outside supplier, as Cordis had done. In 1980 the company hired Larry Stotts, a designer from Motorola Applied Research Group, and set up its own microprocessor design team. "While commercial microprocessors were available," Stotts recalled, "we chose the custom route to meet aggressively low current drain goals to reduce pacemaker battery size, and to have absolute control of the design to assure the highest reliability and quality possible. After seeing the impact that recalls were having in the industry and being a small company, it scared the hell out of us to be dependent on outside suppliers of critical components." The design moved on to production at ZyMos, a custom manufacturer of chips in which Intermedics was the largest shareholder.[47]

Intermedics introduced its Cosmos DDD pacemaker in 1983 to compete with second- and third-generation devices from Medtronic and Cordis. Doctors noticed first that Cosmos was quite small, weighing just 65 grams and occupying a volume of 31 cc. Extensively programmable, it could also store information about past behavior of the heart and the pacemaker that the physician could retrieve via the user-friendly programmer. Hartlaub and others had expressed concern that software-driven pacemakers might contain bugs in the software that would cause malfunctions, but with Cosmos it was possible to modify the original program long after implantation by means of externally programmable patch registers. The designers had included two safeguards against pacemaker-mediated tachycardias. Like other DDD pacers, Cosmos had a programmable atrial refractory period; but if a PMT did occur, the pacer would abort it by counting 15 beats at the upper rate limit, then omitting the next stimulus. Finally, Cosmos had a programmable "sleep" function: it could revert to a low power state, then "wake up" when prompted by a sensed physiological signal such as an R wave. Cosmos and its successor models were distinctive and effective DDD pacemakers. Praised by experts such as cardiologist Peter H. Belott, Cosmos became the Intermedics flagship product and established the company as the leader in the DDD market segment for more than a decade.[48]

The DDD pacemakers of the 1980s had admirable technical features, but introducing these new devices put the manufacturers under severe financial strain, partly because the FDA now required extensive premarket testing that added months of product development time. Each company also had to train its sales force and develop a full-blown marketing plan to persuade doctors to move from single-chamber to dual-chamber pacemakers. Industry observers predicted robust growth in the DDD market segment—but this did not happen. As Medicare adopted the DRG payment system and pressures intensified on doctors not to implant pacers in marginal cases, the annual number of new implantations declined sharply between 1981 and 1985 from about 120,000 to fewer than 90,000; sales growth outside the United States slowed down as well. What most concerned the manufacturers was that, contrary to expectations, physicians proved reluctant to switch from familiar VVI pacing to the new generation of DDD pacemakers.[49]

## PRACTITIONER IMPLANTERS AND DUAL-CHAMBER PACING

All DDD pacers of the early 1980s presented the physician with new choices and complexities. As pacemakers emulated nature more nearly, they had also become more difficult for doctors to comprehend and manage because new, programmable functions for dual-chamber pacing now accompanied the more familiar functions for stimulus rate, amplitude, and duration. For example, the physician could set the sensitivity thresholds for both leads so that they reliably sensed spontaneous P and R waves within the heart. It was necessary to choose an appropriate AV interval (the interval between an atrial and a ventricular pacemaker stimulus), and program the postventricular atrial refractory period to avoid PMTs by rendering the atrial sensing circuit blind for a period. Physicians could individualize the settings and revise them later as necessary. Even in uncomplicated cases, setting up a dual-chamber pacemaker required that the physician combine information from the patient's ECG and the programmer's telemetered data to form a detailed and insightful picture of how the implanted pacemaker and the heart were interacting. In more complex cases, the heart's own erratic behavior plus the pacemaker's stimuli in both chambers could produce ECGs requiring great skill at interpretation.[50]

As late as 1989, three-fourths of the pacemakers sold in the U.S. market were still single-chamber devices that operated in VVI mode, a kind of pacing that had been around for more than 20 years. Thus most patients diagnosed with sinus dysfunction and arguably in need of atrial pacing were still receiving VVI pacemakers. The reasons for doctors' caution are fairly obvious. They could quickly and easily place a ventricular lead in most cases, but it was a tricky and time-consuming business to run a second lead down the vein into the right atrium and position it correctly. Many doctors also — correctly — perceived dual-chamber pacers to be complex, difficult to understand, and likely to require frequent reprogramming. Implanters who conscientiously tried to master the new technology had to understand pacemaker timing cycles and be able to interpret complex ECGs. In sum, many doctors remained unpersuaded that the benefits to the patient of dual-chamber pacing outweighed the troubles it made for both doctor and patient.[51] As a clinching factor, DDD pacers were more expensive than single-chamber VVI pacers.[52]

A turnaround got under way at the very end of the 1980s as U.S. sales of DDD pacers rose from an estimated 25,000 units (one-fourth of the pacers sold in 1989) to about 70,000 (one-half of the pacers sold in 1994). The shift toward dual-chamber pacing in the early 1990s coincided with a growing dominance of cardiologists, including electrophysiologists, in the field of pacing and a corresponding decline in surgeons' participation. An additional factor may have been that the pacemaker companies' programmers, the auxiliary devices through which doctors communicated with implanted pacemakers, were becoming more user-friendly. The newer programmers pointed out unusual patterns in the patient's heartbeat, prompted the doctor with suggested settings, and issued warnings when settings contradicted each other or endangered the patient.[53]

Despite the initial slow acceptance of dual-chamber pacing, the companies competed intensely in this market segment from 1983 on. With the entire pacing industry growing rather slowly in the 1980s, it was highly important to have a position in the segment expected to show the highest rate of future growth. Dual-chamber pacers also carried higher selling prices than single-chamber, so unit-for-unit they generated more revenue for the firm and more income for the salesperson. In the background, too, was always the manufacturers' desire to continue the process of innovation so as to challenge competitors and prevent the commodification of the pacemaker.

## OUT OF NOWHERE: ACTIVITRAX

Pacing specialists had been discussing the possible advantages of dual-chamber pacing for many years before DDD pacers reached the market. At Cordis and Medtronic, designers and consultants had begun to plan the general characteristics of DDD pacemakers by about 1976–77; engineering design work and bench testing were underway by 1978–79. Since the early years of pacing, clinical trials of new devices had become more formal and could take many months; the FDA review added further to the time from first clinical use of a new pacemaker to its being launched into the market. Thus the first DDD pacers were released for general use at the end of 1982, some five to six years after the initial planning. By then, every cardiac pacing insider and every market analyst knew that dual-chamber universal pacers were coming.

Increasingly, companies planned to stage the introduction of new products in an orderly way, every two years or so, with each product advancing one major step beyond the last. Medtronic pacemakers of the late 1970s and early 1980s included Xyrel (1977), which offered lithium batteries and a hermetically sealed titanium shell, and Mirel (1978), which was a smaller, simpler cousin. Xyrel-VP (1978) added rate programmability, Byrel (1979) offered AV sequential pacing, Spectrax (1980) added multiprogrammability to the basic Xyrel package, and Spectrax-SXT (1981) added telemetry. Then came Versatrax (1983), with dual-chamber capabilities, Versatrax II (1983), with a longer atrial refractory period, and Symbios (1984), with programmable atrial refractory, advanced telemetry, and an anti-tachycardia function. New products succeeded old ones with a fair degree of predictability.[54]

In 1980, Medtronic appointed Whitney McFarlin to manage its cardiac pacing business. McFarlin became concerned that the products in the pipeline were a bit *too* predictable. He discussed this with Bobby Griffin, the head of R&D and manufacturing in pacing. McFarlin "came to me one day," according to Griffin, "and he said, 'I want two innovations each year.'" Griffin passed this message on to his engineers, one of whom was Hartlaub, the engineer who had managed the design of Spectrax and who was heavily involved with Versatrax. Hartlaub understood Griffin to be saying that the company must be "bimodal. On the one hand, he was: pay attention to business, pay attention to market share and to doing things well. And on the other hand—" On the other hand, he was offering Hartlaub free rein to get some people together and come up with innovations that would rattle the industry.[55]

Hartlaub approached Ken Anderson, an engineer with whom he had worked on the Spectrax project. Even though Anderson was by no means an expert on pacemaker sensing, he asked Anderson to "look at sensors for . . . pacemaker control." Pacemaker leads had been sensing electrical activity within the heart since the 1960s. Dual-chamber pacers could vary the ventricular rate by sensing the atrial rate, though in 1980 many people felt that atrial sensing was not yet reliable enough. The idea of controlling the ventricular rate by sensing some other physiological change seemed attractive, but as far as Anderson could determine, few researchers had focused on the problem.[56]

Anderson found that another engineer at Medtronic "was working with a piezoelectric element for an ambulatory monitor that measured blood pressure." Under the blood-pressure cuff was a piezoelectric crystal that picked up a signal from the sounds of blood gurgling in the veins.

> I was curious what was in this little thing, this little disk, so I tore it apart. They told me it was a piezoelectric element. And as soon as I saw it, I said, this has applications for our pacemakers. These things are remarkable in the fact that they do not use any current. . . . And of course, that is the number-one problem of pacemaker design: how do you add function to the device without depleting the battery? So I knew there was something here.

Anderson's reasoning was the reverse of the usual process. Most investigators would have said, "What can I come up with that will vary the heart rate?" Anderson "looked at it the other way: how can I use this sensor to solve a medical problem? I started with the sensor and asked, how best can I use it? A little bit different approach: an engineer's approach versus a physician's approach."[57]

At first, Anderson assumed that the piezoelectric sensor could sense the movement of the diaphragm. If a patient's breathing rate increased, the sensor could signal the pacemaker to accelerate the heart rate. But when working with laboratory dogs, he found that the sensor also picked up a great deal of noise that turned out to be pressure waves from the movement of muscles. "An engineer is always fighting signal-to-noise ratio, right? That's his big challenge." The noise in fact was much more noticeable than the signal he wanted from the diaphragm. This was disappointing, but then it occurred to him that the noise could be redefined as a signal. "We redefined . . . the sensor to one which was respiration/activity. . . . We were going to accept signals from both." Respiration yielded a low-amplitude, low-frequency signal—roughly 12 to 15 times a minute. Body activity, the movement of muscles, yielded a very high-frequency and high-amplitude signal. The activity that came through most clearly was walking—the feet hitting the ground. That was an ideal sound to sense, in Anderson's opinion, because pacemaker patients "want to be able to walk. We take [walking] for granted, us healthy people. . . . When you're incapacitated, the thing you want to do more than anything else is to walk."

Anderson planned to add two filters, one for high-frequency signals and one for low-frequency. "When there was a lot of this high-amplitude, high-frequency stuff, that's what you look at—that's what you control the rate with. When that went away, then you look at the respiration." If the patient climbed a flight of stairs, swept the kitchen floor, or went for a stroll, the sensor would pick up footfalls and other evidence of body activity and the pacemaker would raise the heart rate accordingly; and if the patient was at rest or asleep and more or less inactive, the pacer would set the heart rate off the patient's breathing rate. That, at least, was the plan. "That was how we were going to do it. We were just dreaming away." Then Griffin told him, " 'Well, this is a good idea—now go reduce it to a miniature device. . . . You've got six months to work on this. And if you can't come up with something in six months, then we've got other things for you to do.' Now the cheese begins to bind, right? We had to come up with these very fancy filters, and you had this very complicated logic, this dual approach. It just got to be too much for us." The six-month deadline forced the engineers to simplify their grand idea of sensing both activity and breathing. "We eventually abandoned the respiration part, abandoned it entirely." Anderson's circuit engineer, Dennis Brumwell, "came up with a nice circuit which we called a zero-crossing detector for the activity [sensor]. The filters got to be very simple. The counting of those detects got to be very simple. Everything just sort of fell in place. . . . We cut it down to the bare minimum."[58]

What emerged by late 1981 was a multiprogrammable, single-lead pacemaker based on the circuitry of a Medtronic device called Enertrax that had had limited market release. Enertrax had an atrial sensing lead for which Anderson substituted the piezoelectric sensor mounted on the inside of the pacemaker housing, so that instead of varying the ventricular rate based on the atrial rate, the pacer varied it based on the level of body activity. The physician could reprogram the standard parameters such as the amplitude of the pacing stimulus by using the same programmer that communicated with all Medtronic pacers. Four new programmable parameters were added: the absolute maximum and minimum pacing rates above and below which the pacer would not go; the intensity of activity that would cause the pacer to raise or lower its pacing rate; and a setting indicating how much the pacing rate would change at a given activity level.[59]

The new "rate-responsive" pacemaker was first implanted in December 1981. As a single-lead device, it obviously did not fit into the Medtronic product line; in fact, as an alternative to DDD pacers for many patients, it would compete with Medtronic's own Versatrax. Bobby Griffin had pushed it to rapid completion partly to prevent a debate within the company and partly to demonstrate that Medtronic would no longer be an also-ran in technology but an important innovator.[60]

From a marketing point of view, the device — named Activitrax — would appeal to doctors who were interested in giving rate variability to their patients but uneasy about the complexities of dual-chamber pacing. A huge number of physicians fit this description, but because of the very limited market research done before the invention of rate-responsive pacing, nobody at the company was sure how Activitrax would go over with the doctors. Would they treat rate-responsive (VVIR) pacing as a poor man's DDD, or as a vast improvement on VVI?

The answer to this question came when Medtronic unveiled Activitrax in May 1983 at a world convention on pacing in Vienna. A paper about the device, presented by Anderson, Brumwell, and others, won an award as the most important one at the meeting.[61] Nearly a thousand physicians requested additional literature or a visit from a Medtronic representative. Subsequent clinical trials — hundreds of implantations in the United States and elsewhere — reassured the skeptics. According to Anderson, some doctors worried that "it's going to run away, people will have too high rates when you go down a cobblestone road, but . . . patients never complained about that." With FDA approval behind it, Activitrax was introduced to the market in early 1986: it defined an entirely new market segment in cardiac pacing — a segment for rate-responsive pacemakers. By early 1987, it was the most prescribed pacemaker in the world. Since the sensing and circuit technology were patented, Medtronic had this segment all to itself for several years. No doubt the availability of Activitrax also helped slow the acceptance of dual-chamber pacing.[62]

Some physicians disapproved of rate-responsive cardiac pacing on the grounds that it ignored the contribution of the heart's upper chambers and thus did not emulate nature. Many senior specialists, including Parsonnet, preferred dual-chamber pacing, with all its complexities.[63] They had taken the time to master it and believed that others should do the same. If not DDD, they would have preferred a rate-responsive pacer that sensed something other than footfalls. This attitude did not particularly bother Medtronic's Bobby Griffin. It was "fine that they felt that way": the longer people debated the point, the longer Medtronic's competitors would remain uncertain about their own development paths. The debate "gave us a longer lead time." Activitrax might not be the most subtle technology for varying the heart rate, but it worked. At one point, Griffin ordered a study of every possible way to vary the heart rate by sensing some physiological change. "Activity [-sensing] fell about eighth on the list. Why did we choose activity? Because we could do it. And the others weren't doable. We did not have the technological capability to implement them, and that one was easy. So we did it."[64]

Activitrax was arguably the most important single pacemaker model ever launched — certainly the most important since the first VVI pacers in 1967. Rate-responsive pacing hugely changed the pacemaker industry, buoying it at a time of decline in unit sales and postponing the day when pacers would become commodities. The product enabled Medtronic to regain part of its lost market share in pacing: by 1989, the company enjoyed a global market share of about 43 percent, compared with a 1980 figure of 36 percent. Perhaps most significant, rate responsiveness redefined the field of cardiac pacing by bringing the subjective experience of the patient to the fore and legitimizing the idea of pacing not only to safeguard the patient's life but to enhance quality of life. Thus Medtronic's marketing highlighted active people: one, a 57-year-old man in Ohio — a pacemaker patient since 1972 — was able to take up his hobby of bow hunting again; another, a young mother in New York, was enabled by her Activitrax to resume normal activities such as cleaning her apartment, playing ball with her daughter, and doing grocery shopping. As one patient said, "I feel so good I forget I have [a pacemaker]." Ken Anderson knew that many pacemaker patients complained "about not being able to go out and play tennis or not being able to walk down the street without having to stop." Many people on VVI pacers had to stop frequently out of dizziness and tiredness because their heart rates could not rise. Anderson called this the "window-shopping syndrome: the patient had to stop and look in the window, not because he wanted to look in the window, but because his heart rate was too slow."[65]

In the 1980s, DDD pacers were being called "AV universal pacemakers" — but this proved to be only a hopeful nickname. Although they were vastly better for most patients than single-chamber VVI pacers, these devices were far from being universal in their application. DDD pacing provided rate variability for many patients because the pacemaker could sense changes in the atrial heart rate and stimulate the ventricle in tune with the atria. Experience with DDD pacers during the early 1980s made it clear to cardiologists, however, that many patients' sinus nodes were unable to accelerate when activity level or emotional state changed. In a wonderful bit of medical jargon, these hearts were said to be "chronotropically incompetent." Such patients could not realize the full benefit of DDD pacing, but single-chamber, rate-responsive pacemakers gave them back rate variability by ignoring the action of the sinus node and sensing body activity directly. It was an obvious next step to combine DDD with rate-responsive pacing in a DDDR pacemaker, and Medtronic took this step in 1989 with a model called Synergist.[66]

Dual-chamber, rate-responsive pacing arrived in advance of clear medical doctrine on its use. Cardiologists had not determined by 1989 what sorts of heart-rhythm disorders should appropriately be treated with DDDR pacing. The patient with both a chronotropically incompetent sinus node and AV block was a clear candidate, because DDDR could provide both AV synchrony and rate variability. But how many such cases were there, and what other disorders might call for DDDR? Clinical researchers worked through these questions in the 1990s, *after* DDDR devices had come into use and claimed a significant share of the bradycardia pacing market.[67]

Along with dual-chamber and rate-responsive pacing, the device manufacturers and senior pacemaker specialists of the early 1980s were becoming interested in pacing for certain forms of rapid heartbeat (tachycardia). Barouh Berkovits and others designed experimental devices intended to sense the onset of a tachycardia and halt it by throwing in one or more timed stimuli that would interrupt the rhythm. Although these devices were years away from market release, many in the field believed that "the next major evolutionary phase" would be pacing to manage tachycardias. Some forms of tachycardia could degenerate into ventricular fibrillation, from which few would recover. Even as they competed for market share in bradycardia pacing, the device manufacturers were considering future applications for the pacemaker. They were also developing implantable defibrillators.[68]

## CHAPTER 10

PREVENTING SUDDEN CARDIAC
DEATH: THE IMPLANTABLE
DEFIBRILLATOR

       In the 1980s, the pacemaker industry transformed itself into the "cardiac rhythm management industry" by introducing an entirely new kind of implantable machine that halted ventricular tachycardia (VT — a ventricular rate of more than 100 beats per minute) and that most sinister of rhythm disturbances, ventricular fibrillation (VF — random electrical activity in the ventricles with no organized ventricular beat). Thomas Lewis had pointed out in 1915 that "fibrillation of the ventricles is incompatible with existence. . . . If it occurs in man, [it] is responsible for unexpected and sudden death."[1] The implantable cardioverter-defibrillator (ICD) astonished the international community of heart specialists and attracted the gaze of the media as an important "breakthrough" in treatment. The ICD acts in the most dramatic possible way to save the lives of men and women susceptible to a common cause of heart-disease mortality in industrialized countries: sudden cardiac death (SCD). The ICD represents the quandary that new technologies can pose for the health-care system: it is effective for many kinds of patients, but the cost of treatment is not negligible.[2]

       A word about terminology: Until 1983 the device we call the ICD could not cardiovert and was generally known as the AID, or automatic implantable defibrillator. For a time after 1983 it was called the AICD — automatic implantable cardioverter-defibrillator — but the *automatic* dropped off as redundant. *Cardioversion*, a term introduced in the 1960s by Bernard Lown and Barouh Berkovits, referred to timed electrical discharges or shocks that disrupted and terminated tachycardias. Cardioversion was thus somewhat less drastic than defibrillation. ICDs do *cardiovert*, but the word is little known outside the community of heart-rhythm specialists. (The manufacturers now say that *ICD* stands for *implantable cardiac defibrillator*.)

       The invention and industrialization of the ICD offers interesting parallels to the development of implantable pacemakers more than a decade earlier, along with a few striking contrasts. With both inventions, clinical research groups conceived of the idea and led the effort in the early going by tackling disparate problems that ranged from identifying the population of appropriate patients to designing hardware to raising money. In both cases, too, the larger community of heart specialists expressed some initial skepticism. At first, the indications (medically defined circumstances warranting use of the technology) were quite narrow for both inventions, fixed third-degree heart block for the pacemaker and demonstrated high risk of ventricular fibrillation for the defibrillator. But the very existence of the technology and the early reports of miraculous recoveries by patients then prompted researchers to intensify their investigations into other brady- and tachyarrhythmias that might be treatable with the new devices. From the start, electrophysiologists (cardiologists specializing in rhythm disorders of the heart) and medical-device manufacturers were intensely interested in expanding the population of patients. As with the pacemaker a generation earlier, the ICD held the promise of both professional honor and financial gain.

In some respects, the ICD project under Michel Mirowski unfolded differently from the Chardack-Greatbatch invention of a mercury-powered implantable pacemaker two decades earlier. Unlike its cousin the pacemaker, the prototype ICD was the work of a single group of clinical-research physicians and engineers instead of several small teams working independently of each other but sharing ideas back and forth. Development of the ICD involved a more formal mapping out of problems and possible solutions than had been true with the pacemaker. From early in the 1970s, the developers engaged in long-range preparation by submitting more than 50 patent applications and carrying out numerous tests that would lead toward high-quality manufacturing and FDA review. The entire effort cost millions of dollars and engaged the efforts and talents of dozens of people.

In both cases — the pacemaker and the ICD — device-manufacturing corporations took over the effort after the first clinical successes. The competences of an established company were needed to prepare the device for large-scale production, to manage clinical trials with hundreds of human implantations, and to plan a marketing effort that included training sales reps and physicians. In the case of the ICD, the entire ramping-up period — from prototype in the animal lab, to first clinical use, to the manufacture of thousands of safe and reliable units — took place during the era of FDA oversight of implantable devices. This was a fact of life for the device industry by the late 1970s.

## MICHEL MIROWSKI AND THE BEGINNINGS
## OF THE DEFIBRILLATOR PROJECT

The European upheaval of the 1940s profoundly affected the life and career of Mieczyslaw (Michel) Mirowski (1924–90).[3] A Jew born and raised in Warsaw, Mirowski fled his homeland alone as an adolescent in December 1939, after the German invasion and occupation of western Poland. He crossed into the Soviet Union, moved from city to city, and eventually returned to Poland as a junior officer in a Polish army unit that the Soviets had formed. After the war, Mirowski took his medical training at the University of Lyons, receiving the M.D. degree in 1953, then worked in Mexico and the United States. In 1962–63 he was in Baltimore as a research associate of pediatric cardiologist Helen Taussig. He settled in Israel as a practitioner cardiologist.

Between 1953 and 1970, Mirowski coauthored some 30 scientific papers on atrial-rhythm disturbances and other heart problems. Mirowski's biographer, John A. Kastor, remarks that these are "superb examples of inductive or 'armchair' electrocardiographic reasoning" rather than reports on direct experimental work. Mirowski was by inclination an intellectual and a theoretician rather than a gifted experimenter in the laboratory or the clinic.[4]

The sudden death of an old friend in 1967 after several episodes of ventricular tachycardia, and perhaps his own desire to try a new direction, precipitated Mirowski's decision to invent an implantable defibrillator. He later told Kastor that cardiologists with whom he discussed the idea "all told me that defibrillators couldn't be miniaturized." He decided to proceed in part "because people told me it couldn't be done." Recognizing that he would need funding and expert assistance, he accepted an invitation to become chief of the new coronary-care unit at the Sinai Hospital in Baltimore and negotiated a half-time release to pursue his research. Once established at Sinai in the fall of 1968, Mirowski began looking for talented associates. He recruited Morton Mower, a young cardiologist who ran the hospital's heart station and had a good knowledge of medical electronics, and William Staewen, director of the bioengineering lab at the hospital.[5]

The Mirowski team began with the knowledge that ventricular fibrillation was the event behind most sudden cardiac deaths and that it was possible to terminate VF with a strong countershock that depolarized the entire heart. But no one had ever attempted to defibrillate from a battery-powered device, let alone a device small enough to be implanted. To terminate VF, external defibrillators delivered shocks on the order of 400 joules.

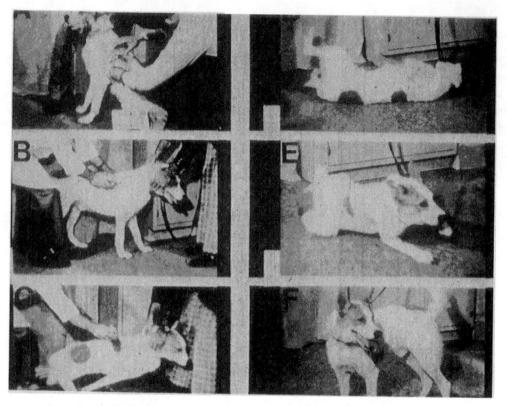

**Dog Being Defibrillated, 1975.** These are selected frames from a movie that Michel Mirowski and colleagues made depicting the use of an implanted defibrillator with a laboratory dog: (A) the group measures the charge time of the defibrillator; (B) alternating current is applied to put the animal into ventricular fibrillation; (C) the heart is no longer pumping and the animal collapses; (D) the defibrillator senses VF and fires its shock; (E) 5 seconds after the shock; and (F) 10 seconds later. (Courtesy of Dr. Morton M. Mower)

The capacitor for such a device, if implanted, would occupy most of the thoracic cavity. Mirowski, however, believed that "most of the energy used for external defibrillation is wastefully dissipated in the tissues surrounding the heart." Put the electrodes close to the heart and it would be possible to defibrillate with a smaller shock. The group set up a makeshift catheter electrode in the superior vena cava of a dog and placed a small plate — a broken defibrillator paddle — under the skin of the dog's chest. In August 1969, less than a month into their collaboration, they put the dog into fibrillation and successfully brought it back to normal rhythm with a single shock of 20 joules.[6]

A fully implanted defibrillator would have to include circuitry able to

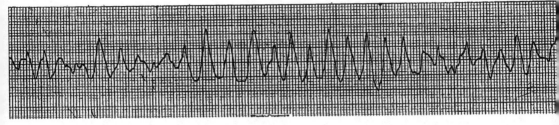

ECG of Ventricular Fibrillation. In this tracing, no P waves or QRS complexes exist. The heart lacks an organized heartbeat; electrical activity is chaotic. (Kevin R. Brown and Sheldon Jacobson, *Mastering Dysrhythmias: A Problem-Solving Guide* [1st ed., © 1987], 137, 146; reprinted courtesy of Delmar, a division of Thomson Learning [fax: 800/730-2215])

sense the onset of VF with a high degree of reliability. As a first attempt, the group developed a sensing circuit that utilized blood pressure measured from a pressure transducer located near the tip of the catheter.[7] Within three months, they put together a working tabletop model from commercially available components and successfully defibrillated dogs via implanted electrodes. At this stage, according to Mower, "the goal was simply to demonstrate that such a device could be reasonably designed and constructed to perform the task of automatic defibrillation." Even so, they ran into opposition. NIH made it clear to Mirowski that the agency would provide no funding, and a preliminary report was rejected by the *New England Journal of Medicine* because of skeptical peer reviews. These difficulties strongly suggested that the "cardiology establishment" in the United States was skeptical of the project. Lacking any outside funding, Mirowski and Mower dug into their own pockets to pay for experimental animals and electrical components.[8]

Mirowski's plan did, however, attract the attention of Medtronic: Earl Bakken visited the lab in Baltimore in spring 1970 to witness a demonstration of the defibrillator. When he asked what would have happened to the dog used in the experiment if the device had not worked, the researchers "disconnected the defibrillator and refibrillated the animal. . . . The dog died, of course." Bakken was impressed. Medtronic agreed to a joint development program and set some engineers to work replicating Mirowski's and Mower's experiments, but using sensing electrodes rather than the pressure catheter. After several months had passed, Mirowski and Mower began to feel that the Medtronic people "weren't moving fast enough." Medtronic conducted a marketing survey and concluded that the time-horizon for the defibrillator project was so distant — an estimated 20 years — that it was not commercially viable. "The company decided we weren't worth the effort," Mower said, "and we parted."[9]

From Medtronic's point of view, the technological problems facing the defibrillator project seemed almost overwhelming in 1970–71. Mirowski and Mower were trying to invent an implantable defibrillator at a time when pacemakers still used discrete circuitry and mercury-zinc batteries potted in epoxy. Cardiologists told Medtronic that "sudden death was a big problem" but pointed out that "there was no way they could identify who was at risk of sudden death and who was not. So they would not know who to put it in. . . . So we asked, when do you think the diagnostic technology will be there to identify who's going to be at risk? And the general view was probably . . . 15 to 20 years. . . . That was sort of their sense — that there was risk stratification work and technological work that was probably going to take a couple of decades."[10]

Bernard Lown, director of the coronary-care unit at Boston's Peter Bent Brigham Hospital, had worked closely with Barouh Berkovits in designing and testing the DC defibrillator and the cardioverter; he was probably the most respected medical expert on defibrillation in hospitals. In an editorial published in *Circulation*, the flagship journal of the American Heart Association, Lown mounted a powerful attack on Mirowski's project. He pointed out several of the technical problems that the development group was already trying to address, but conceded that these were probably soluble. Lown was most concerned about identifying people at high risk of death from VF. What was the point of inventing a device for which it was not possible to select appropriate patients? He also wondered how Mirowski proposed to test the "operational readiness" of the defibrillator once it had been implanted. "Even in this age of derring-do and erosion of ethical constraints, it is unlikely," Lown wrote, "that VF will be induced deliberately to ascertain performance. Barring such a test, how can one be certain that the fixed charge will prove adequate [to defibrillate]?" And what if a patient were found dead? Would this indicate that the device had failed to defibrillate the heart, or perhaps that it had electrocuted the patient?[11]

Medical researchers may sometimes exaggerate the degree of opposition or misunderstanding that at first greeted their work; in Mirowski's case, however, the opposition was real. In 1972 the idea of implanting a defibrillator apparently seemed outlandish to many observers, not only to Lown. Defibrillators at that time were good-sized pieces of equipment that delivered powerful shocks; hospital personnel had to be trained carefully in their safe use. From the era of AC defibrillation, there were old stories of repeated defibrillatory shocks' burning or cooking the heart.[12] Not until

Mirowski and Mower reported in the mid-1970s that defibrillating via a catheter electrode required a much smaller shock than closed-chest defibrillation did this misperception begin to clear up. Then, too, many heart specialists remembered Claude Beck's unfortunate effort to popularize open-chest cardiac massage in the 1950s and the more recent controversies and embarrassments surrounding the efforts to develop an artificial heart.[13] Mirowski's project may have appeared likely to embarrass the profession yet again.

By all accounts, Mirowski had a formidable personality: he was intense, dedicated, and sober. Criticism of the defibrillator project greatly distressed him and drove him to work harder. Stories got back to him in the early 1970s that someone had called the defibrillator "a bomb inside the body" and that another critic had described the idea of inducing tachyarrhythmias to identify high-risk candidates as "Auschwitz revisited." Mirowski took Medtronic's decision to drop the defibrillator project "as a personal affront — he was a pretty emotional man. We didn't believe in his idea, was the way he interpreted it. He was very angry about that. He didn't see it as a business decision." On the other hand, as Kastor has pointed out, the skeptical reception served to discourage other researchers from jumping into the field.[14]

Opinion of the implantable defibrillator began to change after 1975. The group by then had a prototype that they could implant in dogs. They made a short movie showing a dog being put into fibrillation and collapsing unconscious, then standing up and wagging its tail after the defibrillator had fired and terminated the VF. When a viewer asked if the researchers had used Pavlovian methods to train the dog to fall down and jump up, they improved the film by superimposing the dog's ECG. The film persuaded many doctors that Mirowski and Mower were working on an invention that would have practical benefits for patients.[15]

## SUDDEN CARDIAC DEATH

Bernard Lown had warned against inventing new medical devices simply as an exercise in virtuosity, but he failed to notice that new treatments often do not follow in an orderly fashion from new scientific understanding of disease. Technological development is sometimes out in advance of laboratory science and may actually encourage the laboratory scientists to focus on a neglected problem. This had happened just a few years before with the implantable pacemaker: most ideas about indications for pacing had emerged after pacemakers came into clinical use. By 1972, a new understanding or way of thinking about sudden death from ventricular fibrillation or tachycardia was gathering force; and despite Lown's remark about "derring-do," researchers were even beginning to induce VT and VF deliberately in human subjects. They were willing to do so under controlled conditions because the machines to bring the heart back to a normal rhythm stood close at hand — machines that Lown himself had helped introduce a decade earlier. A more sophisticated understanding of sudden cardiac death and new ways of selecting appropriate patients for the ICD were developing in parallel with the ICD itself.[16]

The underlying causes of VF were not at all clear. As we saw in chapter 1, researchers knew that powerful electrical shocks could kill by disrupting normal electrical conduction in the heart and throwing it into VF, but these kinds of events were unrelated to any background disorder of the heart itself. The reframing of VF as one of the prime causes of death in modern societies began with the publication in 1929 of Samuel Levine's *Coronary Thrombosis*, the first book in English to analyze the risk factors, clinical course, and consequences of heart attacks. The author, a Boston cardiologist, used his own and others' clinical cases to show that arrhythmias, including VF, could follow a heart attack within hours or days. The standard treatment for heart attacks emphasized extended bed rest and the use of the drug quinidine to treat ventricular tachycardia when it was detected — which was not often, because hospitals did not put patients on continuous ECG monitoring. For the patient who went into VF after a heart attack, there was hardly time to detect the condition before the patient had died; obviously, no treatment existed.[17]

Thus matters stood until the 1950s, when Electrodyne introduced both a cardiac monitor that displayed the ECG on an oscilloscopic screen and an AC defibrillator that could terminate VF through the closed chest. In 1960, William Kouwenhoven, at Johns Hopkins University, who had been working on the problem of VF for three decades, described a technique of external cardiac massage that any hospital employee could be trained to use. Today we know the Kouwenhoven method as cardiopulmonary resuscitation, or CPR. Berkovits's cardioverter and DC defibrillator (1962) added to the hospital armamentarium. By 1964, Berkovits's company, American Optical, was selling complete arrays of integrated heart-management machines to hospitals. New drug treatments to suppress dangerous arrhythmias also contributed to the growing conviction that VT and VF were treatable.[18]

The problem by the early 1960s was not so much the act of defibrillating the heart as uncertainty over how to deploy the personnel and technology of the hospital to respond quickly enough to save patients routinely. In the

United States, Hughes W. Day, of Bethany Hospital in Kansas City, Kansas, proved the key innovator. In 1962, Day organized the first coronary intensive-care unit (CCU). It had eleven beds: each patient was monitored continuously for several days after admission for a heart attack. The monitor could sense the onset of VT, VF, or coronary standstill and sound an alarm; the screen display, equivalent of the ECG rhythm strip, informed staff at a glance what sort of arrhythmia they had to contend with. Because of the concentration of patients, trained staff, and the new rescue machines in a single location, the hospital was able to respond immediately and appropriately when a patient went into cardiac arrest. For the first 300 patients admitted to the unit, mortality was reduced from 39 percent to 20.6 percent. An analysis of 150 monitored patients in 1966 revealed that one-half had experienced major ventricular arrhythmias in the first 72 hours after their heart attacks. By 1970, the CCU had become central to the acute-care treatment of heart-attack victims.[19]

These innovations also raised the question of cardiac arrest outside the hospital. An important 1967 study of sudden deaths in the United States that used death certificates and defined sudden death as death within two hours of the onset of symptoms found that some 12 percent of all natural deaths were sudden and that 88 percent of all sudden natural deaths — or more than 10 percent of all the deaths in the United States — were sudden cardiac deaths. Thus as Mirowski's group began its work, U.S. cardiologists had been given a practical working definition of SCD and evidence that it was not occasional but commonplace.[20] Epidemiologists' estimates ranged from 300,000 to 600,000 SCDs annually.[21] Later studies showed that three-quarters of SCD victims had had prior heart attacks, many of which had not come to the attention of doctors. A heart attack that had severely damaged the left ventricular wall seemed particularly highly correlated with SCD. The key point was that heart-attack victims remained at risk of ventricular arrhythmias long after discharge from the hospital.[22]

Broadly speaking, those concerned about sudden cardiac death since 1967 have tried four strategies to address the problem of nonhospital cardiac arrest. The first, to identify people most at risk and put them on powerful antiarrhythmic drugs, has proved effective in some cases and is still under investigation. Community-based emergency services combined with broad community training in CPR, the second strategy, has saved thousands of lives by reaching SCD victims and defibrillating them before they have died or suffered irreversible brain damage. But it is an expensive strategy, and less than half of those whom emergency teams reach actually survive to leave the hospital. A third approach has gained momentum in the

1990s with the introduction of automatic external defibrillators that can diagnose an arrhythmia and decide to shock without any operator input. Such units are now stored (in much the same way as fire extinguishers) on airplanes, in the hallways of public buildings, and in sports stadiums. This approach depends on bystanders' recognizing the possibility of SCD and choosing to break the glass and apply the defibrillator paddles, with subsequent prompt arrival of an ambulance or emergency medical team to take over the care of the victim.[23] Mirowski's idea of implanting the life-saving device in the body of the likely victim of SCD was the fourth strategy — and the most reliable.

## FROM TABLETOP TO IMPLANTABLE

After Medtronic dropped Mirowski and Mower, the two cardiologists needed another corporate partner to carry the defibrillator project further. At a cardiology convention in October 1972, Mirowski was introduced to Dr. Stephen Heilman, the founder and president of a small external-device company named Medrad, located in Pittsburgh. Over lunch, Mirowski explained what he was trying to do. Heilman thought that the defibrillator idea seemed "logical and valid"; the device, if ever completed, would have "high impact, both clinically and in a business sense." And Mirowski impressed him as a person. "He was intense. . . . His dedication to the project particularly impressed me," Heilman recalled. "This was important considering the inertia and skepticism I had encountered from the medical profession toward new concepts. . . . I sensed in him a deep intelligence and a dedication that stood out." By the time they had finished lunch, Mirowski and Heilman had agreed to join forces. Medrad soon brought in a young biomedical engineer, Alois Langer, whose Ph.D. thesis dealt with multidimensional electrocardiographic analysis. Early in 1973, with Langer and Heilman flying in frequently from Pittsburgh, the Baltimore group began to attack the key problems in the design of an implantable defibrillator.[24]

In the original model, a sensing electrode detected the onset of VF and triggered a capacitor to charge up from the battery. After the capacitor had fully charged, a matter of a few seconds, the device fired a defibrillating shock of about 25 joules. It could recycle up to three times if the previous discharge was ineffective, with the fourth shock at 30 joules. After four unsuccessful shocks, the defibrillator would recharge and reset itself, then begin again. But the designers faced several technical roadblocks in moving from a tabletop embodiment of the battery-powered defibrillator to an implantable version, and because of the high voltages involved they could borrow very little from the design of pacemakers of the 1970s.

The most basic issue concerned the defibrillatory shock: what should be the shape and duration of the waveform and from what point should it be delivered? The questions were obviously interconnected. The team ran tests in Baltimore on dogs and in Pittsburgh on baboons; the goal of a 100 percent success rate with the first or second shock proved elusive for months. They tested many waveforms, but the design and positioning of the electrodes turned out to be the crucial variable. The investigators eventually decided to place one electrode on a catheter positioned in the superior vena cava close to its junction with the right atrium, with a flexible patch electrode around the lower end of the heart so as to distribute the electrical current through the ventricles.[25]

What made an implantable defibrillator possible at all was the fact that delivering the shock from within the heart would greatly reduce the amount of energy required to defibrillate. Even so, the device would need a battery that both stored enough energy for many shocks and yet was small enough to be implanted. No such battery existed in 1973, but Medrad worked with Honeywell, a large manufacturer of high-tech devices and components, to develop a lithium-vanadium pentoxide battery with enough energy for at least one hundred shocks. In the same period, the engineers selected a capacitor with high energy density; they also designed an efficient technique for charging the ICD's two capacitors from the battery.[26]

The sensing function of the defibrillator posed a third set of problems. How would the device recognize VF when it occurred? In the early 1970s, the electrical behavior of the heart in fibrillation was known solely from ECG tracings taken at the surface of the body, but the defibrillator sensor would be implanted. Sensing VF by tracking the intracardiac blood pressure had been a stopgap; a defibrillator for use with human beings would need more reliable indicators of fibrillation. The ICD could not be allowed simply to monitor the heartbeat and fire in its absence: the absence of detected heartbeats might be caused by a defective component rather than the onset of fibrillation. What was needed was some detectable property unique to VF. In 1976, engineers at Medrad devised a detection algorithm that relied on the fact that in normal ventricular rhythm the intracardiac electrogram has a peak probability at any given instant of being at zero potential, while in VF it varies randomly. The circuit continuously checked the proportion of time that the electrogram spent close to zero potential. "In essence," they wrote, "[VF] is identified by the striking absence of isoelectric potential segments."[27]

Throughout the late 1970s, Mirowski's group tested numerous variations with dogs. In doing so, they were also experimenting with surgical techniques. This continued, as the equipment continued to evolve, through the early 1980s with human subjects. By 1982, surgeons under the leadership of Levi Watkins Jr., at Johns Hopkins University Hospital, had arrived at a minimally invasive technique. First they introduced the catheter electrode from the patient's left internal jugular vein and positioned it in the superior vena cava. They next made a small incision in the middle of the chest to gain access to the lower portion of the heart, inserted the patch electrode, and sutured it to the outside of the ventricles. Having established the electrodes, the surgeons developed an abdominal pocket for the pulse generator, exactly as William Chardack had done in 1960 for the implantable pacemaker. Finally, they tunneled the leads beneath the patient's skin to the generator and attached them.[28]

Medrad also planned how to transform the invention into an industrial product. This did not mean mass production — that would come later, after FDA approval — but quality-control procedures to ensure that no patient received a defective unit. Several experienced quality-control people joined Medrad from Arco Medical, the nuclear pacemaker manufacturer located near Pittsburgh; among these was Steve Kolenik, who assumed overall charge of quality control. Medrad also asked the Applied Physics Laboratory (APL) at Johns Hopkins University to review the design of the device and help to plan its conversion from prototype to manufactured product. On the advice of the APL, Medrad altered the design to use fewer and more easily available components. In 1979, Medrad also created a subsidiary called Intec Systems. The parent company concentrated its efforts on the continued development of the defibrillator; Intec handled relations with the FDA and plans for commercializing the invention.[29]

Preclinical development and testing of the invention had lasted from 1973 until 1979 (see appendix A). Now the development group applied to the Institutional Review Board at Johns Hopkins Hospital for permission to test the device in human subjects. They submitted results from numerous bench and animal tests to back up their claim that the device would be safe. On February 4, 1980, the first human implantation took place. The leads were positioned and connected to the generator and then the surgical team deliberately put the patient, a woman, into VF to test the ability of the device to sense and shock. (This was the practice that Bernard Lown had disparaged in 1972.) The implanters expected that the defibrillator would sense VF within 5 to 15 seconds, but it didn't do so. After 20 seconds they readied an external defibrillator; at 35 seconds ("which seemed like 35 hours," one participant commented) they placed the paddles on the woman's chest. At 40 seconds, just as they were about to shock the heart externally, the ICD delivered its shock and restored a normal heartbeat. The patient left the hospital a few days later and the designers made changes in the sensing algorithm to prevent extended detection periods in the future. Mirowski was confident that because of the years of painstaking preparatory work, the clinical trials under FDA supervision would go well: "We will be more Catholic than the Pope," he told Heilman. Over the next five and a half years, surgeons working with the Medrad team implanted defibrillators in some eight hundred patients who were considered to be, despite drug therapy, at extremely high risk of sudden cardiac death.[30]

Stephen Heilman later commented that at times his relationship with Mirowski had been strained and that he found the inventor "very demanding and frequently overly concerned with insignificant minor matters." Mirowski had insisted until very late in the day that the device should defibrillate from a single catheter lead even though test models had ongoing problems with high and unstable thresholds of defibrillation. Finally, when he realized that Heilman was almost "ready to give it up," Mirowski relented and accepted a design change permitting an apical cup or patch electrode in addition to the catheter in the superior vena cava. Despite these and other conflicts, Heilman commented that "Michel was not always right but, on the other hand, he was right so often that if you disagreed with him, you had to think very seriously about your position."[31]

The implantable defibrillator, like the implantable pacemakers of the 1960s, was intended to protect people at risk of sudden death for whom no comparably effective treatment existed. During its development, the ICD ran into far more opposition than the implantable pacer ever had, perhaps in part because of the size and dramatic effect of its electrical discharge. Unlike the pacer's stimuli that were so tiny as not to be felt, it produced shocks that patients often compared to a punch to the chest or the kick of a mule.[32] And that was all it did. Like the early asynchronous pacers, the first-generation ICD performed a simple task, and years later, people remembered it as a "shock-only device."

## FROM MEDRAD TO CPI

As clinical trials proceeded, Medrad sought outside financial backing. A venture-capital group in Minneapolis got involved, then the pharmaceutical company Eli Lilly got wind of the project and quietly invested in it. In 1984, during the FDA review, Lilly approached Medrad with an offer to acquire the ICD technology for its CPI subsidiary. During the Lilly years, Cardiac Pacemakers, Inc., had languished; it was far behind with DDD pacemakers and its market share was shrinking. On the other hand, CPI was an experienced manufacturer of implantable devices, had a domestic and foreign sales force, and had Lilly's money behind it. Hoping to boost CPI's business in both tachyarrhythmia management and bradycardia pacing, Lilly bought the ICD technology in June 1985. Mower, Heilman, and several other members of the development group joined CPI. As Heilman saw it, "the project was on the verge of becoming a major business. CPI was an established cardiac implantables business and since very substantial amounts of capital were still needed, the Lilly offer could not be refused." Moving to CPI was simply one more step in the long process of bringing the defibrillator into general clinical use.[33]

While Intec continued to manage data collection for the FDA review and Medrad worked on improved versions of the defibrillator, CPI's engineers concentrated on educating themselves about the intricacies of the device and preparing to manufacture it. Their initial goal was simply to make the existing model more reliable and manufacturable. Automating the production process was the key to gearing up production and holding the cost of the device at an acceptable level. CPI at this time was producing only about 400 pacemakers per month, so the new device would mean a substantial expansion. CPI's manufacturing engineers designed and built dozens of new tools and fixtures. Workers at the manufacturing plant next door to company offices near St. Paul were trained on a pilot manufacturing line; a Just-in-Time program was put in place that included dozens of new quality checks.[34]

In approving the defibrillator for general release, FDA required that the manufacturer develop a training program and a training manual for physicians, in effect ratifying the long-standing practice of the heart-rhythm management industry whereby manufacturers took a large role in educating physicians and other medical personnel about the technical aspects of their devices. Thus at the same time it was learning the details of the device technology and preparing to manufacture the device, CPI was bringing in 30 to 50 physicians per month for two-day seminars on the implantable defibrillator. In January 1986, Medicare announced that it would accept ICD implantation as a reimbursable procedure. With production scaling up and more than twelve hundred physicians from four hundred medical centers trained, CPI introduced its Ventak defibrillator late in 1985.[35]

## THE PROBLEM OF PATIENT SELECTION

The ideal patient for a new life-saving technology is one whose prognosis, in the absence of the new technology, would be bleak. For the first implantable defibrillators, investigators and the FDA decided that candidates must have survived at least two episodes of cardiac arrest that had not occurred just after a heart attack and must have had at least one episode of VT or VF while being monitored. The arrhythmias should also have proved refractory to antiarrhythmic drugs. It was ethically justifiable to implant the new machine in such patients because without it they stood little chance of surviving for long. Their exceedingly poor prognosis was reflected in the fact that 11 ICD candidates, about one-fourth of those chosen, died after selection for the program but before implantation. Successfully treating patients like this would dramatically establish the effectiveness of the new device.[36]

Early clinical experience with the invention persuaded Mirowski's development group to give it an important new capability. Research by the early 1980s had made it clear that there were premonitory symptoms before the onset of sudden cardiac death. Holter monitoring (see chapter 5) indicated that hours to minutes before cardiac arrest, the heart showed signs of growing instability, such as an acceleration of the rate and the onset of numerous premature ventricular contractions that could send the heart into ventricular tachycardia. If sustained, VT could cause sudden death or degenerate into VF.[37] The group decided, therefore, to modify the defibrillator to detect tachycardia as well as fibrillation. To do this, in 1982 they added a separate rate-sensing electrode and new circuitry. These changes transformed the implantable defibrillator into a true ICD (implantable cardioverter-defibrillator).[38]

European clinical trials began in 1983 under the management of Ela, the French pacemaker company. By March 1984, 323 patients, all at high risk of sudden death despite drug therapy, had received ICDs in the United States. A typical group of such patients, factoring in their ages and underlying heart diseases, would ordinarily have had a one-year mortality risk of at least 25 percent; for ICD patients who had received the improved versions of the device, the one-year mortality was 3.5 percent. No reports indicated that the defibrillator had injured any patient.[39]

When FDA approved the implantable defibrillator for general market release in October 1985, the agency broadened the candidate population by requiring only one previous episode of cardiac arrest or recurrent unstable ventricular tachyarrhythmias. The FDA also required that every candidate should be proved inducible into sustained VT or VF. This meant, in practice, that every candidate for an ICD must be evaluated by a clinical electrophysiologist, an expert at analyzing disorders of heart rhythm. Electrophysiologists used several diagnostic technologies to estimate a patient's risk of sudden cardiac death. Among these were Holter monitoring and the signal-averaged ECG (useful for identifying those at high risk of a re-entrant ventricular tachycardia). But the most characteristic method has been *invasive electrophysiologic testing*. Here the electrophysiologist introduces a transvenous pacing electrode and tries to induce VT/VF by dropping a series of stimuli into the heartbeat. The artificial stimuli emulate the premature ventricular contractions that often trigger VT. If the heart can easily be sent into a dangerous arrhythmia, this is presumptive evidence that the same kind of event might happen spontaneously.[40] Thus the ICD, like the pacemaker, from its market release was embedded in a larger set of practices centered on machines and the specialists skilled in using them. FDA's decision (with which Medicare agreed) to require an invasive electrophysiologic (EP) workup effectively handed the ICD to the electrophysiologists and defined it as one of their "proprietary" technologies. The growth of demand for ICDs would fuel the expansion of electrophysiology as a subspecialty of cardiology.[41]

During the late 1980s, evidence from hundreds of EP tests showed that many people who have survived episodes of cardiac arrest cannot be artificially induced to go into VT/VF. This suggested that Medicare should loosen the guidelines for implantation of an ICD. The program did so in 1991, by announcing that it would cover ICD implantation for patients who had had a documented episode of VT or VF that had not occurred in the immediate aftermath of a heart attack. This change gave the physician far more discretion in deciding whether a patient should receive an ICD.[42] Six years after FDA had approved the new device for general market release, the momentum was building for a large increase in ICD utilization. Further improvements in the product would soon add to the momentum.

## FURTHER DEVELOPMENT OF THE ICD

Beginning in 1985, CPI would have nearly six years to develop the ICD market as it chose. The company enjoyed formidable patent protection; the most far-reaching of the Mirowski patents would last until late 1990.[43] But viewed in the cold light of day in its original commercial embodiment, the ICD struck many as a fairly limited and primitive technology compared with state-of-the-art pacemakers of the 1980s.

The original Ventak ICD of 1985 had an estimated worklife of only two to three years. It consisted of a bulky pulse generator (250 grams — more than half a pound) based on a printed circuit board assembly with hundreds of discrete components. The patient had to undergo general anesthesia and extensive surgery, just as in the early days of the pacemaker. The Ventak could detect VT as well as VF, but it was a "shock only" device: when it detected either problem, it delivered a powerful shock of up to 30 joules. A full shock of this magnitude would knock the patient, if still conscious, unconscious. It was also a "committed" device, meaning that once activated by a sensed arrhythmia, the Ventak would charge up and fire without confirming its original diagnosis. As a result, it sometimes shocked inappropriately. Even in this early embodiment, the ICD proved safe and generally successful for the candidate population. But as some observers noted, the true effectiveness of the device remained something of a mystery because many ICD patients continued on antiarrhythmia drugs and many never went into VT or VF, with the result that their ICDs were never called on to deliver a shock.[44]

From 1985 through 1990, CPI earned revenues in excess of $300 million on global sales of more than 20,000 ICDs. But this was just scratching the surface. To encourage the growth of the market, CPI would have to improve the device by making it less frightening, easier for the doctor to comprehend and manage, longer-lived, and practical for a wider range of patients. The company used the years from 1985 to 1991 as a period of intense learning, planning, and product development. Then in the 1990s, CPI and new competitors introduced a series of ICD models that made the therapy more acceptable and broadened the population of potential patients, much as VVI pacing, transvenous leads, and lithium batteries had done in cardiac pacing 20 years earlier (see appendix A).

*Tiered-therapy* ICDs, available from CPI and other companies by 1993, offered anti-tachycardia pacing, both low-energy and high-energy shocks, and in some cases bradycardia pacing. The ICD itself selected the appropriate response to a sensed arrhythmia within limits that the electrophysiologist programmed into the device. In reducing the need for high-energy shocks, tiered therapy eased the patient's anxiety and prevented unnecessary pain, while increasing the longevity of the device. By the mid-1990s, ICDs also generally offered biphasic waveforms for their defibrillating shocks. Laboratory and animal studies had shown that these were more effective at defibrillating. This in turn has permitted designers to use simpler electrode systems without compromising effectiveness. Today some manufacturers are moving away from the patch electrode; the ICD housing itself (the "hot can") serves as the anode. By reducing energy requirements, designers are also able to use smaller batteries and downsize the generators. Some ICDs of the late 1990s weighed less than 100 grams and, like pacemakers, could be implanted pectorally under local anesthesia.[45]

Improvements in battery technology and the far greater density of integrated circuits have made possible these improvements, as well as parallel improvements in bradycardia pacemakers. Developments in ICD technology tracked the developments in pacing of the 1970s and 1980s: the ICD, like the pacemaker, moved from a device able to deliver only one crude kind of treatment ("shock only" being the equivalent of asynchronous pacing) to a device able to sense and choose the most effective and least intrusive response. More recently still, manufacturers have begun to introduce dual-chamber ICDs. Yet beneath all the refinements, both the pacemaker and the ICD can behave like their distant ancestors when all else fails: The pacer can fire in asynchronous mode and the defibrillator can deliver a high-intensity electric shock. ICDs have moved in 10 or 12 years as far as the pacemaker did in 30. This is because physicians and company product managers and design engineers now build on decades of experience and have the pacing paradigm before them.[46]

Incremental improvements in the ICD have made it more attractive to implant the devices in a broad spectrum of patients, and already in the early 1990s there was talk of prophylactic implantation in patients who had not experienced even a single episode of sudden cardiac death. Medicare now covers implantation in patients with recurrent episodes of sustained VT who have not had an episode of cardiac arrest and in others who have had nonsustained VT with a weakened left ventricle. The population of sustained VT cases alone was larger than the population of SCD survivors. Industry analysts of the early 1990s found it hard to restrain their enthusiasm as the number of ICD implantations grew at 25 percent annually.[47]

The future of the ICD as an industrial product depended not only on improved technology but on growth in the number of specialists who implanted the devices. It was estimated in 1993 that there were only about eight hundred electrophysiologists in the United States, with just a few dozen doing most of the ICD implantations; but as that number grew, it would inevitably mean heavier ICD use. The manufacturers could act to hasten growth by developing their relationships with electrophysiologists through ongoing programs of physician training and product support.[48] The future of the ICD also depended on electrophysiologists' adopting a strongly interventionist way of thinking about how to manage heart arrhythmias. Mirowski and Mower reminded cardiologists of a saying of Dwight Harken's: "A device is safe when its use is safer for the patient than the prognosis of his disease, and is the best available." Harken had been a heart surgeon; now combating VT and VF required that the electrophysiologist emulate the surgeon's storied boldness and confidence in technology. Here, too, the device companies could encourage the more aggressive attitude through their physician-education programs. As electrophysiologist John Fisher observed in 1990, the familiar notion of "conservative treatment" was being turned inside out: "Conservative today, for a growing number of cardiologists, means choosing the therapy providing the maximum protection, even if it is more invasive than what has traditionally been considered conservative management."[49]

## COMPETITION FOR MARKET SHARE

From the day of the first clinical implantation of an ICD in 1980, all pacemaker companies began to plan their future entry into this entirely new business. The basic Mirowski patent, owned by Lilly/CPI from 1985 on, barred entry until October 1990, but the other manufacturers hoped to design and test their own ICD models, learning from CPI's experience all the while, until the door opened. By 1989, five firms were preparing ICDs that leapfrogged the early shock-only models and included more sophisticated capabilities. An industry analyst commented that these investigational devices "appear remarkably similar to one another" in size and behavior. This suggested that a company's ability to train and support the physician might spell the difference between huge success and a marginal position once the competition got under way.[50]

Industry observers followed two companies, Ventritex and Medtronic, with particular interest. From 1989 on they suggested that once the two firms reached the market with their products, CPI's position would swiftly erode. Ventritex was a new company, the first serious entrant to the heart-rhythm management competition since the early 1970s. This newcomer, which opened its doors in Sunnyvale, California, in 1985, employed two engineers who had worked for Intermedics in the early stages of ICD planning, Ben Pless and Mike Sweeney; Sweeney had also helped design an anti-tachycardia pacemaker for Intermedics. The two designers understood the clinical issues and were able to deal with the high-voltage requirements of an ICD. In planning a radically new technology, a company's existing product-development structure and base of knowledge can at times be more an impediment than a help, as Medtronic had illustrated with the Activitrax pacemaker. Pless and Sweeney constituted a radical engineering cell. They applied for and were granted important patents; by the late 1980s they had come up with a prototype of an ICD that excelled Ventak-P, CPI's second-generation shock-only product.

Ventritex went public in January 1992, raising nearly $60 million from the sale of its stock. The funds went toward developing a sales and marketing capability and to prepare for manufacturing the ICD. A few weeks later, the FDA gave its approval for the company to release its Cadence ICD, with the head of the review panel referring to the device as an "electrophysiologist's dream." Cadence had all the features that had been talked about since the late 1980s: it was highly programmable, offered tiered therapy and a biphasic waveform, and was able to store diagnostic electrograms so that the physician could determine what had been going on in the patient's heartbeat before, during, and after the delivered therapy.[51]

Industry analysts' reports for the investment community tend to base their assessments heavily on product innovations; in this case, observers thought that Cadence was "ahead of" CPI's new Ventak-PRx defibrillator and Medtronic's PCD. Ventritex had certainly done a remarkable thing in developing a safe and reliable ICD within seven years of start-up. But Ventritex had concentrated on its Cadence product and did not have a set of transvenous defibrillating leads to accompany it. The company expected doctors to use leads from Medtronic or CPI with Cadence. Medtronic, however, labeled its leads as usable only with Medtronic ICDs. CPI announced that its Endotak lead could be used with other ICD brands, but doctors proved reluctant to mix products from different companies in the absence of specific FDA approval.[52]

Ventritex had other liabilities, especially a limited capability in the technical support that physicians had come to expect from the device companies and the prospect of years of patent suits. By the end of 1992 it was clear that despite the successful Cadence ICD, Ventritex would have to settle for about 15 percent of the ICD market.

**PATENTS AS BARRIERS TO ENTRY**

The implantable-medical-device industry by the late 1980s had become one of the most litigious among all U.S. industries; the disputes touched on in earlier chapters are only a few of the most prominent.[53] The main underlying reason was that no single company possessed all the knowledge needed to manufacture pacemakers and ICDs that would meet contemporary expectations of safety and effectiveness while satisfying the desire of the physician community for additional features and capabilities. Each company therefore felt a continuing pressure to develop new patents in order to trade its own proprietary technology for access to technology controlled by others. This practice, known as cross-licensing, has been standard in the industry since the mid-1980s. When the patents to be cross-licensed are thought to be very different in value, money may change hands as well.

One indicator of the true value of a device company is the size and significance of its patent portfolio. Ventritex in the early 1990s held 17 patents with 21 applications pending, including important patents on biphasic waveforms and stored electrograms — but it also faced infringement suits from Medtronic, Telectronics, and Lilly/CPI. Intermedics, the former employer of the two key founders of Ventritex, sued in 1990 alleging that trade secrets had been misappropriated and that Cadence infringed on Intermedics patents. As one observer commented, "Ventritex's arrival on the market will mark an open season." But Ventritex was able to arrange a cross-licensing agreement with CPI in 1993 by offering a license on the waveform patent in exchange for a license on CPI's defibrillator patents. We can hazard the suggestion that CPI was willing to ease Ventritex's way into the market because it wanted to add to the competitive pressures on Medtronic.[54]

Medtronic had begun development work on its own tiered-therapy device known as the PCD (pacemaker-cardioverter-defibrillator) by 1982; the intention was to time the FDA's review and approval of the device to coincide with the expiration of CPI's master defibrillation patents. The company received permission from FDA to begin U.S. clinical trials in November 1989, and the PCD cleared the regulatory process in mid-1992. To get this product out the door, Medtronic had undergone a harrowing series of disputes over patents. A market-release date in the early 1990s required that bench, animal, and clinical testing go forward in the 1980s — while the CPI master patents were still in effect. Holding off until the CPI patents expired would have meant that Medtronic could not enter the ICD market until the late 1990s. In 1983, while Medtronic was still at an early stage of developing its product, Eli Lilly filed a patent infringement suit. Lilly/CPI won damages of $26.5 million in federal court in Philadelphia, but an appeals court overturned that award. The case went to the U.S. Supreme Court, and on June 18, 1990, the Court came down in Medtronic's favor, holding that patent protection does not extend to the clinical-testing phase. Lilly could not claim damages for the period before market release of the Medtronic defibrillator.[55]

Even after the expiration of the earliest Mirowski patents, Lilly/CPI held many fundamental patents on technology vital for any implanted defibrillator. Medtronic would have to come to an agreement with Lilly. A year after the Supreme Court decision, the two companies agreed that CPI would license its portfolio of ICD-related patents to Medtronic in exchange for a license on Medtronic's important portfolio of patents in bradycardia pacing, including the patent on rate-responsive pacing and a basic patent on anti-tachycardia pacing. Medtronic also agreed to provide rate-responsive pacemakers to CPI that the latter would sell under its own name until CPI was able to manufacture a rate-responsive pacer of its own. With this agreement, Medtronic secured its legal position as a manufacturer of

ICDs. That Medtronic was willing to give away so much is an indication of how badly it needed to be in the ICD market.[56]

Not only Medtronic but every company in the industry found itself in the same situation. The market for bradycardia pacers was said to be maturing; growth in that area was slowing down. The prospect of enormous future profit from the sale of ICDs plus the long lead time needed to design and test a new implantable product, see it through clinical trials, and prepare to manufacture it on a large scale persuaded the device manufacturers that they simply must gain access to each other's patented technologies, and especially to CPI's master patents, well before 1990 if they were to have products ready to roll out by 1992 or 1993. They must not only match CPI's defibrillators but leap beyond them by beating CPI to the market with tiered-therapy devices.

To the physicians eager to bring implanted defibrillators into widespread clinical use, the companies' propensity for suing each other was almost incomprehensible, particularly when judges issued injunctions preventing doctors from publishing data on the performance of various ICDs. The physicians viewed the litigiousness of the companies as undermining the flow of information between company-employed engineers, clinical investigators, and the medical community at large. Most did not realize that companies could not afford to wait on their ICD development until all key patents had expired. They had to begin work immediately and hope to negotiate licensing arrangements further down the line. It was a risky strategy— but the alternative, not to get into the ICD market, was unthinkable.[57]

## ICD THERAPY: SOCIAL BENEFITS AND COSTS

Today the newest ICDs are the size of pacemaker pulse generators and can be implanted in the patient's upper chest. They deliver their shock therapy through transvenous leads. Their onboard computers store information about heart activity and download it on command so that the doctor can evaluate it. ICDs have prevented thousands of sudden cardiac deaths, and in many cases have added years to the lifespans of their carriers—years as invalids, in some cases; years of satisfying activity, including productive work, in other cases. However, an ICD patient may experience negative side effects from the implantation surgery, inappropriate discharges of the device, or psychological reactions of fear or depression.

Although the majority of ICD carriers are middle-aged or elderly men and women with years of coronary artery disease or other chronic heart problems behind them, the industry prefers to highlight younger patients in order to show that ICD therapy can be appropriate for a wide range of people and heart problems. One such patient was a woman named April who lived in suburban Detroit at the time her case was published in 1997.[58] April was born in 1966, around the time when Michel Mirowski began to think about designing an implantable defibrillator. She had no history of heart disease, but on Super Bowl Sunday in 1990, for no apparent reason, her heart went into VT or VF and she collapsed. April and her husband were visiting her parents in Salt Lake City at the time. Her mother called 911 and an emergency medical team arrived in time to defibrillate her, get her to the hospital, and save her life. When she left the hospital a few days later, April had an implantable defibrillator; she was 23 years old.

As she described it later, she "wasn't excited about having this big box in my stomach. It looked like I swallowed a Sony Walkman." She received a replacement unit in the summer of 1993 — "and I was like, every three and a half years? I might as well have a zipper" — and another replacement in mid-1997. This third ICD was one-third smaller than the first one and was expected to last twice as long. April did not need a shock until one night in 1994 when her heart speeded up dangerously while she was asleep; the ICD sensed the onset of VT and halted it. She was eight and a half months pregnant with her second child. "Had I not had my defibrillator, I would have died in my sleep and therefore lost a baby. Now that it's saved my life, it could be the size of a gallon milk jug."

At a seminar for adolescents and young adults who had received ICDs, April described the sensation. "I was suh-MOH-kin. My heart felt very hot. I felt the electric current go from the box all the way up to my heart and then the initial jolt zapped me. It sort of felt kind of sore and hot. I didn't think, 'Ow, that hurt so bad'; it just really scared me. It was more scary than painful. I freaked out. I thought, 'What have I done to this baby?' . . . I never dreamed it would go off." To her relief, the baby was uninjured. In fact, all three of her children were born after she had received her first ICD. Her obstetrician induced labor about two weeks before each due date so that her electrophysiologist and various technicians could be scheduled to be present. They turned off the ICD for each delivery. An external defibrillator stood by, but April did not need it.

Individual case histories, however heartwarming, cannot be the final word when a major new treatment technology like the ICD is introduced: broader societal issues transcend the good or bad experience of any particular patient. Public uneasiness about high-tech interventional medicine and the ongoing national debate over the high cost of health care (compared with the cost in other countries) have forced manufacturers and doctors to justify new treatments like ICD therapy with more than anecdotal evidence. Assessments of the ICD have been more numerous and a great deal more formal than anything published about the implantable pacemaker during its first two decades — though the ICD was never subjected to the gold standard of a prospective randomized clinical trial before general market release.[59] Assessments follow two other tracks, one devoted to the issues of safety and effectiveness and the other to the question of overall societal benefit for the money spent.

On the question of clinical effectiveness, there is always an underlying question: Effective for what kinds of patients? A treatment might be highly effective for one kind of heart problem but considerably less effective for another. The early clinical trials focused on narrowly defined groups of patients such as those who had been revived from sudden cardiac death. In the 1990s, several large trials studied a variety of patient populations and compared ICD therapy with alternatives such as antiarrhythmic drugs and catheter ablation of the conduction cells that fired spontaneously to trigger VT and VF. The concept is simple enough: Take a population of patients — recent heart-attack victims, for instance — and give some ICD therapy, others the alternative therapy. Then compare the results after some time has elapsed. A number of technical problems rendered these studies hard to interpret. Many ICD patients also took drugs to treat their heart problems; should a successful — or unsuccessful — outcome be ascribed to the drugs, the device, or the combination? Should implanted defibrillators that never fired because their recipients did not happen to go into VT or VF be discarded from the study or be regarded as successes? How should the studies categorize deaths caused by arrhythmias that were not "sudden" by the conventional definition?[60]

Behind these loomed another question. As cardiologist J. Thomas Bigger Jr., put it in 1991, "It is possible that totally eliminating arrhythmic causes of death might not extend life very much because death due to heart failure or [heart attack] might nullify the potential benefit of effective antiarrhythmic treatment." He meant that most ICD patients were, after all, very sick people with serious underlying diseases of the heart. One could implant a defibrillator that worked perfectly, yet the patient might well die of some other heart-related problem. Sophisticated and effective ICD therapy might not significantly extend the patient's life or improve quality of life.

Of all the clinical trials of arrhythmia therapies under way in the 1990s, the one that made the headlines was the AVID trial ("Antiarrhythmics vs. Implantable Defibrillators"), which began in 1993. AVID posed a straight-forward question: In patients with a history of VF or serious VT and therefore clearly at high risk of sudden cardiac death, would ICDs reduce the risk as compared with the newest antiarrhythmic drugs such as amiodarone? The study encompassed 1,016 patients who averaged 65 years of age. These patients were randomly assigned to ICD or drug therapy. Random assignment ensured that the two groups, ICD and drug patients, were essentially identical.

In April 1997, the National Heart, Lung, and Blood Institute called a sudden halt to the clinical trial. A review of the data showed that after one year of treatment, the ICD patients had a death rate that was 38 percent lower than that of the drug patients. Of those who survived the first year, the ICD patients had a 25 percent lower death rate in the second year. There seemed no point in continuing the study: AVID made it clear that for patients of this sort, the ICD should be the default therapy. Cardiologists had generally reached a consensus by 1997, if not before, that ICD implantation should be the "first line of defense" for patients with a tendency to slip into dangerous ventricular tachycardia or fibrillation.[61]

Today the ICD is regarded as safe and effective when appropriately and correctly implanted and managed; but during the lifetime of this medical device (1980 to the present), planners and policymakers in health care raised the stakes. In sharp contrast to the reception that implantable pacemakers got in the 1960s, observers of the 1990s became intensely interested in the question of overall social cost of the ICD in relation to its benefits and in the "rational diffusion" of ICD treatment. They recognized that the same dynamics were at work in the case of the ICD as with the pacemaker a generation earlier. The practice of third-party payment masked the cost of treatment to patients and caregivers, while improvements in the treatment and the accumulation of experience with it encouraged doctors to apply it to an ever-broader population. Patients who carried implanted defibrillators might have serious underlying heart problems that would require a range of maintenance treatments in addition to the ICD itself. The fact that these patients would live longer further added to the cost of their medical care. In sum, a treatment that saved lives and enhanced the quality of life could also add significantly to the cost of health care for the society as a whole.[62]

In an effort to get beyond case reports and vague general statements, health economists developed more formal methods to compare the benefits and costs of medical treatments. The method most heavily used for the ICD is known as cost-effectiveness analysis: the *effectiveness* of a treatment is defined as the number of years that patients live, on average, after receiving the treatment; the *costs* are defined to include the initial cost of treatment (hospitalization, the EP workup, the ICD, the implantation procedure), ongoing costs (checkups, ICD replacement), and the costs of treating any complications and side effects. Few analysts include indirect costs because they are more difficult to pin down. Indirect costs would include those that the patient and the family incur owing to the patient's impaired ability to perform the activities of daily living, and perhaps the costs that the community incurs because of the diminished productivity of the patient and family members.[63]

Cost is stated in dollars, effectiveness in years of life gained. Typically, analysts compare the cost-effectiveness of two alternative treatments such as the ICD and antiarrhythmic drugs, so that the result of the analysis presents the incremental cost of one treatment over another.[64] All this can be summarized in a simple formula: (Cost $A$ − Cost $B$) / (Effectiveness $A$ − Effectiveness $B$), where $A$ and $B$ are the two therapies to be compared. One early analysis compared two "typical" (hypothetical) patients from 1986, when the ICD was just coming into use. Patient A received an ICD and survived for 5.1 years; the total cost of A's treatment was $121,540. Patient B was put on a program of treatment using antiarrhythmic drugs and survived for 3.2 years at a cost of $88,990. The cost-effectiveness of the ICD would be ($121,540 − $88,990) / (5.1 years − 3.2 years), which reduces to $17,100 per year of life gained.[65] The formula thus yields a rough sense of the *incremental cost per life-year saved* (LYS) for a new therapy as compared with an alternative therapy. Of course the analysis is only as solid as the researcher's data and assumptions, and one might wish to refine it in many ways — perhaps by discounting the "years of survival" if the quality of life of many patients is low. But cost-effectiveness analysis has the virtue that it provides a simple, comprehensible way to talk about the comparative merits of different treatments.

Studies of the ICD found that ICD therapy was more effective than alternative treatments in terms of prolonging life, but costlier per LYS than drug therapy. Estimates of the incremental cost ranged from about $7,500 to about $31,000 per LYS. They also indicated that ICD therapy grew less costly in the 1990s — apparently because the devices lasted longer, which meant that fewer replacement procedures were needed, and because transvenous defibrillator leads made the implantation safer and quicker.[66]

In a sizeable population, the cost-effectiveness of a treatment depends heavily on the mix of patients. If many ICD patients have advanced heart disease and are likely to die soon no matter what treatment they receive, then the defibrillator may not extend their lives and the incremental cost of ICD therapy will be high. If the ICD is used prophylactically with patients who may never have had episodes of VT/VF but are thought to be at risk, again the incremental cost of the therapy will be high because similar patients on modestly priced therapies, or no therapy, may live just as long. Improving the technological properties of the device tends to lower the incremental cost per LYS, while broadening the patient population tends to raise it.[67]

As the issue of cost containment in the United States moved to center stage after 1980, economists (in the words of economist Nick Bosanquet) "detected a bias towards high-tech and toward excessive production of expensive care, as opposed to mundane medium-tech services." Cost-effectiveness analysis represented one attempt to reduce this bias. Cost-effectiveness studies appear to constitute a body of neutral, authoritative knowledge — but certain value preferences are built in. Studies of ICD therapy have almost always compared it to other paths of comparable intrusiveness and complexity such as treatment with amiodarone or catheter ablation. The technique was framed in such a way that it precluded comparison between entities that were quite different — the cost-effectiveness of ICD therapy versus some far-reaching alternative such as large-scale preventive programs that might reduce the future incidence of coronary heart disease and hence of sudden cardiac death.[68] Researchers' choice not to include indirect costs and not to adjust life-years saved to account for patients' quality of life during those years tended to bias results in favor of the ICD and other interventional medical technologies. In addition, parties interested in the ICD interpret and use the findings in ways congenial to their own outlooks. One manufacturer stated flatly in a product brochure that the studies had "demonstrated [that] ICD therapy is cost-effective." This seemed an overstatement. On the other hand, many accepted medical interventions are far more costly per LYS, when compared with alternative treatments, than is the ICD when compared to its rival therapies.[69]

In the end, even if methods of analysis were completely standardized and the resulting data unimpeachable, judgments about cost-effectiveness must come down to societal values. "At some point," wrote surgeon John C. Schuder, "as more technological advances become available, it may be necessary to ask whether a dollar spent in improving ICDs will be more effective in preventing suffering and extending life than the same dollar expended in other areas of health care." The cost-effectiveness studies helped to frame the question but did not, in themselves, provide answers.[70]

As happened with the pacemaker during the 1960s, device manufacturers and heart specialists gained effective control of the uses, and hence the very definition, of the implantable defibrillator once the technology had proved its worth. Since 1985, manufacturers and heart specialists have co-operated to develop the ICD to meet the needs of a more diverse patient population. By the mid-1990s, industry analysts were predicting world sales of more than 75,000 ICDs in 2000, and revenues of more than $1.5 billion, with 80 percent of sales being in the United States.[71] Electrophysi-ologists have successfully asserted a claim to manage ICD implantation and follow-up. Surgeons and other kinds of cardiologists may also become in-volved, but the ICD belongs to the electrophysiologist above all because of his ability to interpret ECGs and intracardiac electrograms in complex rapid heart rhythms. The result will be that EP as a subspecialty and ICD implantation will grow in tandem. Neither existed a generation ago, but both have become significant parts of the ongoing redefinition of cardiol-ogy during the 1990s.[72]

The manufacturers understood that they must join the competition in this new business. Their old product the pacemaker was nearing the end of its forty-year run as a star of medical technology. In the 1990s, purchasing departments in hospitals were inclined to treat the dual-chamber pacers from the various manufacturers as more or less equivalent in overall func-tionality. They placed large orders for the devices but tried to contract with one or two vendors; and they expected price concessions. The ICD was different: still in an early stage of its development as a product, it most emphatically sustained life, and as a result the marginal differences between competing brands appeared highly important to physicians. The ICD pro-vided a new arena in which Medtronic, CPI, and the other manufacturers could compete on technology.

## CHAPTER 11

## THE 1990S AND BEYOND: "WHEN LIFE DEPENDS ON MEDICAL TECHNOLOGY"

The use of pacemakers and ICDs has continued to grow during the 1990s (see appendix B); unfortunately, estimates of the number of implantations vary greatly, ranging from 192,000 to 317,000 for 1997. Industry analysts believed that cardiac pacing was growing at about 6 percent a year in units sold, about 8 to 10 percent in revenues as purchasers moved from one product generation to the next. Overall, the market for pacemakers, leads, and related technology in the United States exceeded $1 billion in 1997, but price discounting and the constantly changing product mix mean that any estimate is only a rough one. The U.S. market for defibrillators and their leads was growing strongly at more than 20 percent per year and had probably reached $800 million by 1997.[1]

The number of pacemaker implantations continued to rise in the 1990s for three reasons. Older people are at highest risk of heart-rhythm disorders, and the population over the age of 65 grew 11 percent between 1990 and 1999. Physicians, by now entirely familiar with the symptoms and ECG indications of the common forms of bradycardia, were treating a higher proportion of those with slow heartbeats. Finally, the addition of one accepted new indication for implantation, pacing for AV block caused by a new invasive procedure called catheter ablation, marginally raised the number of implantations each year.[2]

Survey data that Alan D. Bernstein and Victor Parsonnet collected for 1997 indicated that doctors implanted pacers at 4,467 hospitals and surgical centers — an astonishing number implying just 34 primary procedures and 9 generator replacements per facility. Clearly the idea of confining pacemaker work to a limited number of large hospitals had not caught on. In more than 85 percent of new cases, the patients received pacemakers for "traditional" indications: heart block, the sick sinus syndrome, and bradycardia induced by drug treatment for some other heart problem. The more exotic indications that researchers had discussed for some years, including anti-tachycardia pacing, accounted for no more than 4 percent of new implantations, partly because cost-effectiveness analyses either had not been done or did not support pacemaker use. Most doctors, hospitals, and third-party payers clearly viewed the pacemaker as a tool for which the accepted uses were well established.[3]

The 1997 survey by Bernstein and Parsonnet showed that on a large scale, surgeons — who were older, on average, than other pacemaker physicians — had withdrawn from the field of pacing during the 1990s. Cardiologists now constituted considerably more than two-thirds of the implanter community, and in four cases out of five they implanted dual-chamber pacemakers. But except for the surgeons' disappearance, surveys generally depicted heart-rhythm management as a stable field during an era of immense change in the U.S. health-care system.[4] Indeed, steady growth constituted part of that stability. Had someone — some agency of the government perhaps? — exempted rhythm management from exposure to the winds blowing through U.S. medicine?

## THE GROWING PRESSURES FOR COST CONTAINMENT AND QUALITY CONTROL

From World War II to the end of the 1970s, physicians, acting on behalf of their patients, had determined how U.S. health-care dollars would be spent. Insurance companies (and Medicare from its inception in the 1960s) paid the bills that hospitals and physicians submitted and rarely questioned the professional judgment of the doctors. These arrangements encouraged the rapid diffusion of new technology and new treatments from university medical centers to community hospitals and local practices without much attention to whether the new was more cost-effective than the old. After a hiatus in the early 1980s, the pacemaker implantation rate had resumed its growth. By 1989, the estimated rate in the United States (359 per million population, according to Bernstein and Parsonnet) had surpassed rates of the 1970s. Several European countries reported even higher figures (appendix B). These numbers should be treated as rough estimates, but they suggested that pacemaker implantation rates depended on a country's age structure, its general level of affluence, and the acceptance and enforcement of clearly defined patient-selection criteria.[5]

In 1983, as we have seen, Congress had authorized the Health Care Financing Administration (HCFA), the agency that manages the Medicare program and pays the lion's share of costs related to cardiac pacing, to introduce a prospective-payment system for hospital-based procedures and services. Under the new arrangement, Medicare no longer waited for the patient's hospital bill to arrive but announced annually the prices that it would pay during the coming year for each hospital-based procedure. Prospective payment gave hospitals an incentive to economize by allowing them to retain any funds remaining if the cost of delivering a particular service proved lower than the Medicare reimbursement. The coming of prospective payment marked the moment when those who paid for health-care services began to exercise their purchasing power to slow the untrammeled growth of expenditures.[6]

In 1989, six years after the introduction of prospective payment for hospital services, Congress applied the same idea to out-of-hospital physician services under Medicare by authorizing HCFA to move to a complex and controversial new prospective-payment system. HCFA assigned values to 7,000 distinct medical services or procedures based on the nature of the physician's work, typical office expenses, and the cost of malpractice insurance. Beginning in January 1992, HCFA reimbursed physicians who treated Medicare patients — eventually some 500,000 physicians — according to the new Resource-Based Relative Value Scale (RBRVS).

With prospective payment in place, HCFA sought year by year to reduce the overall rate of growth in reimbursements to health-care providers. In annually revising its payment schedules for hospitals and physicians, the agency would often lower its reimbursement for some procedures while holding the line for others. All of this reflected what was happening in other sectors of the U.S. economy: the most powerful customers were asserting their power over the providers of services.[7]

The shift to prospective payment in Medicare and the broader debate over the proportion of the U.S. gross domestic product devoted to health care — from 5.3 percent in 1960 to 13.2 percent of a far larger GDP in 1991 — affected every provider in the United States, from hospital to clinic to individual physician. All industrial democracies had experienced pressure on state budgets since the 1970s because of increases in expenditure for health care: the United States was not unique in that respect. But unlike Germany, France, the United Kingdom, Canada, and Japan, the United States had never had a national health-care system managed and financed by the state; it lacked the institutional means and the consent of the electorate openly to impose limits on health-care expenditure. Washington had begun the process with the Medicare payment reforms, but the broader turn toward cost control and intensified scrutiny of physicians' treatment practices reflected the concerns of large private employers, who feared that rising costs for employee health care was undermining their ability to compete in the global economy.[8]

Employers wished to restrain the growth of health-care costs for employees; their chosen means were prepaid health-maintenance organizations (HMOs) and other kinds of managed care arrangements. Managed care plans differed in many ways, but all encouraged enrollees to use cost-efficient providers and discouraged treatments deemed inappropriate or unnecessary. Typically, the plan would manage and pay for employees' medical services in return for premiums from the employer. The plan contracted with a limited number of physicians and facilities, then created financial incentives to encourage enrollees to use these providers exclusively. This enabled the plan to keep track of treatments and outcomes and determine which treatments were most effective — and cost-effective — and whether physicians were providing the kind of treatment the plan favored. Thus, unlike traditional health-insurance plans, these new organizations actively tried to manage the costs and quality of health-care services.[9]

Plans also discouraged enrollees from seeing cardiologists and other specialists unless a gatekeeper primary-care physician had first given approval; this measure was intended to prevent unnecessary consultations with the most expensive physicians. Attempts to discourage the enrollees' excessive use of services and to induce physicians to use statistically validated best-practice methods lay at the heart of managed care and account for the intense hostility that many patients and doctors expressed toward these new forms of health insurance. Patients and doctors particularly objected when a plan refused financial coverage for a test or procedure that a doctor had recommended; but plan designers viewed this as a way to counteract the technological enthusiasm of many doctors and prevent them from gaming the system. As health economist Alain C. Enthoven observed, plans must "second-guess decisions by physicians to subject patients to needlessly risky surgery and needlessly costly tests. . . . Denials are a necessary feature of a well-run plan." The public and many doctors deeply believed that more care — more tests, more of the latest technology — generally equated with better care. But managed care was founded on the belief that some medical treatments accomplish little, while others may do more harm than good — the unstated premise behind Enthoven's comment. Appropriate reductions in the intensity of medical care would not jeopardize patients' health.[10]

Physician practice groups signed on with managed-care plans in order to get patient referrals, but the contracts often required them to follow standard practice guidelines and to seek HMO approval before proceeding with certain treatments. And as employers pressured the HMOs for still lower rates, the HMOs repeatedly pressured the providers to accept reduced payments. Both Medicare and the HMOs in effect forced physicians and hospitals to assume part of the financial risk: the providers were agreeing to

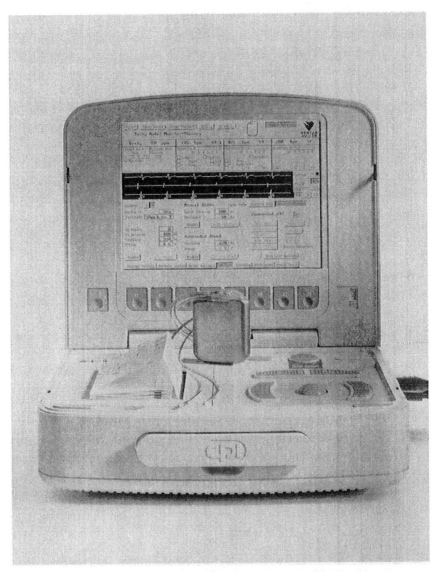

**Guidant Model 2901 Programmer, 1997.** Programmers of the 1990s were essentially sophisticated laptop computers able to download, store, and display or print out real-time and historical data about the behavior of a patient's heart and about the implanted pacemaker or ICD. User-friendly screens displayed the surface ECG and atrial and ventricular electrograms as picked up by the pacemaker leads and prompted the pacing specialist—usually a nurse or technician—with a series of programming choices for setting up the implanted device. As an encouragement to physicians to remain loyal to the company's pacers and ICDs, Guidant—like Medtronic and St. Jude Medical—did not sell programmers but instead gave them away. (Courtesy of Guidant Corp.)

treat a particular condition or perform a certain procedure for a preset fee, but if the case proved complex, the provider might lose money. Doctors and hospitals risked substantial loss if the cost of patient care regularly outstripped the terms of the contract.[11]

These three innovations — insistence on prospective payment arrangements, the unending analysis of data to identify the best treatments and the best providers, and utilization review to eliminate unnecessary treatments and pressure physicians to follow the agreed-upon best practices — together constituted the revolution in U.S. health care of the 1990s. They remade the world in which the pacemaker/ICD manufacturers and physicians worked, but it was the manufacturers who proved most nimble in response.[12]

## PACEMAKER/ICD MANUFACTURERS IN
## THE ERA OF COST CONTAINMENT

Prospective payment schemes did not remove all incentives for hospitals and clinics to buy the latest medical technology; rather, the system encouraged them to emphasize technology that would help reduce their own costs, since a hospital could retain the difference between the prospective payment and the actual cost of caring for the patient. If a new pacemaker model required less time to implant and program or led to fewer postimplantation complications, it might be a money saver for the hospital.[13] But the intense search for cost efficiencies also led hospitals, clinics, and practice groups to merge or form alliances in hopes of extracting volume discounts and other pricing breaks from vendors of medical equipment. This consolidation among medical-device purchasers represented the most important demand-side shift that the companies had faced since the creation of Medicare in the mid-1960s.

Increasingly the manufacturers had to negotiate with regional and national hospital chains and purchasing alliances. Even dealing with a local chain of four or five hospitals in a single metropolitan area presented a new set of circumstances for the manufacturers. Instead of talking directly with the doctors, each of whom would decide what kind of pacemaker to implant, a company often now negotiated with purchasing agents for the entire group of hospitals; and while agents were interested in physicians' preferences, they were interested even more in buying larger quantities of a limited number of different device models.

Some 82 percent of pacemaker recipients are enrolled in Medicare. Under the DRG system, Medicare pays a hospital a flat fee that covers the pacemaker itself and other hospital expenses. In 1996, for example, the base reimbursement for implantation of a dual-chamber pacemaker was

$10,302, but actual payments varied slightly depending on the location of the hospital and other factors. The average list price of a dual-chamber, rate-responsive pacemaker system (leads included) was $7,538, or 73 percent of Medicare's payment. In order to cover its other costs related to the implantation, a hospital group would seek price discounts from the manufacturers in return for volume purchases of pacemakers. By stocking the pacemaker models of a single manufacturer, a large hospital or a hospital chain could also simplify its staff training and standardize its treatment protocols for heart arrhythmias; an HMO could more readily compare outcomes between hospitals and between individual implanters. The same principle applied to angioplasty catheters, heart valves, and a host of other cardiovascular technologies. Purchasing agents also preferred to work with a handful of vendors able to offer an entire "market basket" of equipment for cardiovascular care and prepared to provide follow-up training and service; this implied that they wished to negotiate with large, full-service suppliers that were certain to remain in business for years to come.[14]

Even when manufacturers could offer the products and the support that hospitals wanted, they found that to land these large orders, they were under pressure to give substantial price reductions. To take one example, in 1996 several hospitals around Rochester, New York, worked together to negotiate a three-year contract with two device manufacturers that yielded cost reductions of 14 percent to 30 percent on pacemakers and defibrillators. Physicians were asked to approve the pool of potential vendors before the bidding began, but thereafter they played no part. "Vendors were instructed not to dialogue with the clinicians . . . on this," a hospital officer explained. According to another, "We said to our physicians . . . , 'Tell us what's important to you and trust that we'll negotiate a contract that meets your needs.' " The contract did not specify how the two vendors would share the pie. "What percentage . . . each obtains is totally dependent on their products, their service, their pricing."[15]

The shift toward institutional purchasing did not mean that the traditional relationship between company sales rep and implanting physician had become obsolete. In the late 1990s, Medtronic employed about one thousand sales people, its two main rivals about half that many. The complexity of the technology meant that many doctors believed that they should have a company rep on hand when they implanted a device, and hospitals permitted this as a way to outsource some of their staffing to the pacemaker/ICD suppliers at no charge. The companies, for their part, emphasized that the rep's technical know-how was a "value-added service." For the individual sales rep, the situation had not changed drastically: the rep still wanted to nurture a relationship with physicians who implanted devices. But as a marketing manager put it, "In many instances you must be an approved, contracted vendor. . . . Once you are on contract, you have a 'hunting license' to try to make the clinical sale with the physician."[16]

As hospitals and large clinics grew more assertive in negotiations, the pacemaker/ICD companies for the first time in their history faced the possibility of a "seemingly endless downward price spiral" that was already affecting many other suppliers of health-related products. One observer warned in 1996 that health-care purchasing alliances were beginning to standardize their purchases of angioplasty catheters and pacemakers — "products that until now have escaped the reins of group purchasing because of strong physician preference." The consolidation of their customers and the relentless pressure on prices pushed Medtronic and its rivals to intensify their traditional strategies for survival and growth and to adopt some new strategies as well. By the end of the 1990s, the surviving manufacturers of pacemakers and ICDs were growing rapidly and diversifying their range of products, searching for acquisition targets as a way to add new technologies that promised high growth in the future, and investing in R&D to maintain a stream of innovative new implantable devices. Through these measures, the device makers were able to shelter themselves — at least temporarily — from open price competition and drastically reduced profit margins.[17]

The major firms acquired smaller ones as each tried to get bigger and assemble a complete line of high-tech products for heart care. To Medtronic board member Tom Holloran, "The issue is, can you stay a brady pacing company? Can you stay a brady and tachy company? Or do you need to go beyond that?" Firms specializing in just one or two products, no matter how good those products might be, found that some hospital groups held them to a small cut of the business or even closed them out entirely because they questioned whether small vendors would be able to provide adequate product support or would even be on the scene after a few years. On the other hand, according to CPI president Jay Graf, large and diversified companies could realize scale economies "and offer providers one-stop shops through broadened product lines, and price concessions through increased volume." Large firms could also support larger and more diverse arrays of research talent, so that ideas and knowledge from one product area could energize thinking in others. Smaller companies that specialized in one technology such as brady pacemakers or angioplasty catheters became the prime takeover candidates in the 1990s.[18]

The dominant companies in the industry had often acquired smaller firms before the 1990s in order to gain access to promising new technologies and develop them as clinical products. This practice continued. But the major acquisitions of the decade brought together established companies with hundreds or thousands of employees and significant market shares of main-line products. At the end of the 1990s, only three U.S. manufacturers of implantable heart-rhythm devices remained: Medtronic, Guidant Corporation, with its CPI rhythm-management division, and St. Jude Medical, a heart-valve company that acquired Pacesetter in 1994 and Ventritex in 1996. The world headquarters of two of the three, and the rhythm-management division of the third, are located within a 10-minute drive of each other in the northern suburbs of Minneapolis–St. Paul. Gone were Telectronics and Sulzer Intermedics, bought out for their patent portfolios and other assets and then closed by their new owners.

## MEDTRONIC

Winston R. Wallin, CEO of Medtronic after 1985, explained the medical-device companies' situation memorably: "The thing to remember about the medical device industry . . . [is that] it is still very much of a growth industry. In a growth industry, you either grow or you die, and it's much easier to grow. It's much more appealing than to not do anything and then find that all your competitors grow, and then one day the phone rings and someone wants to buy your company, and that's the end of that."[19]

Wallin was an unlikely choice for CEO of Medtronic in 1985. A Minneapolis native, he had spent many years at Pillsbury, a food-products company. Earnings at Medtronic were in decline during the early 1980s and the stock price had fallen. The company's acquisition program seemed unfocused, and there had been a major product recall on a polyurethane pacing lead. "Medtronic was not the technology leader in the pacemaker industry," Wallin recalled. "It was the biggest, but it did not have as good products as some of the others." With the new Medicare payment system, cardiac pacing seemed unlikely to grow as rapidly in the future. Medtronic was losing market share to rival manufacturers. On the other hand, Medtronic had a great deal of cash, experienced managers in the core pacemaker business, a strong sales force, and a formidable new product in the Activitrax rate-responsive pacemaker.[20]

During Wallin's tenure as Medtronic CEO (1985–91) and that of his successor William W. George (1991–2001), the company rebuilt its market share in rhythm management and refocused its businesses. In his first two years, Wallin doubled research spending. To keep competitors at bay, he diversified the line of pacemakers and pushed the firm toward much shorter product cycles. Medtronic had acquired numerous small companies over the years and had been particularly interested in transcutaneous nerve stimulation, a form of electrotherapy to treat chronic pain. But the acquisitions had not added up to a coherent or successful strategy for diversifying into new technologies and new growth areas. Wallin focused the company's businesses in two ways, first by emphasizing high-tech devices for the treatment of heart disease. He moved Medtronic into the field of mechanical heart valves in 1986 by acquiring two smaller companies, and into tissue valves the following year. He drove the firm to develop an ICD and plan an entire tachyarrhythmia business for the day when CPI's master patents expired or Eli Lilly, CPI's parent firm in the 1980s, agreed to license them to other producers. (Medtronic introduced its first ICD in 1991.) Wallin also made sure that Medtronic did not miss out on the newest major technology area in cardiovascular medicine, angioplasty catheters and stents used by interventional cardiologists.[21]

Second, Wallin redefined Medtronic's focus on nerve stimulation, a field closely related to electrostimulation of the heart. He sold off the transcutaneous nerve-stimulation business in 1992 but pushed Medtronic's Neuro Division into new areas, such as control of urinary incontinence and stimulation of the spinal cord for pain management via implanted electrodes. When Wallin stepped down as chairman of the Medtronic board of directors in 1996, the company was testing neurostimulation devices that reduced essential tremor in Parkinson's disease and managed sleep apnea; these two applications obviously had potential for future growth.[22]

For the last fiscal year before Wallin became CEO, Medtronic had net sales of $379 million, of which 84 percent came from its pacemaker business. When he retired in 1996, after six years as CEO and five as chairman of the board, net sales had risen to $2.17 billion, of which rhythm management accounted for only 68 percent. The company's other heart-related businesses contributed 24 percent of sales, the neuro area 8 percent: Wallin's legacy was a much larger and better balanced firm. Then in 1998, Medtronic went on "an $8 billion shopping spree," buying five companies that manufactured innovative, market-leading products. One of these, Physio-Control, of Redmond, Washington, produced automatic external defibrillators for use in ambulances, airplanes, and public buildings. The other acquisitions bolstered Medtronic's product lines in interventional cardiology and neurostimulation. For fiscal year 2000, Medtronic projected sales of $5 billion, with about 47 percent coming from its heart-rhythm business, 22 percent from neuro, 21 percent from vascular, and 10 percent from cardiac surgery. Fifty years had passed since Earl Bakken and Palmer Hermundslie had opened Medtronic as an equipment-repair business in a north Minneapolis garage.[23]

GUIDANT/CPI
As part of a corporate refocusing, Eli Lilly spun off five of the businesses in its Medical Devices and Diagnostics Division as a new company, Guidant Corp., in September 1994. Guidant could boast major businesses in cardiac-rhythm management (this division still bore the name Cardiac Pacemakers, Inc., or CPI) and vascular intervention (catheters and stents for coronary angioplasty procedures) as well as a smaller business in equipment for minimally invasive cardiovascular surgery. Guidant picked up a number of single-product companies in order to acquire promising new technologies that would fill out its product line. One of these, InControl, a neighbor of Medtronic's Physio-Control in Redmond, Washington, manufactured an implantable atrial defibrillator.[24]

In June 1998, the Swiss conglomerate Sulzer announced that it was putting its Intermedics division up for sale. Three months later, Guidant agreed to buy Sulzer Intermedics for $850 million. To industry observers, this came as no surprise. Guidant was competing with Medtronic for market leadership in ICDs, but it held only about a 10 percent share of the U.S. market for brady pacemakers. Intermedics, on the other hand, had come late to the ICD field and had gained only a tiny market share. It stood little chance of competing with Guidant and Medtronic in that crucial business. The poor showing in ICDs meant that Intermedics was likely to find itself shut out of the broader competition in cardiovascular products. Yet the firm held important assets that Guidant could use: an excellent line of pacing leads, an impressive portfolio of patents in brady pacing, and a strong sales force that had helped Intermedics hold about a 15 percent share of the U.S. domestic market for pacemakers. Guidant appropriated the Intermedics patents, thereby ridding itself of a set of patent suits that the firms had filed against each other. It took over marketing of the Intermedics leads and hired as many of Intermedics' sales reps as it could, while also announcing that it would close down the Intermedics manufacturing plant and lay off most of the Intermedics employees. By adding dozens of experienced sales reps to its own force, Guidant hoped to pick up most of the Intermedics market share in brady pacing and to establish itself more firmly as a full-line manufacturer of cardiovascular products. "Guidant was really getting overwhelmed in the pacemaker market," according to one industry analyst.

"With the acquisition [of Intermedics], they picked up several market share points. And they got it at a very good price, less than three times sales. You could almost say it was a discount."[25]

## ST. JUDE MEDICAL

Founded in 1976 by Manuel Villafaña, St. Jude had established itself by 1990 as the leading manufacturer of artificial heart valves, an industry with worldwide sales of over $330 million and high profit margins. St. Jude's huge market share in the mechanical-valve segment surpassed Medtronic's dominance of the pacemaker industry before 1975. But some of St. Jude's competitors (including Medtronic) were much larger and more diversified. As late as 1993, St. Jude still depended almost entirely on heart valves, had fewer than one thousand employees, and reported net earnings of $110 million. The company's revenues grew only 5 percent in 1993, and industry analysts commented that the heart-valve field was approaching maturity. That year, Ronald A. Matricaria joined the company as CEO with a mandate from the board to diversify the company into technologies that would fuel more rapid growth. A longtime executive at Eli Lilly, Matricaria had once been president of CPI when it was a Lilly subsidiary. He understood the device industry and the larger changes under way in U.S. and global health care. In his first message to shareholders, Matricaria wrote that "we are in the midst of a revolution in the way health care products and services are delivered and purchased. . . . Among the changes already in process are shifts in buying patterns, industry consolidation, [and] increased competition and pricing pressures worldwide."[26]

Fourteen months after Matricaria's arrival, St. Jude purchased the two pacemaker subsidiaries of the German electronics conglomerate Siemens, paying about $500 million in cash for Siemens Pacesetter (in Sylmar, California) and Siemens Elema (in Sweden). The acquisition gave St. Jude the status of second largest pacemaker manufacturer in the United States and positioned it to expand in Europe as well. Two years later, St. Jude added a third crucial product line by buying the defibrillator company Ventritex for $505 million. St. Jude also entered the interventional cardiology market by spending $400 million to acquire Daig Medical, an established firm with an outstanding line of angioplasty catheters.[27]

St. Jude's acquisitions still left it the third-place company in sales revenue from cardiovascular technology. It faced two "monster competitors" in Medtronic and Guidant (see table 9). Perhaps the biggest challenge facing St. Jude lay in the fact that most of its revenues came from rhythm manage-

TABLE 9. THE PACEMAKER/ICD MANUFACTURERS IN 1998*

|  | Medtronic | Guidant | St. Jude |
|---|---|---|---|
| Net sales | $4.13B | $1.897B | $1.016B |
| Non-U.S. sales as % of all sales | 35% | 27% | 41% |
| Net sales, rhythm-management | $2.12B | $825M | $736M |
| R&D | $429M | $276M | $100M |
| R&D as % of net sales | 10.4% | 15% | 9.8% |
| Net income (loss) | $905M† | $359M† | $129M |
| after special charges | | ($2.2M) | |
| Net income before charges as % of net sales | 21.9% | 19% | 12.7% |

| Estimated U.S. market shares, 1998 | | | |
|---|---|---|---|
|  | Pacemakers | ICDs | Combined |
| Medtronic | 51 | 40 | 48 |
| St. Jude/Pacesetter/Ventritex | 22 | 11 | 18 |
| Guidant/CPI/Intermedics | 21 | 48 | 30 |
| Biotronik and others | 6 | 1 | 4 |

*Sources:* Company annual reports, 10-K filings. Market-share estimates from Jill J. Barshay, "Heart-Device Competition Is Heating Up," *Minneapolis Star Tribune,* 22 September 1998, and Karen Padley, "Industry Pulse Quickens with Unveiling of Pacemakers," *St. Paul Pioneer Press,* 14 January 1999.
*CY 1998 for Guidant and St. Jude; FY ending 30 April 1999 for Medtronic.
†Before nonrecurring charges related to acquisitions.

ment but most of its profits from heart valves. As businesses, heart-rhythm management and heart valves differ in important ways. Market shares appear to be more stable in heart valves; physicians tend to remain firmly loyal to a particular valve once they have gained confidence in it. Unlike pacemakers and ICDs, heart valves do not undergo constant modification with new models appearing every year or two. Mechanical valves already have such high success and longevity rates that it would require a huge clinical trial, lasting many years, just to demonstrate that a new design improved significantly on existing ones. St. Jude can count on high profits and a large and stable market share in its valve business without investing heavily in R&D, but rhythm-management companies must invest heavily in R&D and innovate constantly. St. Jude's management faced the challenge of marrying the leading valve company to a couple of rhythm-management companies competing in an entirely different kind of business.[28]

## INTENSIFIED COMPETITION IN
## THE RHYTHM-MANAGEMENT INDUSTRY

In the long-ago era before 1983, few in the health-care field had worried much about the price of new technology. Physicians did not respond in the expected manner to market cues: instead of being attracted to lower-priced pacemakers, they tended to regard low price as a hint of lower quality. It speaks volumes that when Medtronic saw its market share shrink after the Xytron advisories in the mid-1970s, management kept the stockholders happy by raising prices in the confidence that price hikes would not further affect market share.[29]

In the rhythm-management industry, manufacturers have fought for market share not by cutting prices but by investing heavily in new products and promoting them heavily. Hospitals wanted price concessions and did get some, but each pacemaker/ICD manufacturer sought to differentiate its products from those of its competitors and to hold the line on prices by adding new features. It had become gospel in all the companies by 1980 that, at least in the U.S. market, failure to "improve the product" continually would ruin a firm's pricing strategy, undermine its reputation, and erode its market share. In turn, hospitals demanded that new features demonstrably contribute to patients' well-being or make patient care more straightforward and trouble-free.[30]

Product cycles sped up in cardiac pacing and defibrillation as the manufacturers introduced new "families" of devices about every two years or less. In brady pacing, Medtronic and Pacesetter set this process in motion in mid-1989 by introducing dual-chamber rate-responsive (DDDR) pacers. The other firms had spent the 1980s mastering dual-chamber (DDD) pacing and had only recently introduced their first single-chamber rate-responsive (VVIR) pacers. Medtronic and Pacesetter gained a secure position in the most advanced segment of the industry where strong growth in demand could be expected in the future. Intermedics, by contrast, saw its market share in brady pacing decline by about half between 1985 and 1992 because of its lag in rate-responsive pacing, and CPI during those years maintained a precarious position by producing VVIR pacers of older design under license from Medtronic and selling them under its own name.[31]

Medtronic's strategy of emphasizing high-end pacemakers proved effective. The company probably did not gain overall market share in brady pacing between 1989 and 1995, but it profited handsomely because it was well-positioned in DDDR pacers. The *number* of pacemakers sold in the United States grew steadily during the 1990s; but in addition, the *mix* of pacemaker types shifted: DDDR pacers moved from about 10 percent of the total in 1990 to about half in 1995, from 10,000 units to more than 75,000. DDDR pacing represented less of a jump for physicians than the jump from single-chamber to DDD pacemakers a few years earlier. The implantation technique was the same for DDDR as for DDD, and programming and follow-up were becoming easier to manage in the 1990s. For the device companies, DDDR pacers carried higher price tags than older models and were more resistant to pressures for discounting. This was exactly why product innovation remained so important to the industry. Even in an era of cost containment, the firms' revenues and profits in brady pacing continued to grow.[32]

All the manufacturers released a stream of new products in the 1990s; all tried to stake out positions in the DDDR market segment and in ICDs with transvenous leads and dual-chamber pacing/shocking capability. In pacing, Intermedics announced in 1992 a "family" of advanced models called Dash, Dart, Relay, and Stride. Relay, the DDDR pacer of the group, did well in the market, but Intermedics soon introduced still newer devices, Marathon and Momentum. Medtronic came out with its Elite DDDR pacemaker in 1991, Elite II in 1992, Thera in 1994, Thera (*i*-series) in 1995, Kappa 400 in 1996, and Kappa 700 early in 1999. CPI introduced its Prelude DDDR pacer in the early 1990s, a new line of Vigor VVIR and DDDR pacers in 1995, then Discovery and Meridian in 1998 and Pulsar in 1999.

Once a manufacturer had gained command of effective DDDR pacing and transvenous defibrillation, just what did these successive product generations represent in the way of significant innovation? Each new model was physically smaller and longer-lived than the last, in keeping with physicians' well-known preferences. The Intermedics Relay weighed just 27 grams, CPI's Discovery 25 grams. New models also usually included some striking feature that showcased the virtuosity of the manufacturer and permitted the exchange of even greater amounts of telemetered information between physician and implanted pacemaker. Some models took rate-responsive pacing to new levels of sophistication by sensing two separate physiologic indicators such as activity and respiration: Ken Anderson's original idea (chapter 9) had come to pass. In mid-1999, Medtronic released its Jewel AF defibrillator in the United States. This device could discriminate between dangerous ventricular arrhythmias and fast atrial arrhythmias that felt uncomfortable but were not immediately life-threatening, then select an appropriate treatment — pacing to prevent or terminate a tachyarrhythmia or a defibrillatory shock in either chamber.

The microprocessor had added functionality and complexity to pacemakers and ICDs; paradoxically, it also made the devices easier for the physician to use, if not to comprehend. The implanted devices increasingly performed the core of the tasks formerly assigned to physicians: diagnosing the patient's arrhythmia and selecting the appropriate pacemaker or ICD therapy. As a Medtronic press release noted about the Jewel AF, the device could "analyze heart rhythm patterns, much as a physician does when reviewing an electrocardiogram." Physician and support staff still had plenty to do: they selected and implanted the pacer or ICD (this included positioning the leads to sense and stimulate reliably), usually reprogrammed the device from its default settings, dealt with any complications at implantation or later, and periodically checked on the condition and performance of the implanted unit by transtelephone monitoring or at a pacemaker/ICD clinic.[33]

The physician's real point of connection with the pacemaker came through the ancillary piece of equipment known as the programmer. From the earliest models in the early 1970s, programmers had evolved into specialized laptop computers able to receive data from the implanted pacemaker or ICD, display the data for the physician or technician in a variety of tables and graphs (including stored electrograms of key cardiac events such as runs of tachycardia), and transmit reprogramming instructions back to the device. The manufacturers emphasized that the programmers would save staff time and that nurse-specialists and technicians could program a pacemaker appropriately because the pacer and programmer already contained decades of electrophysiological knowledge and could make or recommend appropriate clinical choices in most situations. The programmer had become a significant item in the manufacturers' attempt to present themselves as allies with the hospitals in the search for cost-effective treatment strategies.[34]

In advertising their transvenous leads for pacemakers and ICDs, manufacturers emphasized the same theme as they did with their programmers. The Sweet Tip leads from Guidant/CPI, for example, featured a smooth gel coating that capped the helical screw-in electrode tip. This enabled the implanting physician to push the lead quickly through a vein and into the heart without snagging it. The gel cap would soon dissolve and the physician could screw the exposed electrode tip into the endocardial surface. St. Jude's Pacesetter rhythm-management division introduced a "steerable stylet" that "meets the challenges of torturous [sic] venous pathways" by allowing the implanter to bend or straighten the stylet tip while advancing it down the vein and into the heart without having to withdraw it to reshape the tip. This advance in "intelligent lead technology" enhanced "simplicity, efficiency, and ease" in pacemaker implantation.[35]

As we have seen earlier, the pacemaker companies competed on technology partly to stave off commoditization and because a belief prevailed in the industry that advanced technology impressed the doctors, particularly the academic physicians. Doctors and patients both appeared to regard advanced technology as a token of medical progress. Patients, unable to judge the quality of medical care as accurately as physicians, treated high-tech devices as proxies or indicators of high-quality care. For specialists in a field like cardiology, control of these devices remained an important source of prestige and income. Academic physicians readily accepted new technology and influenced decision making in hospitals. After acquiring skill with a new device (perhaps by participating in the clinical trials that FDA requires), they championed it through presentations at medical meetings and in journal articles. They tested the outer limits of the new technology and shared their knowledge. The manufacturer, by winning the support of well-known specialists, gained prestigious product champions in the medical community.

The stiff new terms of competition in the United States also impelled device manufacturers to look for growth opportunities abroad, particularly in South America, eastern Europe, India, and East Asia. All these areas had growing middle classes that were interested in advanced medical technology. But here, too, size and a wide range of products were vital as a company tried to establish itself in a new market.[36]

## PACEMAKER AND ICD PHYSICIANS IN
## THE NEW WORLD OF COST CONTAINMENT

Physicians who implanted and managed pacemakers and ICDs knew that the devices had at times been overutilized in the past and that numerous studies had revealed large and mysterious variations in Medicare beneficiaries' use of medical and surgical services. This research could be interpreted to indicate that doctors lacked an evidence-based consensus on how to treat various conditions. The professional societies for heart specialists, including NASPE, addressed this problem in the 1980s by asking panels of experts to prepare practice guidelines on a variety of cardiovascular procedures. The written guidelines demonstrated the medical societies' desire to improve the quality of the care and to reduce the number of inappropriate procedures, and probably had some effect on practitioners.[37]

Heart specialists also understood that before Medicare and managed-care plans were likely to embrace a new medical device or procedure, manufacturers and physicians must demonstrate that it would be cost-effective. The cost-effectiveness trials for new devices like the ICD, while complex and burdensome, were important and could be intellectually fascinating. They certainly brought professional honor to the participating physicians — and many of the studies provided strong support for cardiologists' belief that treatment with implantable rhythm devices, when used with appropriate patients, outperformed alternative treatments in cost-effectiveness. But neither the movement for practice guidelines nor the research on cost-effectiveness proved much of a protection from the great wave of cost-control measures that broke over physicians' heads in the early 1990s.

One important goal of the new physician-payment system that HCFA introduced in 1992 was to reduce the overall growth rate in expenditures for physician services under Medicare. To accomplish this, the RBRVS scheme cut back reimbursement levels for many "overpriced" procedures that specialists performed. This meant sharp reductions in reimbursements for several cardiologists' procedures such as cardiac catheterization, bypass surgery, and pacemaker implantation. W. Bruce Fye, a cardiologist and historian, has commented that, by the mid-1990s, RBRVS and other Medicare payment policies were having a "profound (and progressive) negative effect" on cardiologists' incomes.[38]

Under the new payment system, the physician's compensation for implanting a pacemaker declined year by year, reaching $600 for a dual-chamber pacer and leads in 1997. (Because Medicare covers 80 percent of the bill for procedures, the physician could bill the patient an additional $150.) HCFA had already acted in advance of the RBRVS system to cut reimbursement for pacemaker work in the late 1980s; thus, over the decade 1987–97, reimbursement to implanting physicians had declined about 40 percent. Then the Balanced Budget Act (BBA) of 1997 addressed the impending fiscal crisis in Medicare in part by imposing still tighter constraints on payment rates to physicians.[39]

Organized professional groups such as ACC and NASPE found that they could occasionally influence the Medicare fee schedules to a degree. When Congressman Fortney "Pete" Stark sponsored an amendment to the Omnibus Budget Reconciliation Act of 1990 that would have ended reimbursement for ECG interpretation beginning in 1992 on the ground that a computer could be programmed to interpret this "simple diagnostic test," ACC crafted a coalition of physician groups that successfully lobbied to have the provision rescinded. But specialist groups gradually realized that when it came to reimbursement, many other providers — rural hospitals, home-health agencies, nursing homes, a variety of physician-specialist organizations — were also seeking relief from the cost cutting, especially after passage of the BBA in 1997. When HCFA agreed to a more generous schedule for one group, it tightened the schedule for some other group; the overall budget for Medicare did not change. In the view of a NASPE consultant on federal health policy, HCFA was "robbing Peter to pay Paul."[40]

These new cost-containment policies did more than reduce the incomes of U.S. heart specialists; to many, the changes seemed incompatible with the ethical and cultural premises on which modern medicine had been erected. Cardiologists and other specialists often complained that they were finding it progressively more difficult to make autonomous clinical decisions based on their professional judgment and experience. One observer of the health-care scene noted "the enormous resentment and resistance on the part of so many physicians to what sounds like the industrialization of medicine. . . . [The cost-containment revolution] flies in the face of a physician ethic which is very oriented toward individual benefit, toward the art of medicine and not just the science of medicine." He saw this as "an enormous *cultural* problem." Cardiologist Richard Gorlin wrote that "an array of *intruders*" — he mentioned HCFA, insurance companies, and health-care economists, among others — seemed to be "looking at us as the source of the country's economic woes because we order tests and send people to the hospital, strategies that cost money. Whatever doesn't meet a Medicare regulation is viewed as fraud and abuse." Gorlin insisted that cardiologists "are not merely adroit technicians" but experts who understood that each clinical case had its unique features and made treatment decisions based on the fullness of their knowledge and experience.[41]

Gorlin had given voice to the fear of many cardiologists that their professional autonomy was eroding and they were being pressured to use a kind of cookie-cutter approach to patient care. He had a point, for standardization of care had a certain "mass production" feel to it. Doctors were being asked to base their clinical decisions not on personal experience or on what had been "standard practice" but on what the statistical data indicated to be the most cost-effective treatment. It was the loss of control over clinical decision making and the implicit redefinition of medical care as a product that could be standardized that perhaps dismayed them the most.[42]

Advocates of cost containment could argue in reply that postwar scientific medicine had been based, after all, on reductionist understandings of disease and the scientific (including statistical) analysis of courses of treatment to determine which worked best. Cardiologists had flourished under that regime for decades. As medical ethicist Heather Wood Ion observed, "Managed care's utilization would be less painful for many physicians were medical education and the existing system not focused upon the use, documentation, and measurement of success by incidents of technological intervention. This is the world that has rewarded financially more interventions as indicating better care."[43]

Cardiologists and other physicians had long understood themselves as the advocates for their patients, much as a lawyer represents the interests of a client in a lawsuit or criminal case; this tradition was rooted in the ancient Hippocratic oath.[44] They suspected that managed care was recasting them as "double agents" by asking them to consider not only what was best for the patient but what was best for the broader health care system, and in some cases by rewarding physicians financially for limiting services to patients. Gorlin, for one, believed that the double role would undermine patient trust, thereby making it more difficult to forge a physician-patient partnership, to inform and teach the patient, and to enlist the patient's own best efforts in combating an illness or disorder.[45]

The pressures to restrain the growth of their specialty and the claim that new medical technology was the leading culprit in the high cost of health care also ran up against cardiologists' basic belief that "heart disease is still the leading cause of death in the United States and throughout the industrialized world." W. Bruce Fye emphasized that although the wealthy nations needed to focus much more on prevention, ongoing research leading to "more sophisticated tools and techniques and medicines [was] also part of the equation."[46]

While their leaders debated reimbursement rates with HCFA's administrators and tried to get the broader message across to the public, what were practitioners to do? Cardiologist Walter J. Unger, writing in 1994, pointed to one important trend: cardiologists were emulating hospitals and device manufacturers by creating larger practice groups in a quest for economies of scale. Solo or small group practices would not survive in most U.S. metropolitan areas because maintaining and running an office with advanced diagnostic equipment, modern computerized record keeping, and specialists in reimbursement, legal matters, and marketing would crush them.[47]

The Sanger Clinic in Charlotte, North Carolina, founded in the 1940s, provided an example of the trend toward larger practice groups. During the 1990s, the clinic quadrupled in size to 44 heart surgeons and cardiologists. "The issue for us is breadth of service," the clinic administrator explained. The clinic offered the complete range of cardiovascular services through its 11 offices in Charlotte and neighboring communities. Doctors at different facilities could consult with one another and make care decisions through live telemedicine links. Sanger had a pediatric cardiology center and a special arrhythmia clinic; at some sites, it could handle outpatient heart catheterizations. In emergencies, it could even pick up a patient by helicopter. Physicians worked closely with area hospitals to develop "clinical paths," formalized protocols for handling a variety of specific heart problems. All of these capabilities ultimately depended on the clinic's size.[48]

Sanger did nothing but heart care. With its large staff, high volume of patients and procedures, and elaborate computerized record keeping, a clinic such as this one could bring a vast amount of technical knowledge and clinical experience to bear on each patient's specific problems. Staff could constantly seek ways to improve patient outcomes and to speed the flow of patients through its facilities. The clinic approached Regina Herzlinger's model of the focused factory in health care.[49]

A few large cardiology groups formally pursued some version of the efficiency ideology known as Continuous Quality Improvement (CQI), which emphasized setting specific goals and timetables for improving services and measuring the results in repeated iterations. This had the virtue of addressing both sides of the cultural conflict — the demand from payers that cardiologists control costs and make their services more efficient, and the physicians' own desire to maintain patient well-being as the most important validation of their professional authority and standing. The Minneapolis Heart Institute, an association of five physician groups affiliated with Abbott-Northwestern Hospital, set out on the CQI course in 1988 with specific goals such as early removal of chest tubes after open-heart surgery and a same-day admit program for cardiac catheterization. Spokesmen for the group reported in 1995 that their program, which was ongoing, had saved millions of dollars in care expenses, improved the quality of care, and pleased many patients. It was evident, too, that the effort had given physicians and support staff a powerful sense that they had some control over the conditions under which they performed their work. The approach they described could readily be adapted for pacemaker and ICD treatment to reduce common problems such as atrial lead displacement and complications from use of the subclavian puncture technique for venous access.[50]

At decade's end, with the United States enjoying a protracted period of economic growth and prosperity, a backlash against managed care had made the HMO a popular whipping boy. The *Wall Street Journal* reported that "many people today are prepared to spend a much bigger chunk of their discretionary income on health care." Restraining the growth of the system, it seemed, was passé for some.[51]

But even if the specific arrangements so painstakingly built during the 1980s and 1990s — the Resource-Based Relative Value Scale is a prime example — were to be dismantled, returning to a regime of "unfettered applications of procedures and treatments without regard to their true cost and benefit" was out of the question. Cost-conscious behavior had arrived to stay. Employers, as purchasers of health care for their workers, wanted assurances that providers were practicing evidence-based medicine. The federal government would insist that the variations in care practices among different states and regions be equalized. Fye predicted that practice guidelines such as those developed by NASPE "will be widely used, and physicians will be expected to use them and to adhere to them." In the opinion of cardiologist Richard Caso, the future holds "detailed analysis of vast quantities of data to assess treatment outcomes, physician productivity, [and] cost-benefit ratios of various procedures and preventive treatments." For doctors, the future was likely to feel much like managed care.[52]

## ENDURING PATTERNS IN HEART-RHYTHM MANAGEMENT

The growing power of buyers over the providers of medical services and supplies, the growth of managed care, and the FDA's expanding oversight of clinical trials and manufacturing processes greatly troubled executives of medical-device companies in the 1990s. "In an entrepreneurial industry such as this," one of them wrote in 1994, "delay in the acceptance of new technologies (due to an increasingly risk-averse regulatory system) and a reduction in the potential for widespread adoption of new technologies (due to a . . . system that is motivated to minimize the number of procedures) have the potential to destroy the entire industry."[53]

Surprisingly, the makers of pacemakers and ICDs reported accelerating sales growth and robust profits: they repeatedly introduced new kinds of products for a wider spectrum of heart diseases, and their stock prices rose smartly. Medtronic shares in mid-1999 were carrying a price-earnings ratio of 100, an indication of investors' confidence in the future earnings of the company. In a period of turmoil for the American health-care system, how could this be?[54]

An intractable fact of life at the start of the twenty-first century is that populations are aging in North America, western Europe, and Japan. Heart disease is a distinctive problem of the wealthy nations, but in less wealthy ones the middle classes often emulate the Western dietary and lifestyle patterns that contribute to heart disease. As more people survive into their seventies and eighties, the global demand for medical treatment of all forms of heart disease will surely grow. The physical downsizing of pacemakers and ICDs and their far greater sophistication in sensing and diagnosing arrhythmias promise to add to the number of heart patients who can be treated with rhythm-management devices at reduced risk and expense.[55]

The products that Guidant/CPI, St. Jude/Pacesetter/Ventritex, and Medtronic manufacture — their pacemakers, ICDs, ablation catheters, and related technologies — clearly "work" when used properly in appropriate patients, and no comparably effective alternatives exist for treating most heart-rhythm disorders. The candidate population for pacemaker and ICD implantation has grown steadily and will rise precipitously in the early twenty-first century if electrostimulation proves itself as a palliative treatment for congestive heart failure.

In contrast to the 1970s, no major new entrant is likely to come in and gain significant market share from the established pacemaker/ICD companies because this oligopolistic industry today enjoys impossibly high barriers to entry, the result of the structure of interlocking patents and the cost and complexities of FDA premarket approval. As one executive remarked in 1998, "pacing is a *great* business to be in!" Start-up firms such as Ventritex and InControl have sometimes played a creative role by carrying a new product through the early stages of development; for Ventritex, this was a new kind of ICD able to defibrillate with a much less intense shock; for InControl, an implantable atrial defibrillator. But as each product gained FDA approval and approached market release, larger companies stepped in and acquired these small manufacturers.[56]

The cardiac pacemaker is unlikely to become a commodity product like the desktop computer — not just yet, at any rate. The device manufacturers have continued to invest heavily in designing, testing, and marketing new features for their pacemakers. Corporate spokesmen, leading physicians, financial analysts, and business journalists focus intensely on the technological competition among the three leading companies and blow the trumpets over each interesting new device. Many innovations of recent years help to make the pacemaker simpler to understand and easier to use, and this has the effect of empowering smaller hospitals and occasional implanters to continue implanting pacers to manage the more traditional rhythm disorders, heart block and the sick sinus syndrome.[57]

In marketing pacers as "automatic" and easy to use, the device companies could be risking the commoditization of brady pacing. Many doctors concede that they notice "few differences in quality and features offered by various pacemaker vendors" and agree that "selection of one or two device companies through a competitive bidding process" could cut the cost of pacemaker implantation "without compromising quality of care." The brady pacemaker that meets most clinicians' every need, the dual-chamber, rate-responsive pacer with a user-friendly programmer and some automatic features, has been realized. To keep the field growing and to stave off commoditization, the device manufacturers during the 1990s readied a stream of new products to treat for atrial and ventricular tachycardia, atrial fibrillation, and congestive heart failure. These disorders affected a far greater number of Americans than sinus node disorders or heart block. Some of the new technologies are pacemaker-like devices.[58]

For atrial fibrillation and some intractable kinds of tachycardia, cardiologists added an invasive procedure known as *catheter ablation* to their armamentarium at the end of the 1980s. An electrophysiologist advances a catheter up a vein from the patient's groin into the heart and records intra-cardiac electrograms from various sites. Careful study of the electrogram can pinpoint small groups of heart cells that are misfiring and passing on the aberrant heartbeat. Through the catheter, a low-power, high-frequency alternating current is trained on these cells and destroys them by heating them. Since the area affected is small, the heart's overall pumping action is not compromised. But most ablations create permanent heart block by interrupting the conduction pathways, so the typical patient will require a pacemaker. As ablation therapy grew rapidly in the 1990s, reaching about seventy thousand in the United States in 1996, both St. Jude Medical and Medtronic supplied catheters for EP mapping and ablation, although neither had a dominant market share in this business.[59]

Congestive heart failure (CHF), a common condition in which the heart gradually loses the ability to pump an adequate supply of blood to the tissues of the body, also appears to be a new frontier for the rhythm-management industry. At NASPE's Twentieth Annual Scientific Sessions in May 1999, six "Featured Symposia" explored new approaches to CHF, with three of these sessions focusing on pacemaker and ICD therapy. According to data that the American Heart Association has compiled, about 4.6 million Americans suffer from CHF today; the disease is "the single most frequent cause of hospitalization in patients over age 65." Physicians maintained them on diuretics and powerful drugs that aided the pumping action of the heart; a small number might qualify for heart transplants.[60]

No one expected that the pacemaker or the ICD could serve as a complete technological fix for this terrible and widespread disease, but device companies and some cardiologists believed that pacing the left ventricle or both ventricles simultaneously ("cardiac resynchronization") could improve the heart's ability to deliver blood to the lungs and body. Manufacturers also considered biventricular pacemaker/defibrillators because CHF victims had an elevated risk of sudden cardiac death. If pacing for CHF could prove itself, it would far surpass pacing for AV block and the sick sinus syndrome and at least double the sales of pacemakers. In 1999, all three of the U.S. rhythm-management firms announced initial clinical studies of special leads that a cardiologist could pass through a coronary vein leading into the left side of the heart; they announced similar studies of biventricular pulse generators designed to terminate VT and VF in CHF patients as well as deliver stimuli to coordinate the ventricles.

Along with providing an important source of future revenue growth for the manufacturers, the coming of pacemaker-like implantable devices for treating CHF will probably lead to a major renegotiation of the line between cardiologists and other physicians. In the past, general practitioners and internists have often taken care of patients suffering from CHF; but electrophysiologists will almost certainly be the ones who implant and manage the new heart-failure technology.[61]

Cardiologist David L. Hayes, of the Mayo Clinic, the president of NASPE in 1999, once observed that "since the implantation of the first . . . pacing system, clinical cardiac pacing has evolved rapidly. At each step of the way, it has seemed unlikely that further sophistication could occur."[62] Forty years after the earliest implantable pacemakers, the pacemaker was on the verge of yet another reinvention and had a bright future before it. Pacing first appeared as a procedure that would revive desperately sick patients whose hearts had stopped in the hospital; a little later it became a treatment for postsurgical heart block in children; later still, Americans knew the pacemaker as an implanted device that provided a steady heart rate for an elderly person whose natural beat was too slow. In the late 1970s and 1980s, device manufacturers rolled out the pacemaker as a smart machine, an implanted computer that the doctor could interrogate and reprogram and that could diagnose and manage all manner of rhythm disturbances while synchronizing the atrial and ventricular beats and speeding up the heart or slowing it down as required. One might expect that an artifact, the pacemaker, would have a less evanescent character, but the meaning of terms like *pacing* and *pacemaker* has been so thoroughly transformed that what they signify today or may mean tomorrow bears little resemblance to the definitions and assumptions of the 1950s or the 1970s.

Writing in 1987, Seymour Furman and colleagues dismissed the idea that we could ever dissect advanced medical technology away from medical care generally, restricting the one while enjoying continued progress in the other. Consider ventricular tachycardia: Most doctors thought the condition quite rare until Holter monitoring of the heartbeat, a technology invented in the early 1960s, proved that VT was "both common and potentially lethal." The invasive electrophysiological studies that enabled the specialist to figure out just what was going wrong in the patient's heartbeat and every approach to treatment for VT — surgery, catheter ablation, drugs, the ICD — qualified as high-tech. "High technology *is* modern medical therapy," Furman argued. "The two are so intertwined that they cannot be separated."[63]

Furman and his co-authors probably spoke for heart-care doctors generally. Trained on the advanced technology that has come into use since the 1950s, they have seen invasive treatments such as coronary angioplasty, the implantable defibrillator, and catheter ablation restore health and lengthen life for great numbers of patients. Although most doctors are not innovators of high-tech equipment but users of it, they respect the discoverers and innovators in heart medicine; they strongly believe in medical progress. Their professional identities are firmly bound to the technologies they use.[64]

The health-care system pays a price for its commitment to high-tech medicine, as Furman understood perfectly well. This is notably the case when researchers develop a new treatment where no effective treatment had existed before; then "individual and societal costs will not be less, but may possibly substantially increase. Maintaining life in a patient who previously would have perished may result in the need for continuing, expensive custodial care. . . . The widespread application of 'high tech' medicine may lead to expenditures that never previously existed." To push the point a bit further, the very existence of rhythm-management technology has called forth intensive efforts to define new rhythm problems and assign them to the pacemaker and the ICD. Physicians argue correctly that new treatments are often cost-effective on a per-patient basis. But as a treatment broadens into new uses, the overall expense for society grows.[65]

A successful and profitable medical field such as the manufacture and use of implantable technology to "cure" heart-rhythm disorders assumes significant cultural authority. Perhaps what Louis Kahn said of art can be applied to some technologies such as the pacemaker and the defibrillator: a great work of art, he said, does not involve the satisfying of a desire but the creation of a new desire. With each step of their development, the pacemaker and the defibrillator have met doctors' and patients' needs, but at the same time they have opened new possibilities and helped to create new desires for more sophisticated heart-rhythm management devices.

Manufacturers and many physicians disseminate a notion of medical progress that places intervention and advanced technology at the center of the story. To cite a single example, in 1999 Medtronic announced a new corporate motto: When Life Depends on Medical Technology.[66] It is true, of course, that the story of medical progress through endless technological innovation serves the interests of those who create and control the technologies in question; but that circumstance doesn't mean that the story is untrue. The Medtronic motto is making the point that these inventions have extended and greatly improved the lives of many patients. How the American system of health care should allocate these treatments and pay for them is a painful and baffling question precisely because they are so effective at treating disorders of the heartbeat that would otherwise leave many people bedridden, exhausted, frightened, or dead. When I interviewed Paul Zoll in his Boston office in 1990, he wound up our discussion by reminding me, "Be sure to tell them that it really worked." He was speaking of his external pacemaker of 1952, but his statement applies to implanted cardiac pacemakers and defibrillators from the 1960s to the present: They really work.

# DEVICE RELIABILITY, QUALIFICATION
# TESTS, AND IMPROVEMENTS

## STEPS TOWARD PRODUCT RELIABILITY AT MEDTRONIC IN THE 1960S

— In 1966, Medtronic reorganized its operations into five functional departments, two of which — Product Engineering and Production — held primary responsibility for product reliability.

— The company established a reliability assurance program. Specimens of components under consideration for inclusion in implantable pacemakers were dissected and subjected to microscopic examination. Electronic circuits were tested in simulated body fluid. Wires under consideration for leads were subjected to accelerated flex and corrosion testing.

— The company required engineering documentation: detailed engineering drawings specified the placement and tolerances of every component.

— A quality-control program was created: incoming components were checked; for example, batteries were X-rayed to ensure full capacity.

— Employees assembled implantable products in a "clean room" with two-door isolation from the rest of the building (1966). Assemblers scrubbed like surgeons and wore nylon suits and sterile gloves. Air purity in the clean room was checked daily.

— Every assembled unit received a serial number. Assemblers checked and recorded electronic parameters of each pacemaker several times during assembly.

— Medtronic randomly sacrificed 1 percent of newly assembled pacemakers to test them in a bath of simulated body fluid at body temperature.

— Implantable pacemakers were packaged in sealed, sterile cartons ready for use in the operating room. Every finished pacemaker was X-rayed in its carton to verify integrity of the package and condition of the battery.

— Explanted pulse generators were returned to the company for analysis.

*Source:* Medtronic annual reports.

## QUALIFICATION TESTS PERFORMED ON THE MEDRAD/INTEC IMPLANTABLE DEFIBRILLATOR, LATE 1970S

| Type of Test | Specific Test Performed |
|---|---|
| Hybrid microelectronics | 40 hybrid circuits subjected to military-standard stress tests and extended operational life testing at elevated temperatures. |
| Power source tests | 50 battery cells subjected to microcalorimeter analysis, capacity-life testing, reverse charging, thermal shock, mechanical vibration, and other tests. |
| EMI testing | 4 defibrillators (2 with EMI filtered feedthroughs and 2 without) subjected to abnormally high levels of electro-magnetic interference. |
| Environmental testing | 2 defibrillators subjected to physical stress tests: vibration, mechanical shock, free fall to concrete, temperature recycling, etc. |
| Life testing | 36 defibrillators routinely tested for output and charge time over a period of ca. 2 years. |
| Biocompatibility tests | Device materials tested for toxicity in tissue cultures. |
| Chronic animal studies | 13 dogs implanted for a total of 36 implant-months to demonstrate correct detection and treatment of VF and to expose any other defects. |

*Source:* Steve A. Kolenik et al. "Engineering Considerations in the Development of the Automatic Implantable Cardioverter Defibrillator," *Progr Cardiovasc Dis* 36 (September/October 1993): 127.

## CPI'S PRINCIPAL IMPROVEMENTS TO THE ICD, 1985–1995

| Innovation | Product Name, Market Date |
|---|---|
| Hybrid circuitry: smaller device, easier assembly | Ventak 1500, 1988 |
| Energy programmability | Ventak P, 1991 |
| Tiered therapy: device responds to VT with anti-tachycardia pacing, cardio-version, or defibrillation as appropriate | Ventak PRx, 1993 |
| Transvenous defibrillation lead able to be used with all CPI defibrillators | Endotak lead, 1993 |
| Biphasic waveform reduces energy requirements for defibrillation, extends battery life | Ventak PRx II |
| Small generator size permits pectoral implantations | Ventak PRx III, 1994 (180 gm/99 cc) Ventak Mini (some), 1995 (125 gm/68 cc) |

*Sources:* CPI product brochures; industry analysts' reports.

# NUMBER OF IMPLANTATIONS

## PRIMARY (FIRST-TIME) PACEMAKER IMPLANTATIONS PER MILLION POPULATION, 1972

| | |
|---|---|
| British Isles | 22 |
| Finland | 38 |
| Denmark | 56 |
| Netherlands | 59 |
| Norway | 79 |
| France | 100 |
| Federal Republic of Germany | 100–120 |
| Sweden | 123 |
| United States | 110 |

*Source: Cardiac Pacing: Proceedings of the IVth International Symposium on Cardiac Pacing, Groningen, the Netherlands, April 17–18–19, 1973*, ed. Hilbert J. T. Thalen (Assen, Netherlands: Van Gorcum, 1973), 84–87, 90–111.

## THE BERNSTEIN-PARSONNET ESTIMATE OF PACEMAKER AND ICD IMPLANTATIONS, 1997

| | |
|---|---|
| Primary (first time) pacemaker implantations, 1997 | 152,909 |
| New implantations per million population | 571 |
| Pulse generator replacements | 38,946 |
| Primary ICD implantations, 1997 | 28,711 |
| ICD replacements | 6,919 |

*Source:* Alan D. Bernstein and Victor Parsonnet, "Preliminary Results: Survey of Cardiac Pacing and ICD Practice Patterns, 1997 Results" (unpublished; my thanks to the authors for permission to use these data).

**ESTIMATES OF PRIMARY PACEMAKER IMPLANTATIONS PER MILLION POPULATION, VARIOUS COUNTRIES, 1989**

| | |
|---|---|
| Yugoslavia | 67 |
| Japan | 89 |
| United Kingdom | 148 |
| Israel | 184 |
| "Europe" | 235 |
| Sweden | 253 |
| Canada | 279 |
| United States | 359 |
| Federal Republic of Germany | 363 |
| France | 438 |
| Belgium | 538 |

*Source:* Victor Parsonnet and Alan D. Bernstein, "The 1989 World Survey of Cardiac Pacing," *Pacing Clin Electrophysiol* 14 (November 1991, pt. 2): 2074.

# ICHD PACEMAKER
# IDENTIFICATION CODE

| 1st letter<br>*Chamber-paced* | 2nd letter<br>*Chamber-sensed* | 3rd letter<br>*Mode of Response* |
|---|---|---|
| V = ventricle | V = ventricle | I = inhibited |
| A = atrium | A = atrium | T = triggered |
| D = dual chamber | D = dual chamber | D = either |
| O = not applicable | O = not applicable | O = not applicable |

Examples:

• Asynchronous pacing: the ventricular electrode paces but has no sensing function and cannot respond to intrinsic cardiac activity; thus, VOO.

• AV synchronous pacing: sensing in the atrium, pacing in the ventricle; a sensed P wave causes the pacer to fire into the ventricle after an AV delay; thus, VAT.

• Ventricular inhibited ("demand") pacing: the ventricular electrode both senses and paces, and a sensed R wave inhibits it from firing; thus, VVI.

*Source:* Inter-Society Commission for Heart Disease Resources, Pacemaker Study Group, "Implantable Cardiac Pacemakers: Status Report and Resource Guideline," *Am J Cardiol* 34 (1 October 1974): 488.

# ABBREVIATIONS

## INSTITUTIONS AND MEDICAL TERMS

| | |
|---|---|
| ACC | American College of Cardiology |
| AHA | American Heart Association |
| BBA | Balanced Budget Act of 1997 |
| CPI | Cardiac Pacemakers, Inc. |
| DDD | Dual-chamber universal pacing (AV universal pacing) |
| DDDR | Dual-chamber rate-responsive pacing |
| ECG | Electrocardiograph or electrocardiogram |
| EP | Electrophysiology or electrophysiologist |
| FDA | Food and Drug Administration |
| HCFA | Health Care Financing Administration |
| ICD | Implantable cardioverter-defibrillator |
| NASPE | North American Society of Pacing and Electrophysiology |
| NHLBI | National Heart, Lung, and Blood Institute |
| NIH | National Institutes of Health |
| OSRD | Office of Scientific Research and Development |
| OTA | Office of Technology Assessment |
| RBRVS | Resource-Based Relative Value Scale |
| RF | Radiofrequency |
| VOO | Asynchronous or fixed-rate ventricular pacing |
| VVI | Ventricular inhibited or "demand" pacing |
| VVIR | Ventricular inhibited rate-responsive pacing |

## ANNALS AND JOURNALS

The abbreviations used for titles of publications are those to be found in *Index Medicus*.

| | |
|---|---|
| *Acta Chir Scand* | Acta Chiturgica Scandinavica |
| *Acta Med Scand* | Acta Medica Scandinavica |
| *Am Heart J* | American Heart Journal |
| *Am J Cardiol* | American Journal of Cardiology |
| *Am J Med* | American Journal of Medicine |
| *Am J Med Sci* | American Journal of Medical Science |
| *Am J Physiol* | American Journal of Physiology |
| *Ann Cardiol Angéiol* | Annales de cardiologie et d'angéiologie |
| *Ann Intern Med* | Annals of Internal Medicine |
| *Ann N Y Acad Sci* | Annals of the New York Academy of Sciences |
| *Ann Roy Coll Surg Eng* | Annals of the Royal College of Surgeons of England |

| | |
|---|---|
| *Ann Surg* | Annals of Surgery |
| *Ann Thorac Surg* | Annals of Thoracic Surgery |
| *Arch Intern Med* | Archives of Internal Medicine |
| *Arch Surg* | Archives of Surgery |
| *Br Heart J* | British Heart Journal |
| *Br Med J* | British Medical Journal |
| *Bull Hist Med* | Bulletin of the History of Medicine |
| *Cardiol Clin* | Cardiology Clinics |
| *Cardiovasc Clin* | Cardiovascular Clinics |
| *Clin Cardiol* | Clinical Cardiology |
| *Dis Chest* | Diseases of the Chest |
| *IEEE Trans Biomed Eng* | IEEE Transactions in Biomedical Engineering |
| *Int J Tech Assess in Health Care* | International Journal of Technology Assessment in Health Care |
| *Israel J Med Sci* | Israel Journal of Medical Science |
| *J Am Coll Cardiol* | Journal of the American College of Cardiology |
| *J Am Med Assoc* | Journal of the American Medical Association |
| *J Anat Physiol* | Journal of Anatomy and Physiology |
| *J Exp Med* | Journal of Experimental Medicine |
| *J Health Polit Policy Law* | Journal of Health Politics, Policy, and Law |
| *J Hist Med* | Journal of the History of Medicine |
| *J Lab Clin Med* | Journal of Laboratory and Clinical Medicine |
| *J Med Eng Tech* | Journal of Medical Engineering and Technology |
| *J Physiol* | Journal of Physiology |
| *J Surg Res* | Journal of Surgical Research |
| *J Thorac Cardiovasc Surg* | Journal of Thoracic and Cardiovascular Surgery |
| *Lab Clin Med* | Laboratory and Clinical Medicine |
| *MD&DI* | Medical Device and Diagnostic Industry (www.devicelink.com/mddi/) |
| *Med Hist* | Medical History |
| *Med Instrum* | Medical Instrumentation |
| *Med Res Eng* | Medical Research Engineering |
| *Med World News* | Medical World News |
| *Milbank Q* | Milbank Quarterly |
| *N Engl J Med* | New England Journal of Medicine |
| *N Y State J Med* | New York State Journal of Medicine |
| *Pacing Clin Electrophysiol* | Pacing and Clinical Electrophysiology [PACE] |
| *Proc Roy Soc Med* | Proceedings of the Royal Society of Medicine |
| *Proc Soc Exper Biol & Med* | Proceedings of the Society for Experimental Biology and Medicine |
| *Prog Cardiovasc Dis* | Progress in Cardiovascular Diseases |
| *Rev Sci Instrum* | Review of Scientific Instrumentation |
| *Rev Surg* | Review of Surgery |
| *Surg Forum* | Surgical Forum |
| *T&C* | Technology and Culture |
| *Trans Am Soc Artif Int Org* | Transactions of the American Society for Artificial Internal Organs |

## NOTES

### INTRODUCTION

1. Elmer A. Braun, letter to Michael Halonen, 27 February 1990.

2. In the 1990s, a patient with left bundle branch block and episodes of dizziness or fainting would be a candidate for more extensive electrophysiological study to determine whether the heart had a tendency to go into sustained ventricular tachycardia (rapid heartbeat). Such a patient might end up with an implantable defibrillator rather than a pacemaker: S. Serge Barold and Douglas P. Zipes, "Cardiac Pacemakers and Anti-arrhythmic Devices," in *Heart Disease: A Textbook of Cardiovascular Medicine*, ed. Eugene Braunwald, 4th ed. (Philadelphia: Saunders, 1994), 729–30.

3. The pacemaker, a CyberLith IV manufactured by Intermedics, Inc., of Freeport, Texas, was one of the most sophisticated models then available. Although the physician could alter the pacing rate and other parameters through the use of an external programmer, Braun did not recall that anyone had ever reprogrammed it after its initial setup.

4. The second pacemaker was a Cosmos II from Intermedics.

5. Lewis Thomas, "The Technology of Medicine," *N Engl J Med* 285 (9 December 1971): 1366–68. Thomas himself later became a pacemaker recipient: "My Magical Metronome," *Discover* 4 (January 1983): 58–59.

6. U.S. Congress, OTA, *Policy Implications of the Computed Tomography (CT) Scanner*, OTA-H-56 (November 1978); William W. Lowrance, "Summarizing Reflections," in *New Medical Devices: Invention, Development, and Use*, ed. Karen B. Ekelman (Washington, D.C.: National Academy Press, 1988), 164.

7. Arthur Caplan, "The Concepts of Health, Illness, and Disease," in *Companion Encyclopedia of the History of Medicine*, ed. Roy Porter and W. F. Bynum (London: Routledge, 1993), 1:233–48; Walsh McDermott, "Medicine: The Public Good and One's Own," *Perspectives in Biology and Medicine* 21 (winter 1978): 167–87; Walsh McDermott, "Social Ramifications of Control of Microbial Disease," *Johns Hopkins Medical Journal* 151 (December 1982): 302–12; Stanley J. Reiser, "The Machine at the Bedside: Technological Transformations of Practices and Values," in *The Machine at the Bedside: Strategies for Using Technology in Patient Care*, ed. Stanley J. Reiser and Michael Anbar (New York: Cambridge Univ. Press, 1984), 3–19.

8. Shoshana Zuboff, *In the Age of the Smart Machine: The Future of Work and Power* (New York: Basic Books, 1988).

9. Susan Bartlett Foote, *Managing the Medical Arms Race* (Berkeley: Univ. of California Press, 1992), 12.

10. Howard Bird, "An Affair of the Heart," *N Engl J Med* 326 (13 February 1992):

487–88; Mary Moore Free, "The Heart of the Matters of the Heart," *Am J Cardiol* 78 (15 July 1996): 217–18.

11. Free, "Matters of the Heart"; Lynn Payer, *Medicine and Culture* (New York: Holt, 1988), 74–75, 79–85.

12. I use the terms *doctor* and *physician* interchangeably to refer to persons holding the M.D. degree and licensed to practice medicine; thus surgeons are physicians.

13. John B. McKinlay, "From 'Promising Report' to 'Standard Procedure': Seven Stages in the Career of a Medical Innovation," in *Technology and the Future of Health Care*, ed. John B. McKinlay (Cambridge: MIT Press, 1982), 233–70.

14. For an excellent introduction to the history of medical thinking about the heart and treatments for its maladies, see Joel D. Howell, "Concepts of Heart-Related Diseases," in *Cambridge World History of Human Disease*, ed. Kenneth F. Kiple (Cambridge: Cambridge Univ. Press, 1993), 91–102.

15. For a European's analysis of the pacemaker/defibrillator industry, see Patrik Hidefjäll, "The Pace of Innovation: Patterns of Innovation in the Cardiac Pacemaker Industry," Ph.D. diss., Linköping University, Sweden, 1997.

16. Susan E. Bell, "A New Model of Medical Technology Development: A Case Study of DES," *Research in the Sociology of Health Care* 4 (1986): 1–32; Joel D. Howell, "Early Perceptions of the Electrocardiogram: From Arrhythmia to Infarction," *Bull Hist Med* 58 (spring 1984): 83–98; Joel D. Howell, "Diagnostic Technologies: X-Rays, Electrocardiograms, and CAT Scans," *Southern California Law Review* 65 (November 1991): 529–64.

17. Louis Galambos with Jane Eliot Sewell, *Networks of Innovation: Vaccine Development at Merck, Sharp & Dohme, and Mulford, 1895–1995* (Cambridge: Cambridge Univ. Press, 1995). Galambos does not formally define the phrase *cycles of innovation* and it does not appear in his index.

18. James M. Utterback and Fernando F. Suárez, "Innovation, Competition, and Industry Structure," *Research Policy* 22 (1993): 1–21; Michael L. Tushman and Philip Anderson, "Technological Discontinuities and Organizational Environments," *Administrative Science Quarterly* 31 (September 1986): 439–65; Hidefjäll, "Pace of Innovation," esp. 20–23.

19. Gina Kolata comments on the "medical miracle" genre in a review of *Making Miracles Happen*, by Gregory White Smith and Steven Naifeh, *New York Times Book Review*, 7 September 1997. Arne Larsson's web site may be reached through the St. Jude Medical site: www.sjm.com/. In 1994 St. Jude acquired Elema, the Swedish firm that designed and built Larsson's first pacemaker in 1958.

20. Thomas P. Hughes, *Networks of Power: Electrification in Western Society, 1880–1930* (Baltimore: Johns Hopkins Univ. Press, 1983).

21. For overviews, see Michael S. Baram, "Medical Device Legislation and the Development and Diffusion of Health Technology," in *Technology and the Quality of Health Care*, ed. Richard H. Egdahl and Paul M. Gartman (Germantown, Md.: Aspen Systems, 1978), 191–97; U.S. Congress, OTA, *Federal Policies and the Medical Device Industry*, OTA-H-230 (July 1983); and H. David Banta, "The Regulation of Medical Devices," *Preventive Medicine* 19 (November 1990): 693–99.

22. Wiebe E. Bijker, "The Social Construction of Bakelite," in *The Social Construction of Technological Systems*, ed. Wiebe E. Bijker, Thomas P. Hughes, and Trevor Pinch (Cambridge: MIT Press, 1987), 159–87; and Howell, "Diagnostic Technologies."

23. Daniel M. Fox, "The Consequences of Consensus: American Health Policy in the Twentieth Century," *Milbank Q* 64 (1986): 76–99; Daniel M. Fox, *Health Policies, Health Politics: The British and American Experience, 1911–1965* (Princeton: Princeton Univ. Press, 1986).

24. Elmer A. Braun, telephone conversation with author, 13 April 1998. Braun's two most recent implanted devices are dual-chamber, rate-responsive pacemakers (see chs. 10 and 11). Cf. Alden Solovy, "Techno-Treatment and the Body Electric," *Hospitals and Health Networks* 72 (5 July 1998): 42–44, 46.

**CHAPTER ONE. HEART BLOCK AND THE HEART TICKLER**

1. For Zoll's work, see ch. 2, below.

2. William Stokes, "Observations on Some Cases of Permanently Slow Pulse," *Dublin Quarterly Journal of Medical Science* 2 (1846): 73–85, as reprinted in *Classics of Cardiology*, vol. 2, ed. Frederick A. Willius and Thomas E. Keys (Malabar, Fla.: Krieger, 1983), 462–69. Stokes reprints the case report in *The Diseases of the Heart and the Aorta* (Dublin, Hodges & Smith, 1854), 313–16. For earlier descriptions of heart block, see David C. Schechter et al., "History of Sphygmology and of Heart Block," *Dis Chest* 55 (June 1969): 541–54; and J. K. Lewis, "Stokes-Adams Disease: An Account of Important Historical Discoveries," *Arch Intern Med* 101 (June 1958): 130–42.

3. Adams's case from 1819 is reprinted in Stokes, *Diseases of the Heart*, 305–7.

4. Stokes had written the earliest treatises in English on the use of the stethoscope: *An Introduction to the Use of the Stethoscope* (Edinburgh: Machlachlan & Stewart, 1825) and *Two Lectures on the Application of the Stethoscope to the Diagnosis and Treatment of Thoracic Disease* (Dublin: Hodges & McArthur, 1828). See also Stanley J. Reiser, *Medicine and the Reign of Technology* (Cambridge: Cambridge Univ. Press, 1978), chs. 1–2.

5. Robert G. Frank Jr., "The Telltale Heart: Physiological Instruments, Graphic Methods, and Clinical Hopes, 1854–1914," in *The Investigative Enterprise: Experimental Physiology in Nineteenth-Century Medicine*, ed. William Coleman and Frederick L. Holmes (Berkeley: Univ. of California Press, 1988), 212; Dennis M. Krikler, "The Development of the Understanding of Arrhythmias during the Last 100 Years," *Med Hist* (1985): supp. 5, 77–81.

6. Christopher Lawrence, "Moderns and Ancients: The 'New Cardiology' in Britain, 1880–1930," *Med Hist* (1985): supp. 5, 1–33, esp. 4–7; and Steven J. Peitzman, "From Bright's Disease to End-Stage Renal Disease," in *Framing Disease: Studies in Cultural History*, ed. Charles E. Rosenberg and Janet Golden (New Brunswick, N.J.: Rutgers Univ. Press, 1992), 4–19; Frank G. MacMurray, "Stokes-Adams Disease: A Historical Review," *N Engl J Med* 256 (4 April 1957): 643–50; Lewis, "Stokes-Adams Disease."

7. Charles E. Rosenberg, "Disease in History: Frames and Framers," *Milbank Q* 67 (1989): supp. 1, 1–15; Rosenberg, "Framing Disease: Illness, Society, and History," in *Framing Disease*, ed. Rosenberg and Golden, xiii–xxvi; W. Bruce Fye, "Cardiology in 1885," *Circulation* 72 (July 1985): 21–26; Joel D. Howell, "Concepts of Heart-Related Diseases," in *Cambridge World History of Human Disease*, ed. Kenneth F. Kiple (Cambridge: Cambridge Univ. Press, 1993), 91–101.

8. W. H. Gaskell, "On the Innervation of the Heart, with Especial Reference to the Heart of the Tortoise," *J Physiol* 4 (1883): 43–127; quotations at 64, 53; Gerald L. Geison, *Michael Foster and the Cambridge School of Physiology: The Scientific Enterprise in Late Victorian Society* (Princeton: Princeton Univ. Press, 1978), 268–96; Arthur Keith and

Martin Flack, "The Form and Nature of the Muscular Connections between the Primary Divisions of the Vertebrate Heart," *J Anat Physiol* 41 (1907): 172–89; W. Bruce Fye, "A History of Cardiac Arrhythmias," in *Arrhythmias*, ed. John A. Kastor (Philadelphia: Saunders, 1994), 7, 8. The sinus node is also known as the sinoatrial node, or SA node.

9. Gaskell, "Innervation": 64–66; Augustus D. Waller, "A Demonstration on Man of the Electromotive Changes Accompanying the Heart's Beat," *J Physiol* 8 (1887): 229–34; Frank, "Telltale Heart," 248.

10. Frank, "Telltale Heart," 250.

11. Ibid., 250–55; Dennis M. Krikler, "Historical Aspects of Electrocardiography," *Cardiol Clin* 5 (August 1987): 350–51.

12. Frank, "Telltale Heart," 255–56. It was thus Einthoven who introduced and defined the concept of "the normal electrocardiogram." I refer to electrocardiograms as ECGs; the older term *EKG* (from the German *Elektrokardiogramm*) is no longer widely used in the United States.

13. Reiser, *Reign of Technology*, 107; Frank, "Telltale Heart," 258; W. Bruce Fye, "A History of the Origin, Evolution, and Impact of Electrocardiography," *Am J Cardiol* 73 (15 May 1994): 937–49.

14. Wilhelm His Jr., "The Activity of the Embryonic Human Heart and Its Significance for the Understanding of the Heart Movement in the Adult," trans. T. H. Bast and W. D. Gardner, *J Hist Med* 4 (summer 1949): 289–318 (originally published in German, 1893). His's account of his discovery (originally published in German, 1933) also appears in translation: ibid.: 319–33, quotation at 320. Joseph Erlanger, "On the Physiology of Heart-Block in Mammals, with Especial Reference to the Causation of Stokes-Adams Disease," part 2: "On the Physiology of Heart-Block in the Dog," *J Exp Med* (January 1906): 8–58; Fye, "History of Cardiac Arrhythmias": 11–12. Erlanger was awarded the Nobel Prize in medicine in 1944.

15. Tawara's work was first reported in English in L. Aschoff, "A Discussion on Some Aspects of Heart-Block, 1," *Br Med J* ii (27 October 1906): 1103–7. Tawara also showed that the right and left branches of the bundle of His merged into a network of fibers running throughout the ventricles and known as the Purkinje fibers, after their discoverer, the Polish anatomist Johannes Purkinje (1839). Thinking about heart block may also be traced in Joseph Erlanger, "On the Physiology of Heart-Block in Mammals," *J Exp Med* 7 (November 1905): 676–724, and *J Exp Med* 8 (January 1906): 8–58; and Thomas Lewis, "Occurrence of Heart Block in Man and Its Causation," *Br Med J* ii (19 December 1908): 1798–1802, quotation at 1798. It was Erlanger who demonstrated that Stokes-Adams attacks and heart block were associated.

16. G. A. Gibson, "Heart-Block," *Br Med J* ii (27 October 1906): 1113–19; Lewis, "Occurrence of Heart Block."

17. Stokes, "Observations."

18. On the development and diffusion of the ECG machine, see John Burnett, "The Origins of the Electrocardiograph as a Clinical Instrument," *Med Hist* (1985): supp. 5, 53–76; Joel D. Howell, "Diagnostic Technologies: X-rays, Electrocardiograms, and CAT Scans," *Southern California Law Review* 65 (November 1991): 529–64; and Fye, "History of Electrocardiography." Christopher Lawrence and Joel D. Howell have developed the idea that a "new cardiology" rooted in a physiological understanding of the heart gradually supplanted an older way of thinking in the early twentieth century: see Lawrence, "Moderns and Ancients," and Christopher Lawrence, " 'Definite and Material': Coronary Thrombosis and Cardiologists in the 1920s," in *Framing Disease*, ed.

Rosenberg and Golden, 50–82; Howell, " 'Soldier's Heart': The Redefinition of Heart Disease and Specialty Formation in Early Twentieth-Century Great Britain," *Med Hist* (1985): supp. 5, 34–52; and Howell, "Hearts and Minds: The Invention and Transformation of American Cardiology," in *Grand Rounds: One Hundred Years of Internal Medicine*, ed. Russell C. Maulitz and Diana E. Long (Philadelphia: Univ. of Pennsylvania Press, 1988), 243–75.

19. Julius H. Comroe Jr., "The Heart and Lungs," in *Advances in American Medicine: Essays at the Bicentennial*, ed. John Z. Bowers and Elizabeth F. Purcell (New York: Macy Foundation, 1976), vol. 2, 498–99; J. A. MacWilliam, "Electrical Stimulation of the Heart in Man," *Br Med J* i (16 February 1889): 348–50; Joseph Erlanger, "Sinus Stimulation as a Factor in the Resuscitation of the Heart," *J Exp Med* 16 (October 1912): 452–69.

20. David C. Schechter, "Background of Clinical Cardiac Electrostimulation VI: Precursor Apparatus and Events to the Electrical Treatment of Complete Heart Block," *N Y State J Med* 72 (15 April 1972): 954.

21. MacWilliam, "Electrical Stimulation": 348–50, quotations at 349; Peter Bloomfield and N. A. Boon, "A Century of Cardiac Pacing," *Brit Med J* 298 (11 February 1989): 343–44.

22. David C. Schechter, "Background of Clinical Cardiac Electrostimulation, IV: Early Studies on Feasibility of Accelerating Heart Rate by Means of Electricity," *N Y State J Med* 72 (1 February 1972): 399–401. Schechter unearthed a second isolate, a clinical scientist named L. G. Robinovitch, who in 1908, in Paris, used pulsed electrostimulation to revive a patient in drug-induced respiratory depression; Robinovitch noted that her stimulating apparatus, which used external electrodes, also affected the heart: ibid., 402–3.

23. Thomas Lewis, *Clinical Disorders of the Heart Beat* (London: Shaw & Sons, 1912); Fye, "History of Cardiac Arrhythmias." Joel Howell argues that experimental physiology was not closely linked to clinical medicine: "Cardiac Physiology and Clinical Medicine? Two Case Studies," in *Physiology in the American Context, 1850–1940*, ed. Gerald L. Geison (Bethesda, Md.: American Physiological Society, 1987), 279–92.

24. Lidwill's paper, "Cardiac Disease in Relation to Anaesthesia," is reprinted in its entirety in Harry G. Mond, David Smith, and Graeme Gloman, "The First Pacemaker," *Pacing Clin Electrophysiol* 5 (March–April 1982): 278–82.

25. Albert S. Hyman to David C. Schechter, n.d., letter reprinted in Schechter, "Background of Clinical Cardiac Electrostimulation, V: Direct Electrostimulation of Heart without Thoracotomy," *N Y State J Med* 72 (1 March 1972): 609–10. Unless otherwise indicated, all information and quotations in the next five paragraphs come from this source.

26. An ectopic heartbeat is one originating in an area of the heart other than the sinus node. Hyman reported on intracardiac injection in "Resuscitation of the Stopped Heart by Intracardiac Therapy," *Arch Intern Med* 46 (October 1930): 553–68.

27. Albert S. Hyman, "Resuscitation of the Stopped Heart by Intracardiac Therapy, II: Experimental Use of an Artificial Pacemaker," *Arch Intern Med* 50 (August 1932): 286.

28. Ibid.; U.S. patent no. 1,913,595 (June 13, 1933), awarded to Charles H. Hyman and Albert S. Hyman.

29. Wolfgang Schivelbusch, *The Railway Journey: The Industrialization of Time and Space in the Nineteenth Century* (Berkeley: Univ. of California Press, 1986; first published 1977), 99; Schechter, "Background, V": 612; A. R. Clauser, "The Heart Tickler or Pacemaker," letter in *J Am Med Assoc* 100 (20 May 1933): 1628. See also David C.

Schechter, "Early Experience with Resuscitation by Means of Electricity," *Surgery* 69 (March 1971): 360–72; Margaret Rowbottom and Charles Susskind, *Electricity and Medicine: History of Their Interaction* (San Francisco: San Francisco Press, 1984); and Mickey S. Eisenberg, *Life in the Balance: Emergency Medicine and the Quest to Reverse Sudden Death* (New York: Oxford Univ. Press, 1997), 167–77, 139–66.

30. Paul Dudley White, *Heart Disease* (New York: Macmillan, 1934), 669–70; Alfred E. Cohn and Samuel A. Levine, "The Beneficial Effects of Barium Chlorid [*sic*] on Adams-Stokes Disease," *Arch Intern Med* 36 (July 1925): 1–12.

31. See Joel D. Howell, "Early Perceptions of the Electrocardiogram: From Arrhythmia to Infarction," *Bull Hist Med* 58 (spring 1984): 83–98.

32. Aaron E. Parsonnet and Albert S. Hyman, *Applied Electrocardiography* (New York: Macmillan, 1929); W. Bruce Fye, *American Cardiology: The History of a Specialty and Its College* (Baltimore: Johns Hopkins Univ. Press, 1996), 70–75, 125–31, 133; quotation at 131.

33. Mary Shelley, *Frankenstein* (1818), ch. 5; Roslynn D. Haynes, *From Faust to Strangelove: Representations of the Scientist in Western Literature* (Baltimore: Johns Hopkins Univ. Press, 1994), 92–102; S. H. Vasbinder, *Scientific Attitudes in Mary Shelley's Frankenstein* (Ann Arbor, Mich.: UMI Research Press, 1984); Harold Bloom, "Frankenstein, or the New Prometheus," *Partisan Review* 32 (fall 1965): 609–18. For an informative and nonhysterical contemporary account of Hyman's work, see "Device Revives Beats of Heart after Stoppage," *New York Herald Tribune*, 8 December 1932.

34. Letter from Siemens to Dennis Stillings, 1972, as quoted in Stillings, "The Early History of Attempts at Electrical Control of the Heart: Harvey to Hyman," paper delivered at the 24th International Congress of the History of Medicine, August 25–31, 1974, Budapest, Hungary (Minneapolis: Medtronic, 1979), n.p.; Siegfried Koeppen, "Untersuchungen über die Wirksamkeit von Wiederbelebungsmassnahmen bei experimenteller Erstickung," *Klinische Wochenschrift* 14 (1935): 1131–33; Susanne Hahn, *Herz: Das menschliche Herz, der herzliche Mensch* (exhibition catalog, Deutschen Hygiene-Museum, Dresden, 1995), 138–40. My thanks to J. Walter Keller and Dr. Berndt Lüderitz for help in locating these sources.

35. Stillings, "Early History," and Schechter, "Background, VI": 958; Albert S. Hyman, "Cardiac Resuscitation in Sudden Death from Coronary Disease," *Transactions of the American College of Cardiology* 1 (1951): 135–38. A single newspaper story from 1935 also asserts that the invention had succeeded in "bringing a man back to life after his heart had stopped and keeping him alive for twenty-four hours": "Heart Device Aids Revival of 'Dead,'" *New York Times*, 14 February 1935.

36. Seymour Furman has pointed out a possible third case of attempted cardiac pacing in Odessa (Soviet Union) in 1927. This report was published in a French journal but aroused no interest from physiologists or clinicians: Furman, "History of Cardiac Pacing" (MS, April 1998): 3.

37. NASPE, 19th Annual Scientific Sessions: *Final Program* (San Diego, Calif., May 1998), 244, 246.

## CHAPTER TWO. THE WAR ON HEART DISEASE AND
## THE INVENTION OF CARDIAC PACING

1. Walter H. Abelmann, "Paul M. Zoll and Electrical Stimulation of the Human Heart," *Clin Cardiol* 9 (March 1986): 131–35; author's conversations with Arthur J.

Linenthal and Stafford H. Cohen, 14 and 17 October 1999, Boston; Paul M. Zoll, "Resuscitation of the Heart in Ventricular Standstill by External Electric Stimulation," *N Engl J Med* 247 (13 November 1952): 768–71.

2. Lynn Payer, *Medicine and Culture* (New York: Holt, 1988), 124–52; Stanley J. Reiser, "The Machine at the Bedside: Technological Transformations of Practices and Values," in *The Machine at the Bedside: Strategies for Using Technology in Patient Care*, ed. Stanley J. Reiser and Michael Anbar (New York: Cambridge Univ. Press, 1984), 3–19.

3. Rosemary Stevens, *In Sickness and in Wealth: American Hospitals in the Twentieth Century* (New York: Basic Books, 1989), ch. 8; Paul M. Zoll, interviews by author, 5 February 1990 and 19 September 1991, Boston (NASPE, Oral History Archive, Natick, Mass. [hereafter, NASPE]).

4. An electrophysiologist is a cardiologist who specializes in disorders of heart rhythm. Clinical electrophysiology as a recognized subspecialty within cardiology dates from the 1970s, but laboratory research on the electrophysiology of the heart dates back to the nineteenth century. For background on subspecialties in cardiology, see W. Bruce Fye, *American Cardiology: The History of a Specialty and Its College* (Baltimore: Johns Hopkins Univ. Press, 1996).

5. John Parkinson, Cornelius Papp, and William Evans, "The Electrocardiogram of the Stokes-Adams Attack," *Br Heart J* 3 (July 1941): 171–99.

6. Frank G. MacMurray, "Stokes-Adams Disease: A Historical Review," *N Engl J Med* 256 (4 April 1957): 643–50.

7. Charles E. Rosenberg, "Disease and Social Order in America: Perceptions and Expectations," *Milbank Q* 64 (1984): supp. 1, 34–55; Paul M. Zoll et al., "Treatment of Unexpected Cardiac Arrest by External Electric Stimulation of the Heart," *N Engl J Med* 254 (22 March 1956): 541–46; Paul M. Zoll, "Development of Electric Control of Cardiac Rhythm," *J Am Med Assoc* 226 (19 November 1973): 881–86.

8. John A. MacWilliam, "Fibrillar Contraction of the Heart," *J Physiol* 8 (1887): 296–310; John A. MacWilliam, "Electrical Stimulation of the Heart in Man," *Br Med J* i (16 February 1889): 348–50; W. Bruce Fye, "Ventricular Fibrillation and Defibrillation: Historical Perspectives with Emphasis on the Contributions of John MacWilliam, Carl Wiggers, and William B. Kouwenhoven," *Circulation* 71 (May 1985): 858–65. MacWilliam's paper "Some Applications of Physiology to Medicine, II: Ventricular Fibrillation and Sudden Death," *Br Heart J* ii (11 August 1923): 215–19, reviews his own work on VF and comments on the development of knowledge since the 1880s.

9. David C. Schechter, "Flashbacks: Ventricular Fibrillation, Part 1," *Pacing Clin Electrophysiol* 2 (July–August 1979): 490–504; Fye, "Ventricular Fibrillation and Defibrillation"; A. J. Jex-Blake (1913) as quoted by Fye, ibid.: 861.

10. David C. Schechter, "Early Experience with Resuscitation by Means of Electricity," *Surgery* 69 (March 1971): 360–72; Margaret Rowbottom and Charles Susskind, *Electricity and Medicine: History of Their Interaction* (San Francisco: San Francisco Press, 1984), 7–10, 23; Mickey S. Eisenberg, *Life in the Balance: Emergency Medicine and the Quest to Reverse Sudden Death* (New York: Oxford Univ. Press, 1997), 146–51.

11. W. Bruce Fye, "The Delayed Recognition of Myocardial Infarction: It Took Half a Century!" *Circulation* 72 (August 1985): 262–71; Christopher Lawrence, "'Definite and Material': Coronary Thrombosis and Cardiologists in the 1920s: Studies in Cultural History," in *Framing Disease*, ed. Charles E. Rosenberg and Janet Golden (New Brunswick, N.J.: Rutgers Univ. Press, 1992), 50–82.

12. MacWilliam, "Some Applications": 218.

13. D. R. Hooker, William B. Kouwenhoven, and O. R. Langworthy, "The Effect of Alternating Electrical Currents on the Heart," *Am J Physiol* 103 (February 1933): 444–54.

14. L. P. Ferris et al., "Effect of Electric Shock on the Heart," *Electrical Engineering* 55 (May 1936): 514; Fye, "Ventricular Fibrillation and Defibrillation": 862; Carl J. Wiggers and René Wégria, "Ventricular Fibrillation Due to Single, Localized Induction and Condenser Shock Applied during the Vulnerable Phase of Ventricular Systole," *Am J Physiol* 128 (February 1940): 500–505; Robert M. Berne and Matthew N. Levy, *Cardiovascular Physiology*, 6th ed. (St. Louis: Mosby Year Book, 1992), 49.

15. Claude S. Beck, W. H. Pritchard, and H. S. Feil, "Ventricular Fibrillation of Long Duration Abolished by Electric Shock," *J Am Med Assoc* 135 (13 December 1947): 985–86; Claude S. Beck, "Reminiscences of Cardiac Resuscitation," *Review of Surgery* 27 (April 1970): 76–94.

16. Beck, "Reminiscences": 85–86; Hugh E. Stephenson Jr., *Cardiac Arrest and Resuscitation* (St. Louis: Mosby, 1958), ch. 6; Zoll interview by author, 19 September 1991. Beck's technique, never widely accepted, was superseded by closed-chest cardiopulmonary resuscitation (CPR) in the 1960s.

17. Werner Forssmann, *Experiments on Myself: Memoirs of a Surgeon in Germany*, trans. Hilary Davies (New York: St. Martin's, 1974), 75–88; Carl J. Pepine, James A. Hill, and Charles R. Lambert, "History of the Development and Application of Cardiac Catheterization," in *Diagnostic and Therapeutic Cardiac Catheterization*, ed. Pepine, Hill, and Lambert (Baltimore: Williams & Wilkins, 1989), 3–9.

18. André Cournand and H. A. Ranges, "Catheterization of the Right Auricle in Man," *Proceedings of the Society for Experimental Biology and Medicine* 46 (1941): 462–66; Pepine et al., "History of . . . Cardiac Catheterization."

19. André Cournand et al., "Recording of Right Heart Pressures in Man," *Proc Soc Exper Biol & Med* 55 (1944): 34–36; Stephen L. Johnson, *The History of Cardiac Surgery, 1896–1955* (Baltimore: Johns Hopkins Univ. Press, 1970), 129–31.

20. André Cournand, "Cardiac Catheterization: Development of the Technique, Its Contributions to Experimental Medicine, and Its Initial Applications in Man," *Acta Med Scand* 579 (1975): 7–32; Johnson, *History of Cardiac Surgery*, 124–36; Fye, *American Cardiology*, 171–75.

21. Samuel W. Hunter, interview by author, 30 November 1989, Mendota Heights, Minn. (NASPE); Dwight E. Harken, "The Emergence of Cardiac Surgery: Personal Recollections of the 1940s and 1950s," *J Thorac Cardiovasc Surg* 98 (November 1989): 806; Henry Swan, "Cardiac Surgery with Hypothermia," in *History and Perspectives of Cardiology*, ed. H. A. Snellen, A. J. Dunning, and A. C. Arntzenius, 142 (italics in original).

22. Johnson, *History of Cardiac Surgery*, 47–71.

23. Ibid., 78–85; William F. Friedman, "Congenital Heart Disease in Infancy and Childhood," in *Heart Disease: A Textbook of Cardiovascular Medicine*, ed. Eugene Braunwald, 4th ed. (Philadelphia: Saunders, 1994), 935–37.

24. Alfred Blalock and Helen B. Taussig, "The Surgical Treatment of Malformations of the Heart," *J Am Med Assoc* 128 (19 May 1945): 189–202; Johnson, *History of Cardiac Surgery*, 77–85; quotation at 82–83.

25. Dwight E. Harken and Paul M. Zoll, "Foreign Bodies in and in Relation to the Thoracic Blood Vessels and Heart, III: Indications for the Removal of Intracardiac

Foreign Bodies and the Behavior of the Heart during Manipulation," *Am Heart J* 32 (July 1946): 1–19; Dwight E. Harken interview, "Pioneers of Surgery," *Nova* TV program (PBS, n.d. [1980?]).

26. Wilfred G. Bigelow, John C. Callaghan, and John A. Hopps, "General Hypothermia for Experimental Intracardiac Surgery: The Use of Electrophrenic Respirations, an Artificial Pacemaker for Cardiac Standstill, and Radio-Frequency Rewarming in General Hypothermia," *Ann Surg* 132 (September 1950): 531–39; Swan, "Cardiac Surgery with Hypothermia," 142–43.

27. F. John Lewis, Richard V. Varco, and M. Taufic, "Repair of Atrial Septal Defects in Man under Direct Vision with the Aid of Hypothermia," *Surgery* 36 (September 1954): 538–56; Leonard G. Wilson, *Medical Revolution in Minnesota: A History of the University of Minnesota Medical School* (St. Paul, Minn.: Mediwiwin Press, 1989), 498–500.

28. Julius H. Comroe Jr. and Robert D. Dripps, *The Top Ten Clinical Advances in Cardiovascular-Pulmonary Medicine and Surgery, 1945–1975: Final Report*, 2 vols. (Dept. of Health, Education, and Welfare, NIH, January 31, 1977). See also Julius H. Comroe and Robert D. Dripps, "Scientific Basis for the Support of Biomedical Science," *Science* 192 (9 April 1976): 105–11.

29. OSRD, Committee on Medical Research, *Advances in Military Medicine* (Boston: Little, Brown, 1948); Eli Ginzberg, "The Impact of World War II on U.S. Medicine," *Am J Med Sci* 304 (October 1992): 269.

30. See, e.g., Daniel J. Kevles, *The Physicists: The History of a Scientific Community in Modern America* (New York: Knopf, 1978), chs. 21–22; David Dickson, *The New Politics of Science* (Chicago: Univ. of Chicago Press, 1984), ch. 1.

31. President's Commission on the Health Needs of the Nation, *Building America's Health* (Washington, D.C.: GPO, 1952); Donald C. Swain, "The Rise of a Research Empire: NIH, 1930 to 1950," *Science* 138 (14 December 1962): 1233–37; Daniel M. Fox, "The Politics of the NIH Extramural Program," *J Hist Med* 42 (October 1987): 447–66; Daniel M. Fox, "The Consequences of Consensus: American Health Policy in the Twentieth Century," *Milbank Q* 64 (1986): 83–86.

32. Stephen P. Strickland, *Politics, Science, and Dread Disease: A Short History of United States Medical Research Policy* (Cambridge: Harvard Univ. Press, 1972), ch. 5; Paul Starr, *The Social Transformation of American Medicine* (New York: Basic Books, 1982), 335–42.

33. Daniel M. Fox, "Health Policy and Changing Epidemiology in the United States: Chronic Disease in the Twentieth Century," in *Unnatural Causes: The Three Leading Killer Diseases in America*, ed. Russell Maulitz (New Brunswick, N.J.: Rutgers Univ. Press, 1989), 11–31; Daniel M. Fox, *Power and Illness: The Failure and Future of American Health Policy* (Berkeley: Univ. of California Press, 1993), 48–52; James T. Patterson, *The Dread Disease: Cancer and Modern American Culture* (Cambridge: Harvard Univ. Press, 1987), 140–51, 172–85; David J. Rothman, *Beginnings Count: The Technological Imperative in American Health Care* (New York: Oxford Univ. Press, 1997), ch. 2.

34. Kenneth M. Endicott and Ernest M. Allen, "The Growth of Medical Research 1941–1953 and the Role of Public Health Service Research Grants," *Science* 118 (25 September 1953): 341; James A. Shannon, "The Advancement of Medical Research: A Twenty-Year View of the Role of the National Institutes of Health," *Journal of Medical Education* 42 (February 1967): 97–108; James H. Cassedy, "Stimulation of Health Research," *Science* 145 (28 August 1964): 897–902.

35. Fye, *American Cardiology*, 153.

36. Zoll interview by author, 5 February 1990. The phrase I use in the subheading for this section is taken from Eugene Braunwald's title "The Golden Age of Cardiology," in *An Era in Cardiovascular Medicine*, ed. Suzanne B. Knoebel and Simon Dack (New York: Elsevier, 1991).

37. Joel D. Howell, "Machines and Medicine: Technology Transforms the American Hospital," in *The American General Hospital: Communities and Social Contexts*, ed. Diana E. Long and Janet Golden (Ithaca, N.Y.: Cornell Univ. Press, 1989), 109–34; Starr, *Social Transformation*, esp. 347–63.

38. Stevens, *In Sickness*, 216–26; Starr, *Social Transformation*, 347–51.

39. Joel D. Howell, "The Changing Face of Twentieth-Century American Cardiology," *Ann Intern Med* 105 (November 1986): 772–82; Howell, "Machines and Medicine."

40. William H. Sweet, "Stimulation of the Sino-Atrial Node for Cardiac Arrest during Operation," abstract in *Bulletin of the American College of Surgeons* 32 (September 1947): 234; Herman K. Hellerstein and Irving M. Liebow, "Control of Heart Rate with an Intracardiac Thermode," *J Lab Clin Med* 35 (May 1950): 703–7; Herman K. Hellerstein, Denman Shaw, and Irvine M. Liebow, "An Extracorporal Electronic Bypass of the A-V Node," abstract in *Lab Clin Med* 36 (December 1950): 833; Bigelow, Callaghan, and Hopps, "General Hypothermia"; Chester E. Herrod et al., "Control of Heart Action by Repetitive Electrical Stimuli," *Ann Surg* 136 (September 1952): 510–19.

41. Abelmann, "Paul M. Zoll"; Zoll interview by author, 5 February 1990; Arthur J. Linenthal, interview by author, 21 September 1991, Brookline, Mass. (NASPE); Arthur J. Linenthal, *First a Dream: The History of Boston's Jewish Hospitals, 1896 to 1928* (Boston: Beth Israel Hospital and the Francis A. Countway Library of Medicine, 1990), 276–310.

42. Abelmann, "Paul M. Zoll"; Zoll interview by author, 5 February 1990.

43. Zoll interview by author, 5 February 1990.

44. Zoll interview by author, 5 February 1990; Abelmann, "Paul M. Zoll": 132; David C. Schechter, "Background of Clinical Cardiac Electrostimulation, VII: Modern Era of Artificial Cardiac Pacemakers," *N Y State J Med* 72 (15 May 1972): 1166–67.

45. Paul M. Zoll, "The Intermediate History of Cardiac Pacing," in *Boston Colloquium on Cardiac Pacing*, ed. J. Warren Harthorne and Hilbert J. T. Thalen (The Hague: Nijhoff, 1977), 27–28.

46. Zoll interview by author, 5 February 1990.

47. Zoll in Schechter, "Background, VII": 1167; Bigelow et al., "General Hypothermia"; John C. Callaghan and Wilfred G. Bigelow, "An Electrical Artificial Pacemaker for Standstill of the Heart," *Ann Surg* 134 (July 1951): 8–17; Bigelow, "The Pacemaker Story: A Cold Heart Spinoff," *Pacing Clin Electrophysiol* 10 (January–February 1987, pt. 1): 142–50.

48. Schechter, "Background, V": 617; John C. Callaghan, "Early Experiences in the Study and Development of an Artificial Electrical Pacemaker for Standstill of the Heart: View from 1949," *Pacing Clin Electrophysiol* 3 (September–October 1980): 618–19.

49. Zoll, "Development of Electric Control": 882; Zoll interview by author, 5 February 1990; Zoll in Schechter, "Background, VII": 1168.

50. Zoll interview by author, 5 February 1990; cf. Zoll in Schechter, "Background, VII": 1168.

51. Zoll interview by author, 5 February 1990; Zoll, "Resuscitation of the Heart": 769.

52. Zoll, "Resuscitation of the Heart": 769–70; Zoll in Schechter, "Background, VII":

1168; Leona N. Zarsky at Zoll's memorial service, Beth Israel Deaconess Medical Center, 23 February 1999. Zarsky, a surgeon, assisted Zoll with his animal experiments.

53. Zoll, "Resuscitation of the Heart"; Zoll, "Treatment of Stokes-Adams Disease."

54. Zoll, in Schechter, "Background, VII": 1168.

55. Seymour Furman, interview by Earl E. Bakken, 14 August 1980, Minneapolis (*Pioneers in Pacing* video: see bibliographical note); Furman, "Attempted Suicide," editorial in *Pacing Clin Electrophysiol* 3 (March–April 1980): 129.

56. Schechter, "Background, V": 615; Bigelow, "Pacemaker Story."

57. Zoll in Schechter, "Background, VII": 1167.

58. Abelmann, "Paul M. Zoll." In 1973, Zoll received the Lasker Award of the American Medical Association for his contributions to the control of heart arrhythmias.

59. These experiments took place in the surgical recovery room of Toronto General Hospital. See Bigelow and Callaghan in Schechter, "Background, V": 615–17; and Callaghan, "Early Experiences": 619.

60. "Unfortunately, Zoll forgot to indicate in his article [reporting on external cardiac pacing, 1952] the source of the electric circuit diagram that he used in his pacemaker": Bigelow, "Pacemaker Story": 150. For information about Hopps's design, including a photograph, see Callaghan and Bigelow, "An Electrical Artificial Pacemaker." This was the paper that Zoll had heard on October 23, 1950.

61. Callaghan and Bigelow, "An Electrical Artificial Pacemaker": 10–11.

62. Jerry C. Luck and Michael L. Markel, "Clinical Applications of External Pacing: A Renaissance," *Pacing Clin Electrophysiol* 14 (August 1991): 1299–316.

**CHAPTER THREE. HEART SURGEONS REDEFINE CARDIAC PACING**

Part of this chapter was written in collaboration with David J. Rhees and published as David J. Rhees and Kirk Jeffrey, "Earl Bakken's Little White Box: The Complex Meanings of the First Transistorized Pacemaker," in *Exposing Electronics*, ed. Bernard Finn (London: Harwood, 2000). My thanks to David Rhees for permission to reprint some passages from that article.

1. Kenneth M. Ludmerer, *Learning to Heal: The Development of American Medical Education* (New York: Basic Books, 1985; reprint, Baltimore: Johns Hopkins Univ. Press, 1996), 207–13, 217–18, 231–33, 256–64; and Rosemary Stevens, *In Sickness and in Wealth: American Hospitals in the Twentieth Century* (New York: Basic Books, 1989), 201–5.

2. Leonard G. Wilson, *Medical Revolution in Minnesota: A History of the University of Minnesota Medical School* (St. Paul, Minn.: Mediwiwin Press, 1989), 481–85; Stevens, *In Sickness*, 216–26.

3. Samuel W. Hunter, interview by author, 30 November 1989, Mendota Heights, Minn. (NASPE); Norman W. Shumway interview, "Pioneers of Surgery," Nova TV program (PBS, n.d. [1980?]).

4. Stephen L. Johnson, *The History of Cardiac Surgery, 1896–1955* (Baltimore: Johns Hopkins Univ. Press, 1970), 47–72, 137–47.

5. Ibid., 147–49; Wilson, *Medical Revolution*, 486–92; "Mentor of a Thousand Surgeons," *Medical World News* 8 (16 June 1967): 69–76; John W. Kirklin, "The Middle 1950s and C. Walton Lillehei," *J Thorac Cardiovasc Surg* 98 (November 1989): 822–24.

6. Wilson, *Medical Revolution*, 498–500.

7. Ibid., 492; Herbert E. Warden, "C. Walton Lillehei: Pioneer Cardiac Surgeon," *J Thorac Cardiovasc Surg* 98 (November 1989): 833–45.

8. Wilson, *Medical Revolution*, 493, 500–505; C. Walton Lillehei, interview by author, 25 July 1990, St. Paul, Minn. (NASPE). The cross-circulation was "controlled" because the tubes connecting the two bodies passed through a control pump so as to prevent the donor from bleeding excessively into the recipient.

9. C. Walton Lillehei et al., "Direct Vision Intracardiac Surgical Correction of the Tetralogy of Fallot, Pentalogy of Fallot, and Pulmonary Atresia Defects: Report of First Ten Cases," *Ann Surg* 142 (September 1955): 418–42; Vincent L. Gott, "C. Walton Lillehei and Total Correction of Tetralogy of Fallot," *Ann Thorac Surg* 49 (February 1990): 328–32.

10. Wilson, *Medical Revolution*, 508–15. In 1956–57, Vincent L. Gott, a Lillehei surgical trainee, modified the DeWall bubble oxygenator to produce a plastic sheet oxygenator that was inexpensive to manufacture. Both the bubble and the sheet oxygenator were widely adopted for open-heart surgery.

11. Wilson, *Medical Revolution*, 516; Warden, "C. Walton Lillehei"; Vincent L. Gott, "C. Walton Lillehei and His Trainees: One Man's Legacy to Cardiothoracic Surgery," *J Thorac Cardiovasc Surg* 98 (November 1989): 846–51.

12. Lillehei interview by author.

13. Warden, "C. Walton Lillehei": 841; C. Walton Lillehei, interview by Earl E. Bakken, 10 June 1977 (*Pioneers in Pacing* video: see bibliographical note).

14. Warden, "C. Walton Lillehei"; Vincent L. Gott, interview by author, 2 May 1997, Baltimore; Lillehei interview by author.

15. Lillehei interview by Bakken; William L. Weirich, Vincent L. Gott, and C. Walton Lillehei, "The Treatment of Complete Heart Block by the Combined Use of a Myocardial Electrode and an Artificial Pacemaker," *Surg Forum* 8 (1958): 360–63; C. Walton Lillehei et al., "Direct Wire Electrical Stimulation for Acute Postsurgical and Postinfarction Complete Heart Block," *Ann N Y Acad Sci* 111 (11 June 1964): 938–49.

16. Lillehei interview by author.

17. Lillehei interview by Bakken; Earl E. Bakken, *One Man's Full Life* (Minneapolis: Medtronic, 1999), 49; [Northern States Power Co.,] "Report on October 31, 1957 Major Disturbance," typescript, copy in author's possession. My thanks to Ron Hagenson for locating this document for me.

18. Lillehei interview by Bakken.

19. Bakken, *One Man's Full Life*, 50; Lillehei interview by Bakken; Louis E. Garner, "Five New Jobs for Two Transistors," *Popular Electronics* 4 (April 1956): 54–59; Ernest Braun and Stuart Macdonald, *Revolution in Miniature: The History and Impact of Semiconductor Electronics*, 2d ed. (Cambridge: Cambridge Univ. Press, 1982).

20. Earl E. Bakken, interview by William W. Swanson, 22 December 1992 (Bakken Library and Museum, Minneapolis).

21. Earl E. Bakken, interview by author, 23 May 1990, Fridley, Minn.; Bakken, interview by David J. Rhees, 10 January 1997 (Bakken Library and Museum, Minneapolis). At the Karolinska Institute in Stockholm, surgeon Åke Senning and engineer Rune Elmqvist built an external pacemaker of similar design around 1958. Senning had visited Minnesota to study Lillehei's surgical techniques and had observed the use of the myocardial pacing wire. The Swedish external pacemaker was commercialized in 1960 by Elmqvist's employer, the Swedish firm Elema-Schönander: Rune Elmqvist et al., "Artificial Pacemaker for Treatment of Adams-Stokes Syndrome and Slow Heart Rate," *Am*

*Heart J* 65 (June 1963): 731–48; Åke Senning, "Cardiac Pacing in Retrospect," *American Journal of Surgery* 145 (June 1983): 733–39.

22. Bakken interview by Rhees; author's examination of a Medtronic model 5800 pulse generator at the Bakken Library and Museum, Minneapolis.

23. Bakken interview by Rhees.

24. "Inside the Heart: Newest Advances in Surgery," *Time* 69 (25 March 1957): 66–77.

25. On medical devices as tokens of the future, see Nancy Knight, " 'The New Light': X-rays and Medical Futurism," in *Imagining Tomorrow: History, Technology, and the American Future*, ed. Joseph J. Corn (Cambridge: MIT Press, 1986), 10–34.

26. Wilson, *Medical Revolution*, 523; C. Walton Lillehei et al., "Transistor Pacemaker for Treatment of Complete Atrioventricular Dissociation," *J Am Med Assoc* 172 (30 April 1960): 2006–10, quotation at 2007. See also "Heart Timer: Minnesota Reports on Use of Electronic Pacemaker," *New York Times*, 1 May 1960.

27. The Rockefeller Institute, Medical Electronics Center, "Proceedings, Conference on Artificial Pacemakers and Cardiac Prosthesis, September 10, 1958," unpublished mimeographed transcript (New York, N.Y.: Rockefeller Institute, 1958; copy in the Rockefeller Archive Center, North Tarrytown, N.Y.). Unless otherwise referenced, all quotations and other information come from this document. For an abbreviated version of the transcript, see "The Conference on Artificial Pacemakers and Cardiac Prosthesis, 1958," ed. Kirk Jeffrey, *Pacing Clin Electrophysiol* 16 (July 1993, pt. 1): 1445–82.

28. Victor Parsonnet and Alan D. Bernstein, "Transvenous Pacing: A Seminal Transition from the Research Laboratory," *Ann Thorac Surg* 48 (November 1989): 738–40.

29. Robert M. Berne and Matthew N. Levy, *Cardiovascular Physiology*, 6th ed. (St. Louis: Mosby Year Book, 1992), 81–93.

30. Atrial coordinated pacing would later be called P-synchronous or AV-synchronous pacing. In the terminology adopted in 1974 (ch. 9), it would be designated VAT pacing.

31. Herman K. Hellerstein, Denman Shaw, and Irving M. Liebow, "An Extracorporeal Electronic Bypass of the A-V Node," abstract in *J Lab Clin Med* 36 (December 1950): 833; Hellerstein, telephone interview by author, 30 September 1991.

32. The device Briller discussed at the Rockefeller Conference had been formally described in Stanley A. Briller, Nathan Marchand, and Charles E. Kossmann, "A Differential Vectorcardiograph," *Rev Sci Instrum* 21 (September 1950): 805–11.

33. Dryden P. Morse, interview by author, 20 September 1991, Woods Hole, Mass. In interviews and correspondence with the author, Paul Zoll, Earl Bakken, Charles Kossmann, Sam Stephenson, and Herman K. Hellerstein echoed this judgment.

34. Charles Kossman, letter to author, 10 November 1991.

35. Samuel W. Hunter reminiscences in David C. Schechter, "Background of Clinical Cardiac Electrostimulation, VII: Modern Era of Artificial Cardiac Pacemakers," *N Y State J Med* 72 (15 May 1972): 1178; Hunter interview by author, 1989.

36. Samuel W. Hunter, interview by author, 10 January 1997, Mendota Heights, Minn. (NASPE).

37. Norman A. Roth, interview by author, 29 August 1990, Shoreview, Minn. (NASPE).

38. Ibid.

39. Samuel W. Hunter et al., "A Bipolar Myocardial Electrode for Complete Heart Block," *Journal-Lancet* (Minneapolis) 79 (November 1959): 506–8.

40. Roth interview by author; Hunter interview by author, 10 January 1997; Schechter, "Background, VII": 1176–81.

41. Roth interview by author.

42. Ibid.

43. Hunter interview by author, 30 November 1989; see also Schechter, "Background, VII": 1178; and Chris Allen, "St. Paul Man's 1959 Operation Opened Age of the Pacemaker," *Minneapolis Tribune*, 15 April 1979.

44. Hunter interview by author, 10 January 1997.

45. Joel D. Howell, "Machines and Medicine: Technology Transforms the American Hospital," in *The American General Hospital: Communities and Social Contexts*, ed. Diana E. Long and Janet Golden (Ithaca, N.Y.: Cornell Univ. Press, 1989), 109–34.

46. Steven M. Spencer, "Making a Heartbeat Behave," *Saturday Evening Post* 234 (4 March 1961): 50; Mauston obituary, *Minneapolis Tribune*, 7 October 1966.

47. Hunter interview by author, 10 January 1997; Roth interview by author.

48. Roth interview by author; Hunter interviews by author.

49. Hunter interview by author, 10 January 1997; Samuel W. Hunter, "Surgical Electronic Treatment for Stokes-Adams Disease," *Geriatrics* 17 (June 1962): 379–83.

50. Spencer, "Making a Heartbeat Behave."

51. "Toward Man's Full Life" (Minneapolis: Medtronic, 1975), copy at Medtronic Information Resources Center, Fridley, Minn.; Earl E. Bakken, interview by Seymour Furman, 17 May 1996 (NASPE); Robert C. Wingrove, interview by author, 28 April 1998, St. Paul.

52. Kirk Jeffrey, "Pacing the Heart: Growth and Redefinition of a Medical Technology, 1952–1975," *T&C* 36 (July 1995): 583–624; Fye, *American Cardiology*, esp. 250–73, 301–6; American College of Cardiology/American Heart Association Task Force on Practice Guidelines (Committee on Pacemaker Implantation), "ACC/AHA Guidelines for Implantation of Cardiac Pacemakers and Antiarrhythmia Devices: Executive Summary," *Circulation* 97 (7 April 1998): 1325–35.

53. Task Force, "ACC/AHA Guidelines."

**CHAPTER FOUR. THE MULTIPLE INVENTION OF IMPLANTABLE PACEMAKERS**

1. "Sarnoff Predicts 'Disease Machine,'" *New York Times*, 11 November 1959.

2. Robert K. Merton, "Behavior Patterns of Scientists," in Robert K. Merton, *The Sociology of Science: Theoretical and Empirical Investigations* (Chicago: Univ. of Chicago Press, 1973), 331; Murray Eden, "The Engineering-Industrial Accord: Inventing the Technology of Health Care," in *The Machine at the Bedside: Strategies for Using Technology in Patient Care*, ed. Stanley J. Reiser and Michael Anbar (New York: Cambridge Univ. Press, 1984), 49–64; Stuart S. Blume, *Insight and Industry: On the Dynamics of Technological Change in Medicine* (Cambridge: MIT Press, 1992), ch. 1.

3. Wilson Greatbatch, William M. Chardack, and Andrew A. Gage, "Design Considerations Relating to Power Supplies and Physiologic Controls for Implantable Pacemakers," in *Biomedical Instrumentation*, ed. J. Payer, J. Herrick, and T. B. Weber (New York: Plenum, 1967), vol. 3, 281–89; Samuel Ruben, "Sealed Zinc-Mercuric Oxide Cells for Implantable Cardiac Pacemakers," *Ann N Y Acad Sci* 167 (30 October 1969): 627–34;

Ernest Braun and Stuart Macdonald, *Revolution in Miniature: The History and Impact of Semiconductor Electronics*, 2nd ed. (Cambridge: MIT Press, 1982).

4. E.g., Paul M. Zoll in "The Conference on Artificial Pacemakers and Cardiac Prosthesis, 1958," ed. Kirk Jeffrey, *Pacing Clin Electrophysiol* 16 (July 1993, pt. 1): 1454.

5. L. Newton Turk and William W. L. Glenn, "Cardiac Arrest: Results of Attempted Cardiac Resuscitation in 42 Cases," *N Engl J Med* 251 (11 November 1954): 795–803; William Glenn, interview by Earl E. Bakken, 22 March 1979 (*Pioneers in Pacing* video: see bibliographical note).

6. Glenn interview by Bakken, 22 March 1979; A. Mauro, L. Eisenberg, and W. W. L. Glenn, "A Review of Techniques for Stimulation of Excitable Tissue within the Body by Electromagnetic Induction," *Ann N Y Acad Sci* 118 (17 September 1964): 108; William W. L. Glenn et al., "Remote Stimulation of the Heart by Radiofrequency Transmission: Clinical Application to a Patient with Stokes-Adams Syndrome," *N Engl J Med* 261 (5 November 1959): 948–51.

7. W. D. Widmann et al., "Radio-Frequency Cardiac Pacemaker," *Ann N Y Acad Sci* 111 (11 June 1964): 1005.

8. Leonardo Cammilli, Renato Pozzi, and Guiliano Drago, "Remote Heart Stimulation by Radio Frequency for Permanent Rhythm Control in the Morgagni-Adams-Stokes Syndrome," *Surgery* 52 (November 1962): 765–76; Leonardo Cammilli, "Radio-Frequency Pacemaker with Receiver Coil Implanted on the Heart," *Ann N Y Acad Sci* 111 (11 June 1964): 1007–29; Harold Siddons and Edgar Sowton, *Cardiac Pacemakers* (Springfield, Ill.: Charles C. Thomas, 1967), 296–97. Cammilli's group abandoned RF pacemakers when lithium batteries appeared in the early 1970s: Cammilli, letter to author, 22 February 1997.

9. Leon D. Abrams et al., "A Surgical Approach to the Management of Heart-Block Using an Inductive Coupled Artificial Cardiac Pacemaker," *Lancet* i (25 June 1960): 372–74; quotation from L. D. Abrams and J. C. Norman, "Experience with Inductive Coupled Cardiac Pacemakers," *Ann N Y Acad Sci* 111 (11 June 1964): 1032.

10. Siddons and Sowton, *Cardiac Pacemakers*, 90–91; Seymour Furman, "History of Cardiac Pacing" (MS, April 1998), 15; Harold Siddons, "Long-term Artificial Cardiac Pacing: Experience in Adults with Heart Block," *Ann Roy Coll Surg Eng* 32 (1963): 13.

11. "History of Pacemaking," *Medical Electronics* 17 (June 1986): 67–71; Ingvar Karlof and Seymour Furman, "Rune Elmqvist, M.D., 1906 to 1996," *Pacing Clin Electrophysiol* 20 (April 1997, pt. 1): 1002; Patrik Hidefjäll, e-mail message to author, 7 January 1997. Hidefjäll interviewed Rune Elmqvist in October 1993.

12. Åke Senning, "Extracorporeal Circulation," in *History and Perspectives of Cardiology*, ed. H. A. Snellen, A. J. Dunning, and A. C. Arntzenius (The Hague: Leiden Univ. Press, 1981), 149–55; Åke Senning, "Cardiac Pacing in Retrospect," *Am J Surg* 145 (June 1983): 733–34; Clarence Crafoord, "Personal Memories of Cardiac Surgery and the Heart-Lung Machine," ibid., 173–75; Patrik Hidefjäll, "The Pace of Innovation: Patterns of Innovation in the Cardiac Pacemaker Industry," Ph.D. diss., Linköping University, Sweden, 1997, 86–87.

13. David C. Schechter, "Background of Clinical Cardiac Electrostimulation, VII: Modern Era of Artificial Cardiac Pacemakers," *N Y State J Med* 72 (15 May 1972): 1175; Rune Elmqvist, "Review of Early Pacemaker Development," *Pacing Clin Electrophysiol* 1 (October–December 1978): 535–36; Senning, "Cardiac Pacing in Retrospect": 734.

14. The underlying cause of Larsson's heart block remains something of a mystery. Larsson believes that a meal of raw oysters in 1955 had infected him with hepatitis: letter to author, 25 June 1991. Åke Senning, the surgeon, speaks of "hepatitis and myocarditis with a total atrioventricular block due to a virulent infection": Senning, "Cardiac Pacing in Retrospect": 735.

15. Arne Larsson, interview by author, 31 May 1991, Washington, D.C.; Senning, "Cardiac Pacing in Retrospect": 735; Hidefjäll, "Pace of Innovation," 88.

16. Senning, Comment in panel discussion on cardiac pacing, *J Thorac Cardiovasc Surg* 38 (November 1959): 639; Hidefjäll, "Pace of Innovation," 89–90.

17. Senning, "Cardiac Pacing in Retrospect": 735; Elmqvist, "Review of Early Pacemaker Development"; Harold Siddons and O'Neal Humphries, "Complete Heart Block with Stokes-Adams Attacks Treated by Indwelling Pacemaker," *Proc Roy Soc Med* 54 (1961): 237–38; Siddons, "Long-Term Artificial Cardiac Pacing": 12–13; Rune Elmqvist et al., "Artificial Pacemaker for Treatment of Adams-Stokes Syndrome and Slow Heart Rate," *Am Heart J* 65 (June 1963): 731–48; Siddons and Sowton, *Cardiac Pacemakers*, 290–95.

18. Howard A. Frank, interview by author, 16 October 1996, Boston (NASPE); Paul M. Zoll, "Development of Electric Control of Cardiac Rhythm," *J Am Med Assoc* 226 (19 November 1973): 881–86; W. H. Abelmann, "Paul M. Zoll and Electrical Stimulation of the Heart," *Clin Cardiol* 9 (March 1986): 131–35. Frank estimated that the work began in the mid-1950s; see his comments in a comment on a paper by William Chardack, *J Thorac Cardiovasc Surg* 42 (December 1961): 828.

19. Paul M. Zoll as quoted in Dwight E. Harken, "Pacemakers, Past-Makers, and the Paced: An Informal History from A to Z (Aldini to Zoll)," *Biomedical Instrumentation and Technology* 25 (July–August 1991): 319.

20. Clinical investigators thought of threshold in terms of the amplitude of the stimulus in millivolts or milliamperes, and its duration in milliseconds. If the heart is stimulated artificially through an electrode and the current required to trigger a contraction at various durations is measured, the graph of amplitude against duration is called a strength-duration curve. For all combinations of amplitude and impulse width that fall above the curve, the heart is stimulated to contract; below the curve the stimulus will not induce contraction. In general, the curve is hyperbolic in shape: the amplitude required to stimulate declines rapidly with a longer stimulus duration but then reaches a minimum, so that no advantage is gained from further prolonging the stimulus. Zoll was familiar with these concepts. Seymour Furman, David L. Hayes, and David R. Holmes Jr., *A Practice of Cardiac Pacing* (Mount Kisco, N.Y.: Futura, 1986), 44–48.

21. Frank interview by author; Siddons and Sowton, *Cardiac Pacemakers*, 169–74.

22. Paul M. Zoll et al., "Long-Term Electric Stimulation of the Heart for Stokes-Adams Disease," *Ann Surg* 154 (September 1961): 330–31.

23. Ibid.: 331–32.

24. Zoll interview by author, 5 February 1990. The procedure is formally described in Zoll et al., "Long-Term Electric Stimulation": 335. Zoll's daughter Mary Zoll told me (in a conversation, 15 October 1999, Boston) that as a teenager she had suggested to her father that he try Ivory Flakes after having seen advertisements for the product on television. The product slogan affirmed that it was "Ninety-nine and forty-four one-hundredths percent pure."

25. Zoll interview by author, 5 February 1990; Zoll et al., "Long-Term Electric Stimulation": 334–35.

26. Frank interview by author.

27. Paul M. Zoll, Arthur J. Linenthal, and Leona R. Norman Zarsky, "Ventricular Fibrillation: Treatment and Prevention by External Electric Currents," *N Engl J Med* 262 (21 January 1960): 112; Zoll et al., "Long-Term Electric Stimulation": 332–33. The family of the Montefiore patient had asked Zoll to observe the technique and give his opinion. Zoll made it clear to the family and the physicians managing the case (Seymour Furman and John B. Schwedel) that he thought the catheter pacing lead an unsound idea, but he had no alternative to offer: Seymour Furman interview by Earl E. Bakken, 14 August 1980 (*Pioneers in Pacing* video: see bibliographical note). Zoll later recalled, "I was afraid initially that [catheter pacing leads] weren't going to work and that they would break. And that they would move, and so on, but they didn't. That turned out to be a real advance. And I hope I accepted that in good faith—I was wedded before that to the thoracotomy": Zoll interview by author, 5 February 1990. But Zoll continued to express doubts about transvenous pacing into the 1970s. In 1973 he wrote that the transvenous lead was "not entirely satisfactory" and described work on an alternative technique for quickly implanting a myocardial electrode (known as the "harpoon electrode") via a small chest incision: Zoll, "Development of Electric Control."

28. The unit and details of its electrical properties are discussed in Zoll et al., "Long-Term Electric Stimulation": 333–35, and in Paul M. Zoll, Howard A. Frank, and J. Arthur Linenthal, "Implantable Cardiac Pacemakers," *Ann N Y Acad Sci* 111 (11 June 1964): 1068–74.

29. Arthur J. Linenthal, interview by author, 21 September 1991, Brookline, Mass. (NASPE).

30. Wilson Greatbatch, interview by author, 24 June 1997, Tonawanda, N.Y.; see also the interview in Kenneth A. Brown, *Inventors at Work* (Redmond, Wash.: Tempus Books, 1988), 19–44; Wilson Greatbatch, "Biomedical Engineering in the Early U.S. Aerospace Program," *Aviation, Space, and Environmental Medicine* 60 (August 1989): 811–14; and John A. Adam, "Wilson Greatbatch," *IEEE Spectrum* 32 (March 1995): 56–61.

31. Wilson Greatbatch, "Twenty-Five Years of Pacemaking," *Pacing Clin Electrophysiol* 7 (January–February 1984): 143. The blocking oscillator circuit had been invented at the MIT Radiation Laboratory during World War II. Greatbatch recalls that "the 'RADLAB' series was well known to all my generation of engineers and was our 'Bible'": Greatbatch to author, 2 March 1990.

32. What I have given is a plausible reconstruction of how Greatbatch may have come to the idea of an implantable pacemaker. Greatbatch himself has written that the idea of pacing the heart for chronic heart block occurred to him during the summer of 1951, as a pair of brain surgeons described the problem to him over a brown-bag lunch on the Cornell farm campus: Greatbatch, "Twenty-Five Years of Pacemaking": 143. As I showed in chapter 1, physiologists and clinicians had developed a "modern" understanding of the conduction disease known as heart block by World War I. But the problem of actually treating heart block attracted little attention until after Paul Zoll's invention of the external pacemaker in 1952 and the beginnings of open-heart surgery in 1953—and even then, the problem was first conceived in terms of short-term rather than chronic stimulation. The accumulation of knowledge about chronic stimulation of the heart really began only in the years 1958–60.

33. William M. Chardack, letters to author, various dates.

34. William M. Chardack, untitled MS (May 1994) in possession of the author; Wil-

liam M. Chardack, interviews by author, 9 December and 12 December 1996, Gulf-stream, Florida.

35. Chardack MS, May 1994; Chardack interviews by author; William M. Chardack, "Recollections — 1958–1961," *Pacing Clin Electrophysiol* 4 (September–October 1981): 592. I place the pair's first meeting in February 1958. Chardack has been credited with responding to Greatbatch, "If you can do that, you can save ten thousand lives a year": Greatbatch, "Twenty-Five Years of Pacemaking": 143. In light of Chardack's desire to review the literature on heart block, this seems implausible. Moreover, in 1958 no one had even an approximate idea of the annual number of cases of chronic complete heart block in the United States.

36. George B. Penton, Harold Miller, and Samuel A. Levine, "Some Clinical Features of Complete Heart Block," *Circulation* 13 (June 1956): 801–24. Chardack does not recall this paper by name, but it seems probable that he read it. Certainly he recalls the principal findings.

37. Chardack, "Recollections": 592.

38. Ibid.: 593; Chardack MS, May 1994; Chardack interviews by author.

39. A breadboard is a circuit board on which an experimental electrical circuit is assembled. Greatbatch tried germanium transistors in pacemaker models 1 and 3, then stayed with silicon.

40. I will refer to this system as the mercury cell. A battery consists of a number of cells connected together to function as a source of DC (direct current) electrical energy and housed in a common container.

41. Representatives from Mallory assured Greatbatch (and other pacemaker inventors) that the mercury-cell battery would last five years. For Greatbatch's comments about adoption of the mercury cell, I have used Wilson Greatbatch, *Implantable Active Devices* (Clarence, N.Y.: Greatbatch Enterprises, 1983), 4:1.

42. Greatbatch laboratory notebook, 60–62, 123, copy courtesy of William M. Chardack; Greatbatch, *Implantable Active Devices*, 2:2; Chardack, "Recollections": 593; Chardack MS, May 1994; Chardack interviews by author.

43. Greatbatch laboratory notebook, 62. The pacemaker was still functioning when, after 24 hours, the dog died from the effects of the operation to produce heart block.

44. Greatbatch, *Implantable Active Devices*, 2:2.

45. Ibid., 2:3; Chardack, "Recollections": 593; Greatbatch lab notebook, 123.

46. Wilson Greatbatch and William M. Chardack, "A Transistorized, Implantable Pacemaker for the Correction of Complete Atrioventricular Block," *NEREM 1959 Record* [New England Radio and Engineering meeting], 8–9, 113; Greatbatch, "Implantable Pacemakers": 25.

47. Chardack interview by author, 9 December 1996. Chardack added, "That's how we got on [Medtronic's mailing] list."

48. Chardack, "Recollections": 594–95; Greatbatch and Chardack, "A Transistorized Implantable Pacemaker"; Chardack interviews by author. See also "New Pacemaker for Heart Tested," *New York Times*, 6 November 1959.

49. Greatbatch lab notebook, 123; Greatbatch and Chardack, "A Transistorized Implantable Pacemaker"; William M. Chardack, Andrew A. Gage, and Wilson Greatbatch, "Experimental Observations and Clinical Experiences with the Correction of Complete Heart Block by an Implantable Self-Contained Pacemaker," *Trans Am Soc Artif Int Org* 3 (1961): 287.

50. Chardack interview by author, 12 December 1996.

51. Chardack, "Recollections": 595.

52. Ibid.: 595; William M. Chardack, Andrew A. Gage, and Wilson Greatbatch, "A Transistorized, Self-Contained, Implantable Pacemaker for the Long-Term Correction of Complete Heart Block," *Surgery* 48 (October 1960): 643-54; Kirk Jeffrey, "Many Paths to the Pacemaker," *Invention & Technology* 12 (spring 1997): 2-11. A profile of Henefelt appears in Steven M. Spencer, "Making a Heartbeat Behave," *Saturday Evening Post* 234 (4 March 1961): 14, 50.

53. Chardack, Gage, and Greatbatch, "A Transistorized, Self-Contained, Implantable Pacemaker"; Wilson Greatbatch, "Medical Cardiac Pacemaker," U.S. patent no. 3,057,356, application filed 22 July 1960, awarded 9 October 1962. Chardack's name did not appear on the application because he was an employee of the U.S. government as a surgeon for the Veterans Administration.

54. Brown, *Inventors at Work*, 20; Medtronic, *Pulse*, March 1984, cover.

55. William M. Chardack et al., "Clinical Experience with an Implantable Pacemaker," *Ann NY Acad Sci* 111 (11 June 1964): 1088; see also the panel discussion that followed this paper, esp. 1105.

56. Ibid.: 1076; Zoll et al., "Long-Term Electric Stimulation": 338-43; Annetine Gelijns and Nathan Rosenberg, "The Dynamics of Technological Change in Medicine," *Health Affairs* 13 (summer 1994): 31.

57. Alan D. Bernstein and Victor Parsonnet, "Survey of Cardiac Pacing and Defibrillation in the United States in 1993," *Am J Cardiol* 78 (15 July 1996): 190.

**CHAPTER FIVE. MAKING THE PACEMAKER SAFE AND RELIABLE**

1. Kirk Jeffrey and Victor Parsonnet, "Cardiac Pacing 1960-1985: A Quarter-Century of Medical and Industrial Innovation," *Circulation* 97 (19 May 1998): 1979; Parsonnet letter to author, 31 March 1997.

2. Wilson Greatbatch, *Implantable Active Devices. 2: Pulse Generator Design* (Clarence, N.Y.: Greatbatch Enterprises, 1983); William M. Chardack, "Heart-Block Treated with an Implantable Pacemaker: Past Experience and Current Developments," in *Resuscitation and Cardiac Pacing*, ed. Gavin Shaw, George Smith, and Thomas J. Thomson (Philadelphia: Davis, 1965), 213-49.

3. Annetine Gelijns and Nathan Rosenberg, "The Dynamics of Technological Change in Medicine," *Health Affairs* 13 (summer 1994): 31; William M. Chardack, Andrew A. Gage, and Wilson Greatbatch, "A Transistorized, Self-Contained, Implantable Pacemaker for the Long-Term Correction of Complete Heart Block," *Surgery* 48 (October 1960): 645.

4. Victor Parsonnet, "Permanent Transvenous Pacing in 1962," *Pacing Clin Electrophysiol* 1 (April-June 1978): 265; Victor Parsonnet, "Complications of the Implanted Pacemaker: A Scheme for Determining the Cause of the Defect and Methods for Correction," *J Thorac Cardiovasc Surg* 45 (June 1963): 801-12; Victor Parsonnet, "Permanent Pacemaker Insertion: A Five-Year Appraisal," *Ann Thorac Surg* 2 (June 1966): 561-75; Adrian Kantrowitz, "Implantable Cardiac Pacemakers," *Ann N Y Acad Sci* 111 (11 June 1964): 1063-64; "Battery of Problems Complicate Life of Pacemaker," *Med World News* 7 (14 October 1966): 62-63.

5. John B. McKinlay, "From 'Promising Report' to 'Standard Procedure': Seven Stages in the Career of a Medical Innovation," in *Technology and the Future of Health Care*,

ed. John B. McKinlay (Cambridge: MIT Press, 1982), 233–70; Edward B. Roberts, "Technological Innovation and Medical Devices," in *New Medical Devices: Invention, Development, and Use*, ed. Karen B. Ekelman (Washington, D.C.: National Academy Press, 1988), 35–47.

6. William M. Chardack, "A Myocardial Electrode for Long-Term Pacemaking," *Ann N Y Acad Sci* 111 (11 June 1964): 895–96.

7. Chardack interview by author, 12 December 1996, Gulfstream, Florida.

8. Chardack, "A Myocardial Electrode"; Chardack interview by author, 12 December 1996.

9. William M. Chardack, Andrew A. Gage, and Wilson Greatbatch, "Correction of Complete Heart Block by a Self-Contained and Subcutaneously Implanted Pacemaker: Clinical Experience with 15 Patients," *J Thorac Cardiovasc Surg* 42 (December 1961): 814–25; see esp. fig. 9; Chardack, "A Myocardial Electrode": 894; Earl E. Bakken, interview by author, 23 May 1990, Fridley, Minn.

10. Aaron E. Parsonnet and Albert S. Hyman, *Applied Electrocardiography* (New York: Macmillan, 1929).

11. Marvin C. Becker et al., "An Implantable Cardiac Pacemaker for Complete Heart Block," *Journal of the Medical Society of New Jersey* 58 (October 1961): 490–93; Parsonnet, interview by author, 4 December 1995, Newark, N.J.

12. Dryden P. Morse, interview by Parsonnet, 16 February 1997, Moorestown, N.J. (NASPE); Dryden P. Morse, interview by author, 20 September 1991, Woods Hole, Mass.

13. Parsonnet interview by author, 1995; George H. Myers and Victor Parsonnet, "Bio-Medical Engineering, Current Applications at the Newark Beth Israel Hospital," *Journal of the Newark Beth Israel Hospital* 15 (April 1964): 88–92.

14. Victor Parsonnet, "A Decade of Permanent Pacing of the Heart," *Cardiovasc Clin* 2 (1970): 195; Wilson Greatbatch, "Achieving Reliable Pacemakers," in *Proceedings of the Vth International Symposium, Tokyo, March 14–18, 1976*, ed. Yoshio Watanabe (Amsterdam: Excerpta Medica, 1977), 364–68.

15. Samuel Ruben, "Sealed Zinc Mercuric-Oxide Cells," *Ann N Y Acad Sci* 167 (30 October 1969): 627–44; Victor Parsonnet, "Power Sources for Implantable Cardiac Pacemakers," *Chest* 61 (February 1972): 165–73; Victor Parsonnet et al., "The Potentiality of the Use of Biologic Energy as a Power Source for Implantable Pacemakers," *Ann N Y Acad Sci* 111 (11 June 1964): 915–21; Victor Parsonnet, interview by Earl E. Bakken, 21 July 1977 (*Pioneers in Pacing* video: see bibliographical note).

16. Parsonnet interview by Bakken.

17. Victor Parsonnet et al., "A Cardiac Pacemaker Using Biologic Energy Sources," *Trans Am Soc Artif Int Org* 9 (1963): 174–77.

18. Gerhard Lewin et al., "An Improved Biological Power Source for Cardiac Pacemakers," *Trans Am Soc Artif Int Org* 14 (1968): 215–19; I. Richard Zucker et al., "Self-Energized Pacemakers: The Possibilities of Using Biological Energy Sources," *Circulation Supplement* 29 (April 1964): 157–60; Parsonnet interview by Bakken; George H. Myers, comment in panel discussion, *Ann N Y Acad Sci* 167 (30 October 1969): 678.

19. Thomas F. Hursen and Steve A. Kolenik, "Nuclear Energy Sources," *Ann N Y Acad Sci* 167 (30 October 1969): 661–73; Parsonnet interview by author; Victor Parsonnet, "Cardiac Pacing and Pacemakers VII: Power Sources for Pacemakers," *Am Heart J* 94 (October–November 1977): 517–28, 658–64; Edgar Sowton, "Energy Sources for

Pacemakers," in *Proceedings of the Vth International Symposium*, ed. Watanabe, 438–46; Paul Laurens, "Nuclear-Powered Pacemakers: An Eight-Year Clinical Experience," *Pacing Clin Electrophysiol* 2 (May–June 1979): 356–60.

20. Sowton, "Energy Sources," compares the various nuclear units.

21. Harold M. Schmeck Jr., "9 U.S. Nuclear Heart Devices Implanted," *New York Times*, 10 April 1973; Seymour Furman, "Nuclear Pacemakers," editorial in *Pacing Clin Electrophysiol* 2 (March–April 1979): 135–36.

22. *Encyclopedia of Medical Devices and Instrumentation* (1988), s.v., "Pacemakers," by Wilson Greatbatch. For physicians' concerns about safety, see the panel discussion in *Ann N Y Acad Sci* 167 (30 October 1969): 674–78.

23. Furman, "Nuclear Pacemakers": 135.

24. Greatbatch "Pacemakers" article; Seymour Furman, "Controversies in Cardiac Pacing," *Cardiovasc Clin* 8 (1977): 305.

25. William M. Chardack, "Cardiac Pacemakers and Heart Block," in *Surgery of the Chest*, ed. John H. Gibbon Jr., David C. Sabistan, and Frank C. Spencer (Philadelphia: Saunders, 1969), 837, reporting on a study from 1967 that had reviewed many large series.

26. Richard Sutton and Ivan Bourgeois, *The Foundations of Cardiac Pacing, Part 1* (Mount Kisco, N.Y.: Futura, 1991), 177–215.

27. Seymour Furman and George Robinson, "Use of an Intracardiac Pacemaker in the Correction of Total Heart Block," *Surg Forum* 9 (1958): 245–48; Seymour Furman, "Recollections of the Beginnings of Transvenous Cardiac Pacing," *Pacing Clin Electrophysiol* 17 (October 1994): 1697–705.

28. Seymour Furman, interview by author, 8 May 1998, San Diego; Furman, "Recollections": 1698.

29. Furman, "Recollections": 1698.

30. Ibid., 1699; Furman interview by Bakken.

31. Furman interview by Bakken.

32. Seymour Furman and John B. Schwedel, "An Intracardiac Pacemaker for Stokes-Adams Seizures," *N Engl J Med* 261 (5 November 1959): 943–48; "Electrode in Heart Saves Man's Life," *New York Times*, 27 November 1958.

33. Furman interview by Bakken.

34. Furman et al., "Use of an Intracardiac Pacemaker"; Furman letter to author, 14 January 1997.

35. Furman interview by Bakken.

36. Seymour Furman, "Recollections": 1702; Furman interview by author; "Patient Carries Heart Regulator," *New York Times*, 23 June 1959; "Living Minute-to-Minute," *Newsweek* 54 (6 July 1959): 54.

37. John B. Schwedel and Doris J. W. Escher, "Transvenous Electrical Stimulation of the Heart—I," *Ann N Y Acad Sci* 111 (11 June 1964): 972–80; Seymour Furman et al., "Transvenous Pacing: A Seven-Year Review," *Am Heart J* 71 (March 1966): 408–16.

38. Seymour Furman et al., "Implanted Transvenous Pacemakers: Equipment, Technic, and Clinical Experience," *Ann Surg* 164 (September 1966): 465–74; Joel D. Howell, "The Changing Face of Twentieth-Century American Cardiology," *Ann Intern Med* 105 (November 1986): 772–82.

39. D. S. Abelson et al., "Endocardial Pacemaking and Insertion of a Permanent Internal Cardiac Pacemaker," *N Engl J Med* 625 (25 October 1961): 792–94; Victor

Parsonnet et al., "An Intracardiac Bipolar Electrode for Interim Treatment of Complete Heart Block," *Am J Cardiol* 10 (August 1962): 261–65; William M. Chardack, "Heart Block Treated with an Implantable Pacemaker," *Prog Cardiovasc Dis* 6 (May 1964): 517; " 'Intravenous' Cardiac Pacemaking," editorial in *J Am Med Assoc* 184 (18 May 1963): 582–83.

40. Hans Lagergren and L. Johannson, "Intracardiac Stimulation for Complete Heart Block," *Acta Chir Scand* 125 (1963): 562–66; Hans Lagergren et al., "One Hundred Cases of Treatment for Adams-Stokes Syndrome with Permanent Intravenous Pacemaker," *J Thorac Cardiovasc Surg* 5 (November 1965): 710–14; Hans Lagergren, "How It Happened: My Recollection of Early Pacing," *Pacing Clin Electrophysiol* 1 (January–April 1978): 140–43.

41. Harold Siddons and J. G. Davies, "A New Technique for Internal Cardiac Pacing," *Lancet* 2 (7 December 1963): 1204; Alan Harris et al., "The Management of Heart Block," *Br Heart J* 27 (July 1965): 469–82; Rodney Bluestone et al., "Long-term Endocardial Pacing for Heart-Block," *Lancet* 2 (14 August 1965): 307–12; "Medtronic, Inc. Pulse Generator Reference Chart," n.d. [1970s], copy in author's possession.

42. Furman to author, 14 January 1997; Sutton and Bourgeois, *Foundations of Cardiac Pacing*, 184–86. Siddons, Lagergren, and Parsonnet had used the cephalic vein as early as 1961.

43. Furman et al., "Implanted Transvenous Pacemakers": 465–73.

44. Chardack, "Cardiac Pacemakers and Heart Block," 837; Victor Parsonnet, "The Status of Permanent Pacing of the Heart in the United States and Canada," *Ann Cardiol Angéiol* 20 (1971): 287–91, at 289. Parsonnet estimated that in more than 90 percent of primary implantations in 1969, the physician used the transvenous technique.

45. Victor Parsonnet, "A Survey of Cardiac Pacing in the United States and Canada," *Proceedings of the IVth International Symposium on Cardiac Pacing*, ed. Hilbert J. T. Thalen (Assen, Netherlands: Van Gorcum, 1973), 44.

46. Victor Parsonnet and Alan D. Bernstein, "Transvenous Pacing: A Seminal Transition from the Research Laboratory," *Ann Thorac Surg* 48 (November 1989): 738–40; Benjamin J. Scherlag et al., "Catheter Technique for Recording His Bundle Activity in Man," *Circulation* 39 (January 1969): 13–18; Douglas P. Zipes, "The Contribution of Artificial Pacemaking to Understanding the Pathogenesis of Arrhythmias," *Am J Cardiol* 28 (August 1971): 211–22; James A. Reiffel and J. Thomas Bigger Jr., "Current Status of Direct Recordings of the Sinus Node Electrogram in Man," *Pacing Clin Electrophysiol* 6 (September–October 1983, pt. 2): 1143–50; W. Bruce Fye, *American Cardiology: The History of a Specialty and Its College* (Baltimore: Johns Hopkins Univ. Press, 1996), 305–6.

47. Parsonnet et al., "An Intracardiac Bipolar Electrode."

48. Parsonnet, "Permanent Transvenous Pacing in 1962."

49. Seymour Furman et al., "Complications of Pacemaker Therapy for Heart Block" [round-table discussion], *Am J Cardiol* 17 (March 1966): 440.

50. Parsonnet et al., "Complications of the Implanted Pacemaker."

51. Seymour Furman, letter to author, 24 June 1997.

52. Wesley and Helen Johnson, interview by author, 21 January 1991, St. Paul, Minnesota.

53. Samuel W. Hunter in Johnson's medical record, 26 September 1969, used with permission of Wesley Johnson; Louis Lemberg, interview by author, 10 December 1996,

Miami, Fla. Johnson's pacemaker fired at a fixed rate, but his ventricles squeezed more vigorously in times of intense physical activity. He and his wife continued downhill skiing until the early 1990s.

54. Thomas P. Hughes discusses medical procedures as technological systems in "Machines and Medicine: A Projection of Analogies between Electric Power Systems and Health Care Systems," *Int J Tech Assess in Health Care* 1 (1985): 285–95.

55. Victor Parsonnet et al., "A Nonpolarizing Electrode for Endocardial Stimulation of the Heart," *J Thorac Cardiovasc Surg* 56 (November 1968): 710–16; Seymour Furman et al., "Endocardial Threshold of Cardiac Response as a Function of Electrode Surface Area," *J Surg Res* 8 (April 1968): 161–66; Seymour Furman et al., "Energy Considerations for the Cardiac Stimulation as a Function of Pulse Duration," ibid., 6 (October 1966): 441–46.

56. Seymour Furman, letter to author, 27 June 1997; Seymour Furman, Doris J. W. Escher, and Bryan Parker, "The Pacemaker Follow-up Clinic," *Prog Cardiovasc Dis* 14 (March 1972): 515–30.

57. Rodney Bluestone, Alan Harris, and Geoffrey Davies, "Aftercare of Artificially Paced Patients," *Br Med J* i (19 June 1965): 1589–92; Edgar Sowton, "Detection of Impending Pacemaker Failure," *Israel J Med Sci* 3 (March–April 1967): 260–69; Victor Parsonnet et al., "Prediction of Impending Pacemaker Failure in a Pacemaker Clinic," *Am J Cardiol* 25 (March 1970): 311–19.

58. Victor Parsonnet to author, 1 April 1997; Parsonnet et al., "Prediction of Impending Pacemaker Failure."

59. Seymour Furman et al., "Electronic Analysis for Pacemaker Failure," *Ann Thorac Surg* 8 (July 1969): 57–65; Furman et al., "Instruments for Evaluating Function of Cardiac Pacemakers," *Med Res Eng* 6 (third quarter 1967): 29–32; Furman to author, 24 June 1997.

60. Seymour Furman, Bryan Parker, and Doris J. W. Escher, "Transtelephone Pacemaker Clinic," *J Thorac Cardiovasc Surg* 61 (May 1971): 827–34; Furman to author, 24 June 1997.

61. Seymour Furman, "Transtelephone Observation of Implanted Cardiac Pacemakers," *Med Instrum* 7 (May–August 1973): 196–202; Furman to author, 24 June 1997; Parsonnet to author, 1 April 1997; Parsonnet et al., "Prediction of Impending Pacemaker Failure."

62. "Fixed Rate Implanted Pacemaking" [panel discussion], *Israel J Med Sci* 3 (March–April 1967): 270, 273; Parsonnet, "Decade of Permanent Pacing": 187.

63. For a theoretical perspective on such activities, see Edward B. Roberts, "The Development of Biomedical Technologies," in *Critical Issues in Medical Technology*, ed. Barbara J. McNeil and Ernest G. Cravalho (Boston: Auburn House, 1982), 3–22.

64. Edgar Sowton, "Artificial Pacemaking and Sinus Rhythm," *Br Heart J* 27 (May 1965): 311–18.

65. William M. Chardack et al., "The Long-Term Treatment of Heart Block," *Prog Cardiovasc Dis* 9 (September 1966): 123; cf. William M. Chardack et al., "Pacing and Ventricular Fibrillation," *Ann N Y Acad Sci* 167 (30 October 1969): 919–33; Zoll in "Fixed Rate Implanted Pacemaking": 270; I. Richard Zucker et al., "Competitive Idiocardiac and Extrinsic Pacemaker Stimuli in Heart Block," *Am Heart J* 69 (January 1965): 67.

66. Agustin Castellanos Jr. et al., "Repetitive Firing Occurring during Synchronized

Electrical Stimulation of the Heart," *J Thorac Cardiovasc Surg* 51 (March 1966): 334–40; Michael Bilitch, R. S. Crosby, and E. A. Cafferty, "Ventricular Fibrillation and Competitive Pacing," *N Engl J Med* 276 (16 March 1967): 598–604; Leonard S. Dreifus et al., "The Advantages of Demand over Fixed-Rate Pacing," *Dis Chest* 54 (August 1968): 86–89; Chardack et al., "Pacing and Ventricular Fibrillation"; Barouh V. Berkovits, interview by author, 7 May 1993, San Diego, Calif.; Lemberg interview by author.

67. Philip Samet et al., "Hemodynamic Sequelae of Idioventricular Pacing in Complete Heart Block," *Am J Cardiol* 11 (May 1963): 594–99; Sowton, "Artificial Pacemaking and Sinus Rhythm"; "Fixed Rate Implanted Pacemaking"; Victor Parsonnet, interview by author, 5 December 1995, Newark, N.J.; Lemberg interview by author.

68. Louis Lemberg, Agustin Castellanos Jr., and Arthur Gosselin, "Programmed Interval Pacemaking in Atrioventricular (A-V) Block," abstract in *Circulation* 29–30 (October 1964): supp. 3, 111.

69. Barouh V. Berkovits, interview by Bakken, 13 April 1978 (*Pioneers in Pacing*).

70. Barouh V. Berkovits, "Demand Pacemaker," Graduate Symposium on Electrical Control of Cardiac Activity, Buffalo, N.Y., 17–19 May 1965 (unpublished paper in author's possession); George H. Myers and Victor Parsonnet, *Engineering in the Heart and Blood Vessels* (New York, 1969), 34–49; Bryan Parker, "Pacemaker Electronics," in Seymour Furman and Doris J. W. Escher, *Principles and Techniques of Cardiac Pacing* (New York: Harper & Row, 1970), 43–61.

71. Aubrey Leatham, Peter Cook, and J. G. Davies, "External Electric Stimulator for Treatment of Ventricular Standstill," *Lancet* 2 (8 December 1956): 1188; J. Geoffrey Davies, interview by Seymour Furman, 23 June 1996, London (NASPE); Fred Zacouto, "Cardiorythmeur Intracorporel à Inhibition Externe," abstract in *Archives des Maladies du Coeur* 56 (November 1963): 1296; H. J. Sykosch et al., "Zur Therapie mit elektrischen Schrittmachern: Ein implantierbarer, induktiv ausschaltbarer elektrischer Schrittmacher," *Elektromedizin* 8 (September 1963): 139–42; H. J. Sykosch, "Implantierbare Schrittmacher zur permanenten und intermittierenden Stimulierung des Herzens," *Langenbecks Archiv für Klinische Chirurgie* 30 (1964): 288–92. Zacouto's design was awarded a French patent in 1962.

72. *Merriam Webster's Medical Desk Dictionary* (1993), s.v., "Physiological."

73. Barouh V. Berkovits, "Demand Pacing," *Ann N Y Acad Sci* 167 (30 October 1969): 891–95; "A Modern Pioneer's Perspective—an Interview with Barouh V. Berkovits," *Medtronic News* 9 (September 1979): 9; Berkovits interview by author; Seymour Furman, "Physiologic Pacing," editorial in *Pacing Clin Electrophysiol* 3 (November–December 1980): 639–40.

74. Joel D. Howell, "Cardiac Physiology and Clinical Medicine? Two Case Studies," in *Physiology in the American Context, 1850–1940*, ed. Gerald L. Geison (Bethesda, Md.: American Physiological Society, 1987), 279–92; Richard Sutton, John Perrins, and Paul Citron, "Physiological Cardiac Pacing," *Pacing Clin Electrophysiol* 3 (March–April 1980): 207–19.

75. Lemberg interview by author.

76. Lemberg interview by author; Dwight E. Harken to Barouh V. Berkovits, 20 July 1966, copy in author's possession; Walter Zuckerman et al., "Clinical Experiences with a New Implantable Demand Pacemaker," *Am J Cardiol* 20 (August 1967): 232–38; Dwight E. Harken, interview by author, 19 September 1991, Cambridge, Mass.

77. American Optical Co. annual report, 1964, copy at Baker Library, Harvard Busi-

ness School, Cambridge; U.S. patent no. 3,345,990, "Heart Beat Pacing Apparatus," application filed 19 June 1964, awarded 10 October 1997 to Barouh V. Berkovits; "Fail-Safe Pacemaker," *Med World News* 9 (30 August 1968): 42.

78. Louis Lemberg, Agustin Castellanos Jr., and Barouh V. Berkovits, "Pacemaking on Demand in AV Block," *J Am Med Assoc* 191 (4 January 1965): 106–8; Parsonnet, "Status of Permanent Pacing."

79. Dreifus et al., "Advantages of Demand"; Furman and Escher, *Principles and Techniques*, 136–53; Berkovits interview by author.

### CHAPTER SIX. THE INDUSTRIALIZATION OF THE PACEMAKER

1. Patrik Hidefjäll, "The Pace of Innovation: Patterns of Innovation in the Cardiac Pacemaker Industry," Ph.D. diss., Linköping University, Sweden, 1997, 127–28.

2. Michael L. Tushman and Philip Anderson, "Technological Discontinuities and Organizational Environments," *Administrative Science Quarterly* 31 (September 1986): 439–65.

3. Such discussions are remarkable for their absence at national conferences of cardiac-pacing physicians: e.g., "Advances in Cardiac Pacemakers," *Ann N Y Acad Sci* 167 (30 October 1969): 515–1075. By contrast, discussion of national health policy and its implications for cardiac pacing is ubiquitous at medical meetings, in specialty journals, and in textbooks today.

4. Edward B. Roberts, "Technological Innovation and Medical Devices," in *New Medical Devices: Invention, Development, and Use*, ed. Karen B. Ekelman (Washington, D.C.: National Academy Press, 1988), 35–47.

5. Hidefjäll, "Pace of Innovation," app. B.

6. Earl E. Bakken, interview by author, 23 May 1990, Fridley, Minn.; Medtronic company history, 1974 (Medtronic Information Resources Center, Fridley, Minn.). Hermundslie and Bakken were married to sisters. A diabetic, Hermundslie died in 1970.

7. Earl E. Bakken, interview by William W. Swanson, 22 December 1992 (Bakken Library and Museum, Minneapolis); Medtronic company history; Medtronic annual report, 1960.

8. Bakken interview by Swanson; Medtronic company history; Chardack interview by author, 9 December 1996, Gulfstream, Florida; Thomas E. Holloran, interview by author, 4 March 1997, Minneapolis; Medtronic annual report, 1960.

9. Bakken interview by author; Holloran interview by author.

10. Medtronic company history; Bakken interviews by author and by Swanson; Thomas E. Holloran, talk at the Bakken Library and Museum, 17 December 1996; Medtronic annual reports, 1960–63.

11. Medtronic annual report, 1964; company history.

12. Medtronic annual reports, 1962, 1963; Art Detman, "The Reluctant Millionaire," *Dun's* 103 (March 1974): 16; company history; Holloran talk; Holloran interview by author.

13. Michael Toffoli, interview by author and Patrik Hidefjäll, 27 November 1995, Fridley, Minn.

14. Robert Wingrove, interview by author, 28 April 1998, New Brighton, Minn.

15. Wilson Greatbatch, interview by Seymour Furman, 16 February 1996, Bronx, N.Y. (NASPE); Ronald T. Hagenson, telephone interview by author, 12 May 1997.

16. Norman Dann, talk at the Bakken Library and Museum, 20 November 1996.

17. Charles F. Cuddihy Jr., interview by author, 22 July 1996, Minnetonka, Minn.

18. Medtronic annual reports, 1967, 1968; 1974 company history; Holloran talk; Seymour Furman, "Pacemaker Practices in the United States," editorial in *Pacing Clin Electrophysiol* 5 (November–December 1982): 791–92. Chapter 7 and later chapters look further at the role of sales reps in the field of cardiac pacing.

19. Cuddihy interview by author; Earl E. Bakken, "Origin of the Mission Statement," in Thomas E. Holloran, "Earl E. Bakken: Medtronic and Its Mission" (St. Paul: Univ. of St. Thomas Graduate School of Business, May 1991), 12. Honeywell, a corporation based in Minneapolis until the 1990s, was and is a large manufacturer of electrical equipment.

20. "Medtronic Mission," Medtronic 1996 annual report, 54.

21. Daniel R. Denison, *Corporate Culture and Organizational Effectiveness* (New York: Wiley, 1990), 98; Holloran talk.

22. Holloran interview by author; Denison, *Corporate Culture*, 98–100. See also James C. Collins and Jerry I. Porras, *Built to Last* (New York: HarperCollins, 1994).

23. William P. Murphy Jr., interview by Victor Parsonnet, 4 February 1997, Miami (NASPE); Peter Tarjan, interview by author, 11 December 1996, Miami; J. Walter Keller, interview by Earl E. Bakken, 19 October 1976 (*Pioneers in Pacing* video: see bibliographical note).

24. Keller interview by Bakken; J. Walter Keller, interview by author, 11 December 1996, Jupiter Hills, Fla.; Murphy interview by Parsonnet.

25. Keller interview by Bakken; David A. Nathan, interview by author, 10 December 1996, Miami Beach.

26. David A. Nathan and Sol Center, interview by Earl E. Bakken, 17 November 1976 (*Pioneers in Pacing*); Murphy interview by Parsonnet; David A. Nathan et al., "An Implantable Synchronous Pacemaker for the Long Term Correction of Complete Heart Block," *Circulation* 27 (April 1963): 682–85; David A. Nathan et al., "The Application of an Implantable Synchronous Pacer for the Correction of Stokes-Adams Attacks," *Ann N Y Acad Sci* 111 (11 June 1964): 1093–1104.

27. Keller interview by author.

28. Murphy interview, quoting interviewer Parsonnet; A. M. Albisser, "Functional Anatomy of Selected Medical Instruments," in *Medical Engineering*, ed. Charles D. Ray (Chicago: Year Book Publishers, 1974), 1042–43; Andrew G. Morrow, "Electrical Stimulation of the Heart: Technics and Instruments for the Management of Complete Heart Block," *Am J Med* 37 (November 1964): 754–63.

29. David A. Nathan in panel discussion, *Ann N Y Acad Sci* 111 (11 June 1964): 1107; Seymour Furman, "Therapeutic Uses of Atrial Pacing," *Am Heart J* 73 (December 1973): 835–40.

30. *Moody's OTC Industrial Manual* (New York: Moody's Investment, 1971), 321; Medtronic annual report, 1970.

31. Keller interview by author; Tarjan interview by author; William Jakobi, interview by author, 12 December 1996, Miami; Murphy interview, quoting interviewer Parsonnet.

32. Alan Belgard, telephone interview by author, 24 January 1997.

33. Belgard telephone interview by author.

34. Adrian Kantrowitz, "Implantable Cardiac Pacemakers," *Ann N Y Acad Sci* 111 (11 June 1964): 1049–67; Adrian Kantrowitz, interview by author, 14 May 1999, Toronto (NASPE); Morrow, "Electrical Stimulation of the Heart": 758.

35. Kantrowitz, "Implantable Cardiac Pacemakers": 1053.

36. American Optical annual reports, 1958–62 (Baker Library, Harvard Business School, Cambridge); Bernard Lown, " 'Cardioversion' of Arrhythmias," *Modern Concepts of Cardiovascular Disease* 33 (July–August, 1964): 863–68, 869–73.

37. Robert Witt, "The Medical Stocks," *Corporate Report of the Ninth Federal Reserve District*, May 1973: 11–12; Michael P. Moore, "The Genesis of Minnesota's Medical Alley," *University of Minnesota Medical Bulletin*, winter 1992: 7–13; David J. Rhees, " 'Medical Alley': The Origins of the Medical Device Industry in Minnesota," paper presented at the annual meeting of the Society for the History of Technology, London, 4 August 1996; Holloran talk.

38. "Medical Electronics" (Palo Alto, Calif.: Creative Strategies, industry report, 1973), 46; Medtronic annual report, 1973; Earl E. Bakken, "To Our Shareholders, Employees and Friends," Medtronic annual report, 1968.

39. Keller interview by author.

40. Tarjan interview by author.

41. "Cardiac Pacemaker Industry" (Minneapolis: Dain, Kalman, & Quail, industry report, 3 June 1975); Thomas E. Holloran, "Pacemaker Market," Medtronic annual report, 1975, 21–22.

42. Calculated from data in Medtronic annual reports, 1974 and 1975.

43. These numbers represent revisions of estimates I published in "Pacing the Heart: Growth and Redefinition of a Medical Technology, 1952–1975," *T&C* 36 (July 1995): 609; for implants in 1975, "Report on the Pacemaker Industry" (New York: Bernstein, industry report, 18 September 1979), 19. This figure is compatible with my estimate that Medtronic sold 44,800 pacemakers in the United States in that year and held a 60 percent share of the domestic market.

44. W. Bruce Fye, *American Cardiology: The History of a Specialty and Its College* (Baltimore: Johns Hopkins Univ. Press, 1996), ch. 8; "Does the U.S. Need 80,000 Coronary Angiograms a Day?" *Med World News* 15 (20 September 1974): 14–16; Eugene D. Robin, "The Cult of the Swan-Ganz Catheter: Overuse and Abuse of Pulmonary Flow Catheters," *Ann Intern Med* 103 (September 1985): 445–49.

45. Victor Parsonnet, "The Status of Permanent Pacing of the Heart in the United States and Canada," *Ann Cardiol Angéiol* 20 (1971): 290.

46. Rosemary Stevens, *In Sickness and in Wealth: American Hospitals in the Twentieth Century* (New York: Basic Books, 1989), 256–67.

47. Judith M. Feder, *Medicare: The Politics of Federal Hospital Insurance* (Lexington, Mass.: Lexington Books, 1977); Paul Starr, *The Social Transformation of American Medicine* (New York: Basic Books, 1982), 363–78; Fox, *Health Policies, Health Politics*, 201–6. In the 1970s, the mean age of new pacemaker patients was about 72: Seymour Furman, "Controversies in Cardiac Pacing," *Cardiovasc Clin* 8 (1977): 313.

48. Starr, *Social Transformation*, 374–78, 383–88; Stevens, *In Sickness*, 281–83; Edward D. Berkowitz, *America's Welfare State from Roosevelt to Reagan* (Baltimore: Johns Hopkins Univ. Press, 1991), 166–80; Martin Feldstein, *The Rising Cost of Hospital Care* (Washington, D.C.: Information Resources Press, 1971). A jump in demand for hospital and medical services was evident within months of the implementation of Medicare. "Medicare is breaking down financial barriers to doctors and hospitals," commented John H. Knowles, general director of Massachusetts General Hospital, in "What Medicare Has Brought in Its First Year," *Med World News* 8 (30 June 1967): 33.

49. Louise B. Russell, *Technology in Hospitals* (Washington, D.C.: Brookings Institution, 1979), 133, 156, passim.

50. David J. Rothman, *Beginnings Count: The Technological Imperative in American Health Care* (New York: Oxford Univ. Press, 1997), ch. 3.

51. M. Iréné Ferrer, "The Sick Sinus Syndrome in Atrial Disease," *J Am Med Assoc* 206 (14 October 1968): 645–46.

52. By 1973, widely accepted indications for pacemaker implantation included several impairments of AV conduction, including fixed or intermittent complete heart block, heart block that resulted from digitalis therapy, second-degree heart block if it resulted in patient symptoms, and right bundle branch block. Since 1968, physicians had also begun to use pacemakers to treat sinus arrest and the bradycardia-tachycardia syndrome; both these abnormalities were symptoms of sinus node disease. In a survey conducted in 1972, Victor Parsonnet found that "less than half of all new patients have complete heart block": "A Survey of Cardiac Pacing in the United States and Canada," *Proceedings of the IVth International Symposium on Cardiac Pacing*, ed. Hilbert J. T. Thalen (Assen, Netherlands: Van Gorcum, 1973), 43. The quoted phrase comes from "Cardiac Pacemaker Industry," 4.

53. Victor Parsonnet et al. [Pacemaker Study Group of the Inter-Society Commission for Heart Disease Resources, American Heart Association and American College of Cardiology], "Implantable Cardiac Pacemakers: Status Report and Resource Guideline," *Circulation* 50 (1 October 1974): A21–A35; Starr, *Social Transformation*, 384–87; H. David Banta, "Major Issues Facing Biomedical Innovation," in *Biomedical Innovation*, ed. Edward B. Roberts et al. (Cambridge: MIT Press, 1981), 352–78.

54. Medtronic annual reports; J. Gary Burkhead and Theo A. Kolokotrones, "Medtronic" (New York: Smith, Barney, 26 November 1971).

55. "Fail-Safe Pacemaker," *Med World News* 9 (30 August 1968): 42; Walter Zuckerman et al., "Clinical Application of Demand Pacing," *Ann N Y Acad Sci* 167 (30 October 1969): 1055–59. The generic descriptor that pacing specialists eventually adopted was *ventricular inhibited* because a normally conducted heartbeat inhibited the pacemaker from firing. I have used the term *noncompetitive pacing* to emphasize the goal of avoiding competition between a natural and a paced stimulus.

56. Medtronic annual report, 1968, and typed historical list of Medtronic devices, untitled, copy in Medtronic Information Resources Center, Fridley, Minn.; Wilson Greatbatch, Implantable Active Devices (Clarence, N.Y.: Greatbatch Enterprises, 1983), 2:8 through 2:12; Greatbatch interview by Furman; Victor Parsonnet et al., "Clinical Use of an Implantable Standby Pacemaker," *J Am Med Assoc* 196 (30 May 1966): 104–6; Doris J. W. Escher et al., "Standby Pacing in Multiple Arrhythmias," *Israel J Med Sci* 3 (March–April 1967): 221–29.

57. William M. Chardack in panel discussion, *Ann N Y Acad Sci* 167 (30 October 1969): 901. Berkovits and other critics argued that triggered pacing was wasteful of the pacemaker battery and that it was suspect because it did not emulate the natural action of the heart as demand pacing did. For discussion see J. Walter Keller, "Atrial and Ventricular Synchrony: The Engineering-Physiology Interface," *Ann N Y Acad Sci* 167 (30 October 1969): 869–85; and Agustin Castellanos Jr. et al., "Pacemaker-Induced Cardiac Rhythm Disturbances," ibid.: 903–10. Berkovits and the Miami physicians who had worked with him on demand pacing identified what they took to be several flaws in the

Cordis triggered pacer; this attack may have contributed to the subsequent triumph of demand over triggered pacing in the market. My term *noncompetitive pacing* encompasses both the ventricular inhibited and the ventricular triggered mode.

58. Holloran interview by author; Medtronic annual report, 1972. The license fee for 1971–72 was equivalent to 18 percent of Medtronic's net income.

59. See ch. 5.

60. U.S. patent no. 3,478,746, "Cardiac Implantable Demand Pacemaker," application filed 12 May 1965, awarded 18 November 1969 to Wilson Greatbatch.

61. Holloran interview by author.

62. Burkhead, "Medtronic"; Dann talk. Dann was vice president for marketing and sales.

63. Medtronic annual report, 1974; "Medtronic Settles Patent Suit," *Minneapolis Tribune*, 22 February 1974.

64. Wilson Greatbatch, letter to author, 7 May 1997; Wilson Greatbatch, "Pacemaker Patents — Use and Misuse," *Pacing Clin Electrophysiol* 12 (January 1989, pt. 1): 115–16.

65. Medtronic annual report, 1973.

66. The demand pacemaker from American Optical was considered to have a relatively short pacing life.

67. American Optical annual reports, 1962–66; Dann talk.

68. "TeleTrace Telephone EKG System," Medtronic product brochure, October 1973.

**CHAPTER SEVEN. THE PACEMAKER BECOMES A FLEXIBLE MACHINE**

1. H. David Banta, "Embracing or Rejecting Innovations: Clinical Diffusion of Health Care Technology," in *The Machine at the Bedside: Strategies for Using Technology in Patient Care*, ed. Stanley J. Reiser and Michael Anbar (New York: Cambridge Univ. Press, 1984), 65–91; Edward B. Roberts, "Technological Innovation and Medical Devices," in *New Medical Devices: Invention, Development, and Use*, ed. Karen B. Ekelman (Washington, D.C.: National Academy Press, 1988).

2. M. Iréné Ferrer, "The Sick Sinus Syndrome in Atrial Disease," *J Am Med Assoc* 206 (14 October 1968): 645–46; J. Thomas Bigger, "Sick Sinus Syndrome Label for Many Cardiac Problems," editorial in *J Am Med Assoc* 239 (13 February 1978): 597.

3. M. Iréné Ferrer, *The Sick Sinus Syndrome* (Mount Kisco, N.Y.: Futura, 1974), 91–93; David B. Shaw, "The Etiology of Sino-Atrial Disorder," *Am Heart J* 92 (October 1976): 539–40; Henri E. Kulbertus, "Experience with Permanent Pacing in the Sick Sinus Syndrome," *Cardiovasc Clin* 14 (1983): 189–94.

4. Antonio Raviele et al., "Sick Sinus Syndrome: Modern Definition and Epidemiology," in *Proceedings of the International Symposium on Progress in Clinical Pacing*, ed. M. Santini, M. Pistalese, and A. Alliegro (Amsterdam: Excerpta Medica, 1990), 279–88.

5. Ferrer, *Sick Sinus Syndrome*, 97, 100, 107 (emphasis added). Elsewhere Ferrer qualified this statement: if the sinus node was "sluggish, but not dangerous . . . for these patients it would not be fair to implant a pacemaker": Ferrer, "Pacing and Sick Sinus Syndrome," part 2, interview in *Medtronic News* 6 (August 1976): 3–4. See also Michael Bilitch, "Sick Sinus Node Syndrome," in *Modern Cardiac Pacing: A Clinical Overview*, ed. Seymour Furman and Doris J. W. Escher (Bowie, Md.: Charles Press, 1975), 40–44; and

Hilbert J. T. Thalen, "Cardiac Pacing in Sick Sinus Syndrome," in *To Pace or Not to Pace: Controversial Subjects in Cardiac Pacing*, ed. Hilbert J. T. Thalen and J. Warren Harthorne (The Hague: Nijhoff, 1978), 61–72.

6. Victor Parsonnet, "The Status of Permanent Pacing of the Heart in the United States and Canada," *Ann Cardiol Angéiol* 20 (1971): 288; Victor Parsonnet, "A Survey of Cardiac Pacing in the United States and Canada," *Proceedings of the IVth International Symposium on Cardiac Pacing*, ed. Hilbert J. T. Thalen (Assen, Netherlands: Van Gorcum, 1973), 43.

7. These estimates are based on scattered data about numbers of implants; see esp. Victor Parsonnet, "Permanent Pacing of the Heart: 1952 to 1976," *Am J Cardiol* 39 (February 1977): 253. From available data, it is not possible to distinguish between individual physicians and groups of two or three physicians who worked together on pacemakers.

8. Ibid.; Parsonnet, "Survey of Cardiac Pacing."

9. Richard Sutton and Ivan Bourgeois, *Foundations of Cardiac Pacing, Part 1* (Mount Kisco, N.Y.: Futura, 1991), 185–200; Jeffrey A. Brinker, e-mail communication, 23 May 2000.

10. Parsonnet, "Survey of Cardiac Pacing." The *Proceedings of the IVth International Symposium* also contains reports for other countries that are discussed in the text.

11. Ibid.

12. On critical roles in the management of medical technologies, see Edward B. Roberts, "The Development of Biomedical Technologies," in *Critical Issues in Medical Technology*, ed. Barbara J. McNeil and Ernest G. Cravalho (Boston: Auburn House, 1982), 3–22.

13. Michael Bilitch, Ethel E. Cassady, and John S. Lloyd, "Physician Follow-up of Patients with Permanent Cardiac Pacemakers," in *Proceedings of the IVth International Symposium*, ed. Thalen, 443–48.

14. David H. Gobeli, "The Management of Innovation in the Pacing Industry," Ph.D. diss., University of Minnesota, 1982, table A-1.

15. Ibid., 32.

16. "Cardiac Competitors Continue Crowding," *Bio-Medical Insight* 4 (21 November 1973): 187–90; Patrik Hidefjäll, "The Pace of Innovation: Patterns of Innovation in the Cardiac Pacemaker Industry," Ph.D. diss., Linköping University, Sweden, 1997, app. B.

17. Doris J. W. Escher, Seymour Furman, and John D. Fisher, "New, Increased, Logistical, Technical Demands on the Pacemaker Follow-up Center," in *Proceedings of the VIth World Symposium on Cardiac Pacing*, ed. Claude Meere (Montreal: PACESYMP, 1979), n.p.

18. Victor Parsonnet, "Cardiac Pacing and Pacemakers VII: Power Sources for Implantable Pacemakers," part 1, *Am Heart J* 94 (October 1977): 518–19; National Heart and Lung Institute, "A Brief Report on the Status of Long-Life Cardiac Pacemaker Development" (DHEW Pub. No. [NIH] 74-371, August 1973); Thomas P. Hughes, "The Evolution of Large Technological Systems," in Wiebe E. Bijker, Thomas P. Hughes, and Trevor Pinch, eds., *The Social Construction of Technological Systems* (Cambridge: MIT Press, 1987), 73–76. For physicians' opinions, see "Demand Pacers Get a New Rating: Below Expectations," *Med World News* 11 (13 March 1970): 15–16.

19. Sutton and Bourgeois, *Foundations of Cardiac Pacing*, 68–69; "Power Source Report," *Medtronic News* 5 (August 1973): n.p.; Hidefjäll, "Pace of Innovation," 112.

20. Victor Parsonnet et al., "A Nonpolarizing Electrode for Endocardial Stimulation of the Heart," *J Thorac Cardiovasc Surg* 56 (November 1968): 710–16; Seymour Furman et al., "Endocardial Threshold of Cardiac Response as a Function of Electrode Surface Area," *J Surg Res* 8 (April 1968): 161–66; J. Walter Keller Jr., "Improving Pacemaker Electrodes," in *Proceedings of the IVth International Symposium*, ed. Thalen, 275–78; Peter Tarjan, e-mail to author, 3 December 1996.

21. Wilson Greatbatch, "X-ray Techniques for Evaluation of . . . Implantable Mercury Batteries," *Medical Electronomics and Biological Engineering* 3 (1965): 305–6; Wilson Greatbatch, William M. Chardack, and Andrew A. Gage, "Design Considerations Relating to Power Supplies and Physiologic Controls for Implantable Pacemakers," in *Biomedical Instrumentation*, ed. J. Payer, J. Herrick, and T. B. Weber (New York: Plenum, 1967), 281–89; Wilson Greatbatch, interview by Seymour Furman, 16 February 1996 (NASPE); Greatbatch, interview in Kenneth A. Brown, *Inventors at Work* (Redmond, Wash.: Tempus Books, 1988), 19–44, esp. 32–33.

22. Wilson Greatbatch, "Implantable Power-Sources: A Review," *J Med Eng Tech* 8 (March–April 1984): 56; Greatbatch interview by Furman.

23. Wilson Greatbatch et al., "The Solid-State Lithium Battery: A New Improved Chemical Power Source for Implantable Cardiac Pacemakers," *IEEE Trans Biomed Eng* BME-18 (September 1971): 317–24; A. A. Schneider and F. Tepper, "The Lithium-Iodine Cell," in *To Pace or Not to Pace*, ed. Thalen and Harthorne, 116–21.

24. Greatbatch interview by Brown in *Inventors at Work*, 34–35; Wilson Greatbatch, "Achieving Reliable Pacemakers," in *Proceedings of the Vth International Symposium, Tokyo, March 14–18, 1976*, ed. Yoshio Watanabe (Amsterdam: Excerpta Medica, 1977), 364–68; David L. Bowers, "New Pacemaker Devices from a Technical Point of View," in *To Pace or Not to Pace*, ed. Thalen and Harthorne, 126–30; Victor Parsonnet, "Survey of Pacemaker Practices in the United States, 1978," in *Proceedings of the VIth International Symposium*, ed. Meere, n.p.

25. Ronald Hagenson, interview by author, 19 December 1989, Fridley, Minn. Another manager at Medtronic said much the same thing: "[Lithium research] was going on at Medtronic. You end up with one faction that says, this is the wave of the future, we've got to go with it — but you don't have the data; and you've got another faction that says, we're not going to put that across the line until we really understand it: Michael Toffoli, interview by author and Patrik Hidefjäll, 27 November 1995, Fridley, Minn. Siemens acquired Elema-Schönander in 1959 and changed the Elema name to Siemens-Elema in 1972.

26. Thomas E. Holloran, interview by author, 4 March 1997, Minneapolis.

27. The lithium battery had a high internal impedance that rose from 1,000 ohms in a fresh battery to nearly 50,000 ohms near the end of its worklife, requiring a redesign of pacemaker circuitry to accommodate this characteristic.

28. G. E. Antonioli, "Lithium Pacemaker: The First Clinical Experience," *Pacing Clin Electrophysiol* 13 (March 1990): 363–70.

29. Terry Fiedler, "Straight from the Heart," *Minnesota Business Journal* 9 (February 1985): 24–35; Manuel A. Villafaña, interview by author and Patrik Hidefjäll, 1 December 1995, Plymouth, Minn.

30. Wilson Greatbatch, *Implantable Active Devices* (Clarence, N.Y.: Greatbatch Enterprises, 1983), 4: 14; Toffoli interview by author and Hidefjäll; Villafaña interview by author and Hidefjäll.

31. Parsonnet, "Cardiac Pacing: Power Sources": 523–24; Edgar Sowton, "Energy

Sources for Pacemakers," in *Proceedings of the Vth International Symposium*, ed. Watanabe, 438–46; Michael Bilitch et al., "Use Patterns of Permanent Cardiac Pacemakers," in *Proceedings of the VIIth World Symposium on Cardiac Pacing, Vienna, May 1st to 5th, 1983*, ed. K. Steinbach (Darmstadt: Steinkopf, 1983), 975–80; Wilson Greatbatch, "Pacemaker Technology: Energy Sources," *Cardiovasc Clin* 14 (1983): 243.

32. Joann S. Lublin, "Cardiac Pacemakers Monopoly on Lithium Units Comes to an End," *Minneapolis Tribune*, 15 February 1978; "Cardiac Pacemakers Markets" (Wethersfield, Conn.: Theta Technology Corp., industry report, May 1979), sec. 9, 4; Hidefjäll, "Pace of Innovation," 219.

33. Medtronic annual reports, 1973, 1975; "Xyrel Lithium Pacemaker Now in Full-Scale Production," *Medtronic News* 7 (February 1977): 9.

34. Bryan Parker, "Pacemaker Electronics," in Seymour Furman et al., *Principles and Techniques of Cardiac Pacing* (New York: Harper & Row, 1970), 43–61. Medtronic developed a technique of supporting the circuit elements with the conductor interconnects and doing away with the circuit board: David L. Bowers, "Pacemaker Technology," in *Fundamentals of Cardiac Pacing*, ed. Hilbert J. T. Thalen and Claude C. Meere (The Hague: Nijhoff, 1979), 182–83.

35. Ernest L. Braun and Stuart Macdonald, *Revolution in Miniature: The History and Impact of Semiconductor Electronics*, 2nd ed. (Cambridge: MIT Press, 1982), 99; Jerry Hartlaub, "Pacemaker of the Future: Microprocessor Based or Custom Circuit?" in *The Third Decade of Cardiac Pacing: Advances in Technology and Clinical Applications*, ed. S. Serge Barold and Jacques Mugica (Mount Kisco, N.Y.: Futura, 1982), 417; Lawrence Lessing, "The Transistorized M.D.," *Fortune* 68 (September 1963): 131–34, 204, 206, 208.

36. Bowers, "Pacemaker Technology," 182–85; Jacques Buffet, "Technological Progress in Pacemaker Design: Hermetic Sealing," *Medical Progress through Technology* 3 (1975): 133–42; Greatbatch, "Achieving Reliable Pacemakers."

37. Victor Parsonnet et al., "A Permanent Pacemaker Capable of External Non-Invasive Programming," *Trans Am Soc Artif Int Org* 19 (1973): 224–28; Peter Tarjan, "History of Cardiac Pacing — 1," letter in *Am J Cardiol* 41 (March 1978): 614–15; Keller interview by author.

38. Tarjan e-mail to author, December 1996.

39. Robert Wingrove, interview by author, 28 April 1998, New Brighton, Minn.; Keller interview by author; Robert D. Gold, "Cardiac Pacing — from Then to Now," *Med Instrum* 18 (January–February 1984): 17; Peter Tarjan e-mail to author, 21 May 1997; William P. Murphy Jr., interview by Victor Parsonnet, 4 February 1997 (NASPE). CMOS transistors were especially appropriate for low-current applications in battery-powered devices. See Bowers, "New Pacemaker Devices."

40. Parsonnet et al., "A Permanent Pacemaker"; David C. MacGregor et al., "The Utility of the Programmable Pacemaker," panel discussion in *Pacing Clin Electrophysiol* 1 (April–June 1978): 254–59.

41. Braun and Macdonald, *Revolution in Miniature*, ch. 8; Peter Tarjan, e-mail to author, 10 July 1997.

42. U.S. patent no. 3,557,796, "Digital Counter Driven Pacer," application filed 10 March 1969, awarded 26 January 1971 to J. Walter Keller et al.; U.S. patent no. 3,805,796, "Implantable Cardiac Pacer Having Adjustable Operating Parameters," application filed 22 January 1973, awarded 23 April 1974 to Reese S. Terry and Gomer L. Davies.

43. Peter Tarjan, interview by author, 11 December 1996, Miami.

44. MacGregor et al., "Utility of the Programmable Pacemaker"; Doris J. W. Escher, "Future Development of Artificial Pacing," in *Artificial Cardiac Pacing: Practical Approach*, ed. Edward K. Chung (Baltimore: Williams & Wilkins, 1978), 371–86, esp. 376–78.

45. E.g., Arthur W. Silver et al., "Externally Rechargeable Cardiac Pacemaker," *Ann Thorac Surg* 1 (July 1965): 380–88.

46. Robert E. Fischell et al., "A Long-Lived, Reliable, Rechargeable Cardiac Pacemaker," in *Advances in Pacemaker Technology*, ed. Max Schaldach and Seymour Furman (Berlin: Springer-Verlag, 1975), 357–82; Alfred Mann, interview by Seymour Furman, 7 May 1998, San Diego (NASPE).

47. "New Pacemaker Is Rechargeable," *Electronics* 46 (16 August 1973): 35; "A Weekly Heart Throb," *Med World News* 14 (24 August 1973): 5–6; Sowton, "Energy Sources for Pacemakers"; Greatbatch, "Implantable Power Sources: A Review": 58.

48. Mann interview; "Pacemaker Parade," *Bio-Medical Insight* 12 (June 1981): 95. *Bio-Medical Insight* gave the Phoenix Award to Mann and Pacesetter.

49. CPI and Intermedics each had about a 12 percent market share, Pacesetter about 4 percent.

50. "Microlith-P Programmable Pacing System from CPI," product brochure, 1978; "From CPI, the New Microlith Family," advertisement, 1978, copy courtesy of J. Walter Keller.

51. "CPI Sues Cordis," *Minneapolis Star*, 5 December 1977; Albert W. Preston Jr. and Thomas A. Preston, "The Patent System and Cardiac Pacing: Is the System Serving Its Users?" *Pacing Clin Electrophysiol* 8 (July–August 1985): 476–83; Hidefjäll, "Pace of Innovation," 175; Keller interview by author. The Cordis programmer transmitted pulses of uniform width; CPI and other companies used a binary code (two pulse widths): Peter Tarjan, e-mail to author, 2 August 1997. The cost of further litigation was also a factor in the Cordis decision.

52. "Xyrel Lithium Pacemaker Now in Full-Scale Production," *Medtronic News* 7 (February 1977): n.p.; "Rate Programmability Added to the Xyrel Line," *Medtronic News* 8 (March 1978): 10.

53. Dryden Morse and Robert M. Steiner, *The Pacemaker and Valve Identification Guide* (Garden City, N.Y.: Medical Examination Publishing, 1978), 59–61; Intermedics, "You've Already Earned Your Patient's Confidence. We'd Like to Earn Yours," advertisement, Intermedics, n.d. [1978?], copy courtesy of J. Walter Keller.

54. "Microlith-P Programmable Pacing System"; "CyberLith IV Physician's Manual" (Freeport, Tex.: Intermedics, 1981), 12–15; Sutton and Bourgeois, *Foundations of Cardiac Pacing*, 98–100.

55. U.S. patent no. 4,026,305, "Low Current Telemetry System for Cardiac Pacers," application filed 26 June 1975, awarded 31 May 1977 to Robert R. Brownlee and Frank O. Tyers; Frank O. Tyers and Robert R. Brownlee, "A Multiparameter Telemetry System for Cardiac Pacemakers," in *Cardiac Pacing: A Concise Guide to Clinical Practice*, ed. Philip Varriale and Emil A. Naclerio (Philadelphia: Lea & Febiger, 1979), 352; Jason Sholder et al., "Bidirectional Telemetry and Interrogation in Cardiac Pacing," in *Third Decade*, ed. Barold et al., 145–66; Theta Technology, "Cardiac Pacemakers Markets."

56. Quoted in Hidefjäll, "Pace of Innovation," 178.

57. Tyers and Brownlee, "A Multiparameter Telemetry System," 367.

58. Bakken interview by Furman; Hidefjäll, "Pace of Innovation," 178.

59. "Medical Electronics" (Palo Alto, Calif.: Creative Strategies, industry report, 1973); Yehezkiel Kishon and Henry N. Neufeld, "Current Problems in Cardiology," *Med Prog Technol* 5 (1977): 13–19.

60. In July 1993, the author observed pacemaker implantations in Rochester, Minnesota, at St. Mary's Hospital, an affiliate of the Mayo Clinic. The cardiologist who did the implantation decided what kind of pacemaker was needed—a single-chamber, rate-responsive pacemaker, for example—but a nurse-specialist selected the specific brand and model at random from a list of all such pacemakers in stock at the hospital. This degree of evenhandedness has been uncommon.

61. Parsonnet et al., "A Permanent Pacemaker"; Jerry C. Griffin and J. A. Quitman, "Pacemaker Programmability: The Role of Non-Invasive Pacing System Revision," in *Proceedings of the VIth World Symposium*, ed. Meere, n.p.; MacGregor et al., "Utility of the Programmable Pacemaker"; Victor Parsonnet and Todd Rodgers, "The Present Status of Programmable Pacemakers," *Prog Cardiovasc Dis* 4 (May–June 1981): 401–20.

62. Victor Parsonnet, Candice C. Crawford, and Alan D. Bernstein, "The 1981 United States Survey of Cardiac Pacing Practices," *J Am Coll Cardiol* 3 (May 1984): 1322.

63. Escher, Furman, and Fisher, "New, Increased Demands"; Victor Parsonnet, "The Proliferation of Cardiac Pacing: Medical, Technical, and Socioeconomic Dilemmas," *Circulation* 65 (May 1982): 841–45.

64. Parsonnet, Crawford, and Bernstein, "1981 United States Survey"; "Resources Required for Pacemaker Implantation," ed. Robert G. Hauser, symposium in *Pacing Clin Electrophysiol* 6 (January–February 1983): 139–55; Seymour Furman, "Pacemaker Practices in the United States," editorial in *Pacing Clin Electrophysiol* 5 (November–December 1982): 791–92.

65. J. Warren Harthorne, "Pacemaker Implantation in a Medical Center," *Pacing Clin Electrophysiol* 6 (January–February 1983): 140–41.

66. Jerry Hartlaub, interview by author, 2 July 1997, Columbia Heights, Minn.

67. Hughes, "Evolution of Large Technological Systems."

**CHAPTER EIGHT. SLOWING THE PACE**

1. "Pacers' Makers Drag Feet on 'Recalls,'" *Med World News* 16 (21 April 1975): 21–24; FDA, "Responses to Questions in Senator Abraham Ribicoff's Letters of March 11 and March 20, 1975," typescript in author's possession; "General Accounting Office Investigates FDA's Handling of G.E. Pacemaker Recalls," *Newsletter of Biomedical Safety & Standards* 5 (31 March 1975): 34–35.

2. "Pacemaker Parade," *Bio-Medical Insight* 13 (July 1981): 93.

3. "Makers Drag Feet"; "Firms under Fire," *Med World News* 16 (26 May 1975): 6–7.

4. "Makers Drag Feet": 24.

5. "Court Defers FDA Pacemaker Suit," *Med World News* 17 (26 April 1976): 5.

6. E.g., "FDA to Investigate Nader Charges," *Minneapolis Tribune*, 30 March 1973; "Makers Drag Feet."

7. Lewis Cope, "New Pacemaker Is Lighter, Smaller, Longer-Lasting than Previous Models," *Minneapolis Tribune*, 24 January 1975; Bobby I. Griffin, interview by author and Patrik Hidefjäll, 30 November 1995, Fridley, Minn.; Hidefjäll, "The Pace of Innovation: Patterns of Innovation in the Cardiac Pacemaker Industry," Ph.D. diss., Linköping University, Sweden, 1997, 155; Dryden Morse and Robert M. Steiner, *The Pacemaker and Valve Identification Guide* (Garden City, N.Y.: Medical Examination Pub., 1978), 80, 82.

8. Bobby I. Griffin interview as quoted in Hidefjäll, "Pace of Innovation," 566; Jerry Hartlaub, interview by author, 2 July 1997, Columbia Heights, Minn. Medtronic was manufacturing its own hybrid circuits through a subsidiary, Micro-Rel, that the company had created in 1973 after its usual supplier, Motorola, dropped Medtronic as a customer. Xytron was the first product that carried hybrid circuits manufactured at Micro-Rel. New pacemakers are immersed in a test tank that emulates the conditions within the human body; some Xytrons had shorted out at this early stage of testing.

9. "Medtronic Says Some Pacemakers Could Fail," *Minneapolis Tribune*, 3 August 1976; Lewis Cope, "Second Pacemaker Alert Issued," ibid., 23 November 1976; "Some Medtronic Xytron Pacemakers May Be Dangerous," ibid., 28 October 1977. A product advisory letter warns physicians of a problem that may appear in a product and advises them to review the status of the product in each of their patients; it is up to the physician to decide whether the device should be replaced.

10. Bobby Griffin interview as quoted in Hidefjäll, "Pace of Innovation," 156; Hidefjäll, "Pace of Innovation," 156.

11. Charles F. Cuddihy Jr., interview by author, 22 July 1996, Minnetonka, Minn.; Earl E. Bakken interview by Seymour Furman, 17 May 1996 (NASPE).

12. U.S. Congress, OTA, *Federal Policies and the Medical Devices Industry*, OTA-H-230 (July 1983), ch. 5; Michael S. Baram, "Medical Device Legislation and the Development and Diffusion of Health Technology," in *Technology and the Quality of Health Care*, ed. Richard H. Egdahl and Paul M. Gartman (Germantown, Md.: Aspen Systems, 1978), 191–97; Susan Bartlett Foote, *Managing the Medical Arms Race* (Berkeley: Univ. of California Press, 1992), ch. 5.

13. During the 1960s and the early 1970s, the agency attempted with some success to construe its mandate more broadly; it intervened vigorously when evidence began to pile up that implanted IUDs were causing injuries, infections, and deaths: Foote, *Managing the Medical Arms Race*, 116–20.

14. Theodore Cooper, "Device Legislation," *Food, Drug, Cosmetic Law Journal* 26 (April 1971): 165–72; Foote, *Managing the Medical Arms Race*, 114–16; OTA, *Federal Policies*, 97–98; *United States v. An Article of Drug . . . Bacto-Unidisk*, 394 U.S. 784, 798 (1969); *AMP v. Gardner*, 389 F2d 825 (2d Cir. 1968).

15. U.S. Dept. of Health, Education, and Welfare, Cooper Committee, "Medical Devices: A Legislative Plan" (Washington, D.C.: GPO, 1970); OTA, *Federal Policies*, 98; Cooper, "Device Legislation." For the 1973 hearings, see House Committee on Interstate and Foreign Commerce, Subcommittee on Public Health and the Environment, *Medical Devices: Hearings*, 93rd Cong., 1st sess., 23–24 October 1973.

16. Cooper, "Device Legislation"; 170; "Device Makers Charged with Poor Performance," *Electronics* 43 (23 November 1970): 34.

17. Committee on Government Operations, *Regulation of Medical Devices (Intrauterine Contraceptive Devices): Hearings*, 93rd Cong., 1st sess., May 30–June 1, June 12–13, 1973.

18. Pacemaker recalls figured prominently in the final hearings on the bill; see House Committee on Interstate and Foreign Commerce, Subcommittee on Health and the Environment, *Medical Device Amendments of 1975, Hearings*, 94th Cong., 1st sess., 28–31 July 1975. See also "New Bill Would Ban from Sale Any Device that Does Not Comply with Regulations," *Med World News* 16 (30 June 1975): 30–31; "Device Legislation Finally on Its Way," ibid. 17 (3 May 1976): 6; Foote, *Managing the Medical Arms Race*, 120. HIMA proposed several changes to the House and Senate bills but acknowledged

that FDA regulation in some form was inevitable. HIMA's lobbying probably encouraged Congress to add section 501(k) to the Device Amendments, however.

19. Medical Device Amendments of 1976, U.S. Code 21, secs. 360c–360k (1982). For a synopsis, see OTA, *Federal Policies*, 98–102.

20. House Committee, *Medical Device Ammendments of 1975*, 374–87; OTA, *Federal Policies*, 104. The Safe Medical Devices Act (1990) closed the 501(k) loophole and required FDA to order clinical trials for new devices even if they were substantially equivalent to ones already on the market; the 1990 law also changed the procedures for recall of defective devices already in use.

21. OTA, *Federal Policies*, 122, 125. Of the firms responding to the 1981 survey, 46 percent reported that meeting regulatory requirements had become a major burden for them, and 21 percent said that regulation was their single most important problem.

22. Earl E. Bakken interview by David J. Rhees, 10 January 1997 (Bakken Library); William P. Murphy Jr., interview by Parsonnet, 4 February 1997, Miami (NASPE); Seymour Furman, "Regulation of Pacemaker Industry and Practice," editorial in *Pacing Clin Electrophysiol* 1 (July–September 1978): 291.

23. David Mechanic, *From Advocacy to Allocation: The Evolving American Health Care System* (New York: Free Press, 1986), 79; "Medtronic, Inc." (New York: Smith, Barney, industry report, 1971), 15; Ronald E. Hagenson, interview by author, 19 December 1989, Fridley, Minn.; R. D. Peterson and C. R. MacPhee, *Economic Organization in Medical Equipment and Supply* (Lexington, Mass.: Heath, 1973), ch. 4, esp. 40–41.

24. David H. Gobeli, "The Management of Innovation in the Pacing Industry," Ph.D. diss., University of Minnesota, 1982, app. B; "The Leader in Lithium," product brochure, CPI, 1974, copy courtesy of J. Walter Keller.

25. "Microlith-P Programmable Pacing System from CPI," product brochure, 1978; "The CyberLith Programming System," product brochure, Intermedics, n.d. [1978?].

26. "Medtronic, Inc."; Medtronic annual report, 1975; Rader as quoted in Martin Campbell-Kelly and William Aspray, *Computer: A History of the Information Machine* (New York: Basic Books, 1996), 135.

27. Albert Beutel as quoted in Hidefjäll, "Pace of Innovation," 159; Victor Parsonnet, Candice C. Crawford, and Alan D. Bernstein, "The 1981 United States Survey of Cardiac Pacing Practices," *J Am Coll Cardiol* 3 (May 1984): 1329.

28. Medtronic annual report, 1975; "Cardiac Pacemakers Markets" (Wethersfield, Conn.: Theta Technology, industry report, 1979), sec. 10, 39. Intermedics selling expenses in 1977 represented 34 percent of company sales.

29. After Bakken moved from CEO to board chairman, Medtronic had two successive CEOs from consumer-products industries: Dale Olseth (1976–85), from Tonka Toys, and Winston Wallin (1985–91), from Pillsbury.

30. U.S. Senate Special Committee on Aging, *Fraud, Waste, and Abuse in the Medicare Pacemaker Industry: An Information Paper*, 97th Cong., 2nd sess., 1982, 23.

31. Quoted in Hidefjäll, "Pace of Innovation," 159.

32. Cuddihy interview by author; Bobby Griffin interview by author and Hidefjäll; Richard Sanders, interview by author, 16 March 1997, Anaheim, Calif.; Hidefjäll, "Pace of Innovation," 159. Beutel died in a helicopter crash in 1979.

33. "Pacemaker Parade," *Bio-Medical Insight* 12 (June 1981): 92.

34. "Hot Operators: Sales to the Heart Surgeons," *Philadelphia Inquirer*, 19 April 1981, as reprinted in Senate Special Committee, *Fraud, Waste, and Abuse*, 89–90.

35. Senate Special Committee, *Fraud, Waste, and Abuse*, 23–24, 30–31.

36. Senate Special Committee, *Fraud, Waste, and Abuse*, 20–46 ff.; "Report Cites Kickbacks in Sales of Pacemakers," *New York Times*, 6 Sept 1982. Intermedics was not the sole offender but came in for more attention than the other firms. Senior figures in the implanter community, including Seymour Furman, Victor Parsonnet, Jerry Griffin, Robert Hauser, and Michael Bilitch, firmly opposed these excesses and cooperated fully with the Senate Select Committee.

37. Senate Special Committee on Aging, *Pacemakers Revisited: A Saga of Benign Neglect: Hearings*, 99th Cong., 1st sess., 10 May 1985, 4–6, 94–96.

38. "Cordis Staves Off FDA's Pacemaker Crackdown," *Med World News* 16 (22 September 1975): 28; "Court Defers FDA Pacemaker Suit," *Med World News* 17 (26 April 1976): 5.

39. Murphy interview by Parsonnet.

40. For extensive documentation on the Cordis difficulties over meeting FDA standards for "Good Manufacturing Practice," see Senate Special Committee, *Pacemakers Revisited*, 558–905; see also Hidefjäll, "Pace of Innovation," 221–23.

41. Gobeli, "Management of Innovation," table A-1; Parsonnet, Crawford, and Bernstein, "1981 United States Survey": 1323.

42. Victor Parsonnet and Marjorie Manhardt, "Permanent Pacing of the Heart, 1952 to 1976," *Am J Cardiol* 39 (February 1977): 253. For implantation rates in various countries, see Bernard S. Goldman and Victor Parsonnet, "World Survey on Cardiac Pacing," *Pacing Clin Electrophysiol* 2 (September–October 1979): W3–W4.

43. Parsonnet, Crawford, and Bernstein, "1981 United States Survey": 1322.

44. On technological momentum in health care, see Thomas P. Hughes, "Machines and Medicine: A Projection of Analogies between Electric Power Systems and Health Care Systems," *Int J Tech Assess in Health Care* 1 (1985): 285–95.

45. "Cardiac Pacemakers Markets," sec. 9, 6.

46. Thomas A. Preston, "Pacemaker Utilization: The Need for Information," *Pacing Clin Electrophysiol* 4 (March–April 1981): 235.

47. "Cardiac Pacemakers Markets"; "Cardiac Pacemaker Industry" (Minneapolis: Dain, Kalman, & Quail, industry report, 1975).

48. Allan C. Mazur, "Controlling Technology," in *Technology and the Future*, ed. Albert H. Teich (5th ed.; New York: St. Martin's Press, 1990), 214. The earliest use of the phrase "Toward Man's Full Life" occurs in 1974. Medtronic registered the phrase as a trademark.

49. Rosemary Stevens, *In Sickness and in Wealth: American Hospitals in the Twentieth Century* (New York: Basic Books, 1989), 322–27.

50. Public Citizen Health Research Group, "Permanent Pacemakers in Maryland" (Washington, D.C., July 1982); Leonard Scherlis and Donald H. Dembo, "Problems in Health Data Analysis: The Maryland Permanent Pacemaker Experience in 1979 and 1980," *Am J Cardiol* 51 (1 January 1983): 131–36. For efforts to limit the number of implantations in a single hospital, see Atul B. Chokshi et al., "Impact of Peer Review in Reduction of Permanent Pacemaker Implantations," *J Am Med Assoc* 246 (14 August 1981): 754–57. At Brooklyn Hospital, 8 pacers were implanted in 1972, 48 in 1976. A review committee established in 1977 brought the number down to 22 in 1978.

51. Formal criteria for implantation were introduced in 1984 but have been revised and expanded: Joint American College of Cardiology/American Heart Association Task

Force on Assessment of Cardiovascular Procedures (Subcommittee on Pacemaker Implants), "Guidelines for Permanent Cardiac Pacemaker Implantation," *J Am Coll Cardiol* 4 (August 1984): 434–42.

52. U.S. Congress, OTA, *Medical Technology and Costs of the Medicare Program*, OTA-H-227 (July 1984); T. Joseph Reeves, "Advances in Cardiology and Escalating Costs to the Patient," editorial in *Circulation* 71 (April 1985): 637–41; W. Bruce Fye, *American Cardiology: The History of a Specialty and Its College* (Baltimore: Johns Hopkins Univ. Press, 1996), 318–22.

53. U.S. Congress, OTA, Health Technology Case Study #1: *End-Stage Renal Disease*, OTA-HC5–1 (April 1981), 5, table 3; Parsonnet, Crawford, and Bernstein, "1981 United States Survey": 1329. Cf. William S. Stoney et al., "Cost of Cardiac Pacing," *Am J Cardiol* 37 (January 1976): 23–25.

54. E.g., Doris J. W. Escher, "Pacemakers of the 1980s," *Med Instrum* 18 (January–February 1984): 29–34.

55. E.g., Jon B. Christianson, "The Concern about Containment of Health Care Expenditures," in Jon B. Christianson and Kenneth R. Smith, eds., *Current Strategies for Containing Health Care Expenditures* (New York: Spectrum, 1985), 6–8.

56. Social Security Amendments of 1983, U.S. Code, vol. 42, sec. 1395x(v)(1)(A); OTA, *Federal Policies*, 48–50.

57. Victor Parsonnet and Alan D. Bernstein, "Cardiac Pacing Practices in the United States in 1985," *Am J Cardiol* 62 (1 July 1988): 71–77.

58. Jennifer Seery, "DRGs and Pacing—Challenge and Opportunity," *Medtronic News* 13 (December 1983): 3–7; Robert P. Kearney, "Heart Trouble," *Corporate Report Minnesota*, November 1984, 52–55, 57–58.

**CHAPTER NINE. COMPETITION THROUGH INNOVATION**

1. Victor Parsonnet, "A Survey of Cardiac Pacing in the United States and Canada," *Proceedings of the IVth International Symposium on Cardiac Pacing*, ed. Hilbert J. T. Thalen (Assen, Netherlands: Van Gorcum, 1973), 41–48.

2. Paul Starr, *The Social Transformation of American Medicine* (New York: Basic Books, 1982), 370; W. Bruce Fye, *American Cardiology: The History of a Specialty and Its College* (Baltimore: Johns Hopkins Univ. Press, 1996), 218–24, 275–76; Seymour Furman, letter to author, 27 June 1997.

3. Pacemaker Study Group of the Inter-Society Commission for Heart Disease Resources [American Heart Association and American College of Cardiology], "Implantable Cardiac Pacemakers: Status Report and Resource Guideline," *Am J Cardiol* 34 (1 October 1974): 488. As pacemakers became more flexible in the 1980s, others revised the code by adding a fourth and a fifth position so as to indicate new capabilities.

4. President's Commission on Heart Disease, Cancer, and Stroke [DeBakey Commission], *Report to the President: A National Program to Conquer Heart Disease, Cancer, and Stroke* (Washington, D.C.: GPO, 1964). As actually implemented, the regional medical program was watered down considerably from the strongly hierarchical structure that the commission had proposed: "Regional Program Brings Research Results to the Bedside," *Med World News* 8 (8 September 1967): 60–67; Fye, *American Cardiology*, 221–22.

5. Victor Parsonnet, "The Proliferation of Cardiac Pacing: Medical, Technical, and Socioeconomic Dilemmas," *Circulation* 65 (May 1982): 841–45.

6. Victor Parsonnet, "Transtelephone Monitoring," in *Modern Cardiac Pacing*, ed.

Seymour Furman and Doris J. W. Escher (Bowie, Md.: Charles Press, 1975), 273–77; Victor Parsonnet et al., "Followup of Implanted Pacemakers: An Evaluation of Surveillance Methods," *Cardiovasc Clin* 6 (1975): 431–46; "TeleTrace Telephone EKG System" (product brochure, Medtronic, October 1973).

7. Doris J. W. Escher, Seymour Furman, and John D. Fisher, "New, Increased, Logistical, Technical Demands on the Pacemaker Follow-up Center," in *Proceedings of the VIth World Symposium on Cardiac Pacing*, ed. Claude Meere (Montreal: PACESYMP, 1979), n.p.; Victor Parsonnet, Candice C. Crawford, and Alan D. Bernstein, "The 1981 United States Survey of Cardiac Pacing Practices," *J Am Coll Cardiol* 3 (May 1984): 1327–28, 1330–31; Parsonnet, "Proliferation."

8. E.g., Seymour Furman, "Certification of Special Competence in Cardiac Pacing," editorial in *Pacing Clin Electrophysiol* 9 (January–February 1986, pt. 1): 1; Victor Parsonnet, "Cardiac Pacing as a Subspecialty," *Am J Cardiol* 59 (15 April 1987): 989–91. The original Pacemaker Study Group, now including Michael Bilitch, presented an updated set of guidelines and recommendations in "Optimal Resources for Implantable Cardiac Pacemakers," *Circulation* 68 (July 1983): 227A–244A.

9. E.g., Seymour Furman, "Professional Financial Relationships with Industry," editorial in *Pacing Clin Electrophysiol* 6 (January–February 1983): 1; Seymour Furman, "Industrial Control of Medical Research," editorial in *Pacing Clin Electrophysiol* 8 (January–February 1985): 1–3.

10. J. Warren Harthorne, interview by Seymour Furman, 26 January 1997, Boston (NASPE).

11. Ibid.; Dryden P. Morse, interview by Victor Parsonnet, 16 February 1997, Moorestown, N.J. (NASPE).

12. Harthorne interview by Furman; Fye, *American Cardiology*, ch. 9.

13. J. Warren Harthorne et al., "North American Society of Pacing and Electrophysiology [NASPE]," editorial in *Pacing Clin Electrophysiol* 2 (September–October 1979), 521–22; see also Michael Bilitch, Ethel E. Cassady, and John S. Lloyd, "Physician Follow-Up of Patients," *Proceedings of the IVth International Symposium*, ed. Thalen, 443–48; Bernard Dodinot et al., "Professional Qualifications for Pacemaker Implantation," panel discussion in *Pacing Clin Electrophysiol* 1 (July–September 1978): 381–85; J. Warren Harthorne and Victor Parsonnet, "Task Force VI: Training in Cardiac Pacing," *J Am Coll Cardiol* 7 (June 1986): 1213–14.

14. Harthorne interview by Furman; Victor Parsonnet, interview by author, 5 December 1995, Newark, N.J.; Carol McGlinchey, interview by Seymour Furman, 7 May 1998, San Diego (NASPE).

15. Harthorne et al., "North American Society of Pacing and Electrophysiology (NASPE)"; Furman, "Certification."

16. The reports of these policy conferences were published, e.g., in *Pacing Clin Electrophysiol* 6 (January 1983, May 1983), 9 (April 1986), 7 (May 1984), and 11 (July 1988).

17. Information supplied by NASPE, May 1998.

18. Robert G. Hauser, "Summary and Conclusions," in "Resources Required for Pacemaker Implantation," ed. Robert G. Hauser, *Pacing Clin Electrophysiol* 6 (January–February 1983): 154–55; see also Robert A. Chase, "Proliferation of Certification in Medical Specialties: Productive or Counterproductive," *N Engl J Med* 294 (26 February 1976): 497–99.

19. Pacemaker Study Group, "Optimal Resources"; Joint ACC/AHA Task Force on

Assessment of Cardiovascular Procedures, Subcommittee on Pacemaker Implantation, "Guidelines for Permanent Cardiac Pacemaker Implantation," *J Am Coll Cardiol* 4 (August 1984): 434–42.

20. Victor Parsonnet and Alan D. Bernstein, "Pacing in Perspective: Concepts and Controversies," *Circulation* 73 (June 1986): 1092.

21. Furman, "Certification"; Seymour Furman and Michael Bilitch, "NASPExAM," *Pacing Clin Electrophysiol* 10 (March–April 1987): 278–80; "NASPExAM and NASPExAM/AP," pamphlet, May 1996; Parsonnet, "Cardiac Pacing as a Subspecialty."

22. J. Warren Harthorne and Victor Parsonnet, "Physician Competence to Perform Permanent Cardiac Pacemaker Implantations" (unpublished MS, 1992, copy courtesy of Victor Parsonnet).

23. Telectronics Pacing Systems, "Quality Engineered for over 30 Years" (1996).

24. Jason Sholder, Paul A. Levine, et al., "Bidirectional Telemetry and Interrogation in Cardiac Pacing," in *The Third Decade of Cardiac Pacing: Advances in Technology and Clinical Applications*, ed. S. Serge Barold et al. (Mount Kisco, N.Y.: Futura, 1982), 145–66; Hidefjäll, "The Pace of Innovation: Patterns of Innovation in the Cardiac Pacemaker Industry," Ph.D. diss., Linköping University, Sweden, 1997, 181–82.

25. Theta Technology, "Cardiac Pacemakers Markets" (Wethersfield, Conn.: Theta Technology Corp., industry report, May 1979), sec. 9, pages 4 and 14; Terry Fiedler, "Straight from the Heart," *Minnesota Business Journal* 9 (February 1985): 24–35; Hidefjäll, "Pace of Innovation," 178–79, 183–84.

26. Medtronic completed a long-range product planning exercise in 1979 that I have not seen, but I have been told it embodied the kind of analysis I present here.

27. Medtronic annual report, 1977; "Xyrel Family Expanded to Give Physicians Broad Selection," *Medtronic News* 8 (June 1978): 12.

28. "Cardiac Pacemakers Markets."

29. J. Walter Keller, "Atrial and Ventricular Synchrony: The Engineering-Physiology Interface," *Ann N Y Acad Sci* 167 (30 October 1969): 869–85; Seymour Furman, "Therapeutic Uses of Atrial Pacing," *Am Heart J* 73 (December 1973): 835–40. This pacing mode was designated VAT under the ICHD Pacemaker Code.

30. Miami cardiologist Louis Lemberg had suggested the concept of AV sequential pacing to Berkovits in a letter of 10 May 1965, copy in author's possession. See also Philip Samet, Cesar Castillo, and William H. Bernstein, "Hemodynamic Consequences of Sequential Atrioventricular Pacing: Subjects with Normal Hearts," *Am J Cardiol* 21 (February 1968): 207–12. "Basically," Berkovits later said, "we followed the reports that Samet had published that P-wave triggered pacemakers improve cardiac output and that AV interval had a significant contribution to the ventricular output. And also, I like always to imitate nature; and if nature provides an atrium and a ventricle, let's make both of them functional.": Barouh V. Berkovits, interview by Seymour Furman, 21 January 1996, Natick, Mass. (NASPE).

31. Cesar A. Castillo et al., "Bifocal Demand Pacing," *Chest* 59 (April 1971): 360–64. This pacing mode, designated DVI in the ICHD Pacemaker Code, is also known as AV sequential pacing: Seymour Furman et al., "Atrioventricular Sequential Pacing and Pacemakers," *Chest* 63 (May 1973): 783–89; Berkovits interview by Furman; Harthorne interview by Furman.

32. Victor Parsonnet, "Survey of Pacemaker Practices in the United States, 1978," in *Proceedings of the VIth International Symposium*, n.p.; J. Warren Harthorne, "Indications

for Various Types of Pacemakers," in *Boston Colloquium on Cardiac Pacing*, ed. J. Warren Harthorne and Hilbert J. T. Thalen (The Hague: Nijhoff, 1977), 33–47; Arthur B. Simon and Allan E. Zloto, "Symptomatic Sinus Node Disease: Natural History after Permanent Ventricular Pacing," *Pacing Clin Electrophysiol* 2 (May–June 1979): 306; Furman, "Therapeutic Uses of Atrial Pacing"; Richard Sutton, John Perrins, and Paul Citron, "Physiological Cardiac Pacing," *Pacing Clin Electrophysiol* 3 (March–April 1980): 207–19; Seymour Furman and Jay Gross, "Dual-Chamber Pacing and Pacemakers," *Current Problems in Cardiology* 15 (1990): 117–79. Victor Parsonnet showed that in 1973, when about one-third to one-half of the primary implants were for conditions other than complete heart block, nearly nine-tenths of the pacemakers were ventricular inhibited devices: Parsonnet, "Survey of Cardiac Pacing Practices," 43, 45. For a later review of this question, see John Camm et al., "Ventricular Pacing for Sick Sinus Syndrome — a Risky Business?" *Pacing Clin Electrophysiol* 13 (June 1990): 695–99.

33. Kalman Asubel and Seymour Furman, "The Pacemaker Syndrome," *Ann Intern Med* 103 (September 1985): 420–29 at 425; Richard Sutton and Ivan Bourgeois, *The Foundations of Cardiac Pacing, Part 1* (Mount Kisco, N.Y.: Futura, 1991), 126–31.

34. Nicholas P. D. Smyth et al., "Permanent Pervenous Atrial Sensing and Pacing with a New J Shaped Lead," *J Thorac Cardiovasc Surg* 72 (October 1976): 565–70; Francis Robicsek et al., "Self-Anchoring Endocardial Pacemaker Leads," *Am Heart J* 102 (October 1981): 775–82.

35. Hidefjäll, "Pace of Innovation," 189–90; Smyth et al., "Permanent Pervenous Atrial Sensing and Pacing."

36. K. Stokes, K. Cobian, and T. Lathrop, "Polyurethane Insulators, a Design Approach to Small Pacing Leads," in *VIth World Symposium*, ed. Meere, n.p.; Seymour Furman, "Atrial Pacing," editorial in *Pacing Clin Electrophysiol* 3 (July–August 1980): 393–94; Victor Parsonnet, "Routine Implantation of Permanent Transvenous Pacemaker Electrodes in Both Chambers: A Technique Whose Time Has Come," editorial in *Pacing Clin Electrophysiol* 4 (January–February 1981): 109–12.

37. M. E. Leckrone et al., "A Microprocessor-Based, Two-Chamber Physiologic Pacemaker," in *Third Decade*, ed. Barold and Mugica, 167–89; Harthorne interview by Furman; Hidefjäll, "Pace of Innovation," 190–94.

38. Michael L. Hardage and Don J. Stillwell, "Computers in Cardiac Pacing," in *Modern Cardiac Pacing*, ed. S. Serge Barold (Mount Kisco, N.Y.: Futura, 1985), 355–73.

39. Robert D. Gold, "Cardiac Pacing — from Then to Now," *Med Instrum* 18 (January–February 1984): 17, 19; Hidefjäll, "Pace of Innovation," 192; Peter Tarjan, interview by author, 11 December 1996, Miami, Fla.

40. Jerry Hartlaub, "Pacemaker of the Future: Microprocessor Based or Custom Circuit?" in *Third Decade*, ed. Barold and Mugica, 417–28; Jerry Hartlaub, interview by author, 2 July 1997, Columbia Heights, Minn. At the end of the 1970s, Medtronic had inaugurated but abandoned a program to design custom microprocessors.

41. "Medtronic Introduces Its Versatrax A-V Universal Pacemaker," *Medtronic News* 12 (December 1982): 20–21; "Dual Sensing and Pacing Make Versatrax the Jogger's Pacemaker," *Medtronic Pulse*, January 1983, 12–13; Victor Parsonnet and Alan D. Bernstein, "Cardiac Pacing in the 1980s: Treatment and Techniques in Transition," *J Am Coll Cardiol* 1 (1983): 339–54.

42. Seymour Furman and John D. Fisher, "Endless Loop Tachycardia in an AV Universal (DDD) Pacemaker," *Pacing Clin Electrophysiol* 5 (July–August 1982): 486–89.

43. Peter H. Bellot, "Clinical Experience with Over 250 DDD Pacemakers," in *Modern Cardiac Pacing*, ed. Barold, 439–81; "Symbios Pacing System Soon to be Commercially Available," *Medtronic News* 14 (March 1984): 17–18.

44. Michael Toffoli, interview by author and Patrik Hidefjäll, 27 November 1995, Fridley, Minn.

45. Michael R. Leckrone, "An Analysis of Product Development Problems at Cordis Corporation" (unpublished paper, Nova University, Miami, February 1982; copy courtesy of Peter Tarjan); Hidefjäll, "Pace of Innovation," 221–22. For a review of DDD products from all the pacemaker firms, see Bellot, "Clinical Experience"; and Hidefjäll, "Pace of Innovation," 194–213.

46. Confidential industry sources.

47. Larry Stotts, as quoted in Hidefjäll, "Pace of Innovation," 208–9.

48. Dryden Morse, Robert M. Steiner, and Victor Parsonnet, *A Guide to Cardiac Pacemakers: Supplement, 1986–1987* (Philadelphia: Davis, 1986), S86–S87; Belott, "Clinical Experience," 474–77; Hidefjäll, "Pace of Innovation," 209–10.

49. Victor Parsonnet, Alan D. Bernstein, and Donna Galasso, "Cardiac Pacing Practices in the United States in 1985," *Am J Cardiol* 62 (1 July 1988): 71–77; Hidefjäll, "Pace of Innovation," 215.

50. David L. Bowers, "New Pacemaker Devices from a Technical Point of View," in *To Pace or Not to Pace*, ed. Hilbert J. T. Thalen and J. Warren Harthorne (The Hague: Nijhoff, 1978), 126–30; Parsonnet and Bernstein, "Cardiac Pacing in the 1980s"; J. Warren Harthorne, "Programmable Pacemakers: Technical Features and Clinical applications," *Cardiovasc Clin* 14 (1983): 135–47; H. J. Sykosch, "The New Generation of Pacemakers," *Progress in Artificial Organs* 1 (1983): 317–23.

51. Alan D. Bernstein and Victor Parsonnet, "Survey of Cardiac Pacing in the United States in 1989," *Am J Cardiol* 69 (1 February 1992): 334; Alan D. Bernstein and Victor Parsonnet, "Survey of Cardiac Pacing and Defibrillation in the United States in 1993," *Am J Cardiol* 78 (15 July 1996): 191; "Hospital Supply Industry: Update on Angioplasty Catheters and Pacemakers" (New York: PaineWebber, industry report, 1989), 6. See also Seymour Furman, "Comprehension of Pacemaker Cycles," in Seymour Furman, David L. Hayes, and David R. Holmes Jr., *A Practice of Cardiac Pacing* (Mount Kisco, N.Y.: Futura, 1986), 159–218; Jeffrey A. Brinker, "VVI vs. DDD—New Twists to the Ongoing Controversy," editorial in *Intelligence Reports in Cardiac Pacing and Electrophysiology* 9 (June 1990): 1, 4.

52. William B. Stason, Charles A. Saunders, and Hugh C. Smith, "Cardiovascular Care of the Elderly: Economic Considerations," *J Am Coll Cardiol* 10 (August 1987): 18A–21A; S. Serge Barold and Richard S. Sanders, "Rate-Adaptive Cardiac Pacing: Cost versus Technology versus Patient Benefit," *Am Heart J* 125 (June 1993): 1828–34.

53. Bernstein and Parsonnet, "Survey of Cardiac Pacing, 1993": 191; "Guidant (GDT): Re-initiating Coverage" (New York: Morgan Stanley, industry report, 1995), 11–12; cf. "American Heart Association Meeting" (New York; Cowen, industry report, 1996), 69 (www.cowen.com, visited 18 August 1999).

54. Medtronic annual reports, 1978–84; Hidefjäll, "Pace of Innovation," 209; "How an Idea Becomes a Product," *Medtronom* no. 2 (February–March 1976): 5–10.

55. Bobby I. Griffin, interview by author and Patrik Hidefjäll, 30 November 1995, Fridley, Minn.; Jerry Hartlaub, interview by author, 2 July 1997, Columbia Heights, Minn.

56. Kenneth Anderson, interview by author, 23 July 1998, Fridley, Minn.; Hidefjäll, "Pace of Innovation," 228. Leonardo Cammilli's group at the University of Florence in Italy had described an experimental pacer that detected the changes in blood pH level occurring with increased activity: Leonardo Cammilli et al., "A New Pacemaker Auto-Regulating the Rate of Pacing in Relation to Metabolic Needs," in *Proceedings of the Vth International Symposium, Tokyo, March 14–18, 1976*, ed. Yoshio Watanabe (Amsterdam: Excerpta Medica, 1977), 414–19. Around the time Anderson was working on activity sensing, several other groups were investigating ways of altering a pacemaker's stimulus rate based on other indicators of increased metabolic demands such as central venous oxygen content, respiratory rate, or changes in the Q-T interval of the heartbeat.

57. Anderson interview by author.

58. Anderson interview by author.

59. Anderson interview by author; "Activitrax Becomes World's Most Prescribed Pacemaker," *Medtronic News* 16 (winter 1986/87): 16–18.

60. Bobby Griffin interview by author and Hidefjäll.

61. D. P. Humen et al., "A Pacemaker Which Automatically Increases Its Rate with Physical Activity," in *Proceedings of the VIIth World Symposium on Cardiac Pacing, Vienna, May 1st to 5th, 1983*, ed. K. Steinbach (Darmstadt: Steinkopf, 1983), 259–64.

62. "Fresh Spirit of Worldwide Teamwork Forges Medtronic's Victory at Vienna," *Medtronic Pulse*, June 1983, 1–4; "Activitrax Becomes World's Most Prescribed Pacemaker"; Hidefjäll, "Pace of Innovation," app. A; Suzanne Loeffelholz, "Medtronic Outpaces the Pack," *Financial World*, 15 November 1988, 12; Rachael M. Scherer, "Medtronic" (New York: Dain Bosworth, industry report, 1990), 6.

63. See the exchange between Richard Sutton and Anthony Rickards in "Physiology of Dual-Chamber Pacing," *Pacing Clin Electrophysiol* 6 (March–April 1983, pt. 2): 355–56.

64. Bobby Griffin interview by author and Hidefjäll.

65. "The Sky's the Limit!" *Medtronic Pulse*, fall 1985, 2–9; Anderson interview by author.

66. David L. Hayes, "Advances in Pacing Therapy for Bradycardia," *International Journal of Cardiology* 32 (1991): 183–96, quotation at 184–85; Scherer, "Medtronic," 6–8.

67. Hayes, "Advances": 189; Seymour Furman, "Rate Modulated DDD (i.e., DDDR) Pacing," editorial in *Pacing Clin Electrophysiol* 15 (March 1992): 247.

68. "A Brief Look at Pacing in the 1980s with Dr. Seymour Furman," interview in *Medtronic News* 10 (December 1980): 9; J. Warren Harthorne, "The Future of Cardiac Pacing," in *Modern Cardiac Pacing*, ed. Barold, 949–58.

**CHAPTER TEN. PREVENTING SUDDEN CARDIAC DEATH**

1. Thomas Lewis, *Lectures on the Heart* (New York: Paul B. Hoeber, 1915); W. Bruce Fye, "Disorders of the Heartbeat: A Historical Overview from Antiquity to the Mid-20th Century," *Am J Cardiol* 72 (1 November 1993): 1065–67.

2. There appears to be no canonical definition of sudden cardiac death, but the phrase points generally to the spontaneous and rapid onset — within hours or even minutes — of a rhythm disorder that renders the heart unable to pump. Typically this means ventricular fibrillation or some varieties of ventricular tachycardia, but standstill is a possibility, too. Thus a Stokes-Adams attack that did not spontaneously revert to a heartbeat would be considered a form of sudden cardiac death. When the heart fails to pump, circulatory

collapse follows: the victim will lose consciousness within seconds and die within minutes. In most cases, the victim suffers from an underlying form of heart disease such as coronary artery disease (including recent heart attacks), congestive heart failure, hypertrophic cardiomyopathy, or a congenital rhythm disorder. In general, predictive factors for coronary artery disease and heart attack such as age, obesity, cigarette smoking, and elevated cholesterol levels are also predictors of sudden cardiac death: Robert J. Myerburg and Agustin Castellanos, "Cardiac Arrest and Sudden Cardiac Death," in *Heart Disease: A Textbook of Cardiovascular Medicine*, ed. Eugene Braunwald, 4th ed. (Philadelphia: Saunders, 1994), 756–89; John P. DiMarco and David E. Haines, "Sudden Cardiac Death," *Current Problems in Cardiology* 15 (April 1990): 187–232, esp. 187–92.

3. John A. Kastor, "Michel Mirowski and the Automatic Implantable Defibrillator," *Am J Cardiol* 63 (15 April and 1 May 1989): 977–82, 1121–26. A complete bibliography of Mirowski's publications may be found in *Pacing Clin Electrophysiol* 14 (May 1991, pt. 2): 866–71.

4. Kastor, "Michel Mirowski": 977–81; John A. Kastor, "Abnormal Atrial Rhythms, an Early Interest of Michel Mirowski," *Pacing Clin Electrophysiol* 14 (May 1991, pt. 2): 916–19.

5. Kastor, "Michel Mirowski": 981–82; Albert I. Mendeloff, "Michel Mirowski and the Department of Medicine at the Sinai Hospital of Baltimore," *Pacing Clin Electrophysiol* 14 (May 1991, pt. 2): 873–74; Morton M. Mower, "Building the AICD with Michel Mirowski," ibid.: 928.

6. Kastor, "Michel Mirowski": 1122; Mower, "Building the AICD": 928–30. The best account from an engineering standpoint is Steve A. Kolenik et al., "Engineering Considerations in the Development of the Automatic Implantable Cardioverter Defibrillator," *Prog Cardiovasc Dis* 36 (September–October 1993): 115–36.

7. Mower, "Building the AICD": 929.

8. Kastor, "Michel Mirowski": 1123; Morton M. Mower and Robert G. Hauser, "Developmental History, Early Use, and Implementation of the Automatic Implantable Cardioverter Defibrillator," *Prog Cardiovasc Dis* 36 (September–October 1993): 89. In 1970, a second group working independently also reported a tabletop prototype of an implantable defibrillator but did not develop it into a practical device: see John C. Schuder, "Completely Implanted Defibrillator," *J Am Med Assoc* 214 (9 November 1970): 1123. A third team, physician Leo Rubin and electrical engineer Peter Hudson, also pursued the idea at the U.S. Army's Fort Monmouth Lab in the early 1970s and conducted experiments with dogs at Newark Beth Israel Medical Center.

9. Kastor, "Michel Mirowski": 1123; Morton M. Mower, interview by author, 15 October 1998, Baltimore.

10. Michael Toffoli, interview by author and Patrik Hidefjäll, 27 November 1995, Fridley, Minn.

11. Bernard Lown and Paul Axelrod, "Implanted Standby Defibrillators," editorial in *Circulation* 46 (October 1972): 637–39; Kastor, "Michel Mirowski": 1122–23; Mower interview by author.

12. Dwight E. Harken, interview by author, 19 September 1991, Cambridge, Mass.

13. Thomas A. Preston, "The Artificial Heart," in *Worse than the Disease: Pitfalls of Medical Progress*, ed. Diana B. Dutton (Cambridge: Cambridge Univ. Press, 1988), 91–126.

14. Kastor, "Michel Mirowski": 1123, passim; Mower interview by author; quotation is from Toffoli interview by author and Hidefjäll. For Mirowski's personality, see *Pacing Clin Electrophysiol* 14 (May 1991, pt. 2), a special Mirowski memorial issue.

15. Kastor, "Michel Mirowski": 1124.

16. John A. Kastor et al., "Clinical Electrophysiology of Ventricular Tachycardia," *N Engl J Med* 304 (23 April 1981): 1007–8.

17. Samuel A. Levine, *Coronary Thrombosis: Its Various Clinical Features* (Baltimore: Williams & Wilkins, 1929); W. Bruce Fye, "Acute Myocardial Infarction: A Historical Summary," in *Management of Acute Myocardial Infarction*, ed. Bernard J. Gersh and Shahbudin H. Rahimtoola (New York: Elsevier Science, 1990), 3–13; W. Bruce Fye, "A Historical Perspective on Atherosclerosis and Coronary Artery Disease," in *Atherosclerosis and Coronary Artery Disease*, ed. Valentin Fuster, Richard Ross, and Eric J. Topol (Philadelphia: Lippincott-Raven, 1996), 1–12. Thrombosis refers to the blockage of a blood vessel (in this case, a coronary artery) by a thrombus, or blood clot; heart attacks can also be brought on by the narrowing of a coronary artery (ischemia) because of the gradual buildup of fatty deposits (atherosclerosis). For VF that began in the operating room, Clude Beck's open-chest cardiac massage and defibrillation (1947) could sometimes save the patient; see ch. 2.

18. Paul M. Zoll et al., "Treatment of Unexpected Cardiac Arrest by External Electric Stimulation of the Heart," *N Engl J Med* 254 (22 March 1956): 541–46; Paul M. Zoll, Arthur J. Linenthal, and Leona R. Norman Zarsky, "Termination of Ventricular Fibrillation in Man by Externally Applied Electric Countershock," ibid. (19 April 1956): 727–32; Paul M. Zoll et al. "Ventricular Fibrillation: Treatment and Prevention by External Electric Currents," *N Engl J Med* 262 (21 January 1960): 105–12; William B. Kouwenhoven, James R. Jude, and Guy Knickerbocker, "Closed-Chest Cardiac Massage," *J Am Med Assoc* 173 (9 July 1960): 1064–67; Bernard Lown, Raghavan Amarasingham, and Jose Neuman, "New Method for Terminating Cardiac Arrhythmias: Use of Synchronized Capacitor Discharge," *J Am Med Assoc* 182 (3 November 1962): 549–55; "Single Shock Reverses Cardiac Arrhythmias," *Med World News* 3 (25 May 1962): 22–23; Fye, "Acute Myocardial Infarction": 10; W. Bruce Fye, "Ventricular Fibrillation and Defibrillation: Historical Perspectives with Emphasis on the Contributions of John MacWilliam, Carl Wiggers, and William B. Kouwenhoven," *Circulation* 71 (May 1985): 858–65.

19. Hughes W. Day, "An Intensive Coronary Care Area," *Dis Chest* 44 (October 1963): 423–27; Paul M. Zoll, "The Cardiac Monitoring System," interview in *J Am Med Assoc* 186 (2 November 1963): 34–36; "When Every Second Counts," *Med World News* 7 (18 February 1966): 104–12; W. Bruce Fye, *American Cardiology: The History of a Specialty and Its College* (Baltimore: Johns Hopkins Univ. Press, 1996), 176–81.

20. Lewis Kuller, Abraham Lilionfeld, and Russell Fisher, "An Epidemiological Study of Sudden and Unexpected Deaths in Adults," *Medicine* 46 (July 1967): 341–61; Lewis Kuller, "Sudden Death in Arteriosclerotic Heart Disease: The Case for Preventive Medicine," *Am J Cardiol* 24 (November 1969): 617–28.

21. Philip R. Reid, "The Automatic Implantable Defibrillator: Current Status and Future Development," in *Modern Cardiac Pacing*, ed. S. Serge Barold (Mount Kisco, N.Y.: Futura, 1985), 745; Robert J. Myerburg, "Epidemiology of Ventricular Tachycardia/Ventricular Fibrillation and Sudden Cardiac Death," *Pacing Clin Electrophysiol* 9 (November–December 1986, pt. 2): 1335.

22. Myerburg, "Epidemiology": 1334–38, reporting on a number of studies. See also Robert J. Myerburg, Kenneth M. Kessler, and Agustin Castellanos, "Pathophysiology of Sudden Cardiac Death," *Pacing Clin Electrophysiol* 14 (May 1991, pt. 2): 935–43.

23. DiMarco and Haines, "Sudden Cardiac Death": 197–203 (antiarrhythmic drugs), 203–5 (community emergency services); Mickey Eisenberg, *Life in the Balance: Emergency Medicine and the Quest to Reverse Sudden Death* (New York: Oxford Univ. Press, 1997); D. W. Weaver et al., "Use of the Automatic External Defibrillator in the Management of Out-of-Hospital Cardiac Arrest," *N Engl J Med* 310 (15 September 1988): 661–66.

24. M. Stephen Heilman, "Collaboration with Michel Mirowski on the Development of the AICD," *Pacing Clin Electrophysiol* 14 (May 1991, pt. 2): 910–15.

25. Kolenik et al., "Engineering Considerations": 116–19. In 1985, a second patch electrode was added to lower the defibrillation threshold.

26. Kolenik et al., "Engineering Considerations": 122–24; Mower and Hauser, "Developmental History": 89–90.

27. Michel Mirowski et al., "Automatic Implantable Defibrillator," in *Artificial Cardiac Pacing: Practical Approach*, ed. Edward K. Chung, 2d ed. (Baltimore: Williams & Wilkins, 1984), 300; Kolenik et al., "Engineering Considerations": 119–21, quotation at 119.

28. Levi Watkins Jr. and Eric Taylor Jr., "The Surgical Aspects of Automatic Implantable Cardioverter-Defibrillator Implantation," *Pacing Clin Electrophysiol* 14 (May 1991, pt. 2): 953–59.

29. Kolenik et al., "Engineering Considerations": 129; Robert G. Hauser and M. Stephen Heilman, "The Industrialization of the ICD," *Pacing Clin Electrophysiol* 14 (May 1991, pt. 2): 906; Heilman, "Collaboration": 911.

30. Kolenik et al., "Engineering Considerations": 129; Kastor, "Michel Mirowski": 1124; Abhijit Acharya, "Regulatory Issues in the Development and Future of the AICD," *Pacing Clin Electrophysiol* 14 (May 1991, pt. 2): 880–82.

31. Heilman, "Collaboration": 912–13.

32. Alan D. Bernstein et al., "Patients' Attitudes toward Implanted-Defibrillator Therapy," abstract in *Pacing Clin Electrophysiol* 19 (April 1996, pt. 2): 605.

33. Mower interview; Patrik Hidefjäll, "The Pace of Innovation: Patterns of Innovation in the Cardiac Pacemaker Industry," Ph.D. diss., Linköping University, Sweden, 1997, 264; Heilman, "Collaboration": 913; confidential industry sources. Under contract with Medrad/Intec, CPI had already developed the rate-sensing lead for the ICD. Intec was liquidated in 1986 but Medrad continued as an independent company. Today it is a supplier of diagnostic imaging devices.

34. Hauser and Heilman, "Industrialization of the AICD": 907–8.

35. Ibid.

36. Mirowski et al., "Automatic Implantable Defibrillator": 302–3.

37. Myerburg, "Epidemiology."

38. Philip R. Reid et al., "Clinical Evaluation of the Internal Automatic Cardioverter-Defibrillator in Survivors of Sudden Cardiac Death," *Am J Cardiol* 51 (June 1983): 1608–13; Kolenik et al., "Engineering Considerations": 128–35.

39. Masood Akhtar et al., "Sudden Cardiac Death: Management of High-Risk Patients," *Ann Intern Med* 14 (March 1991): 504–6; Mower and Hauser, "Developmental History": 92; Acharya, "Regulatory Issues": 880–81.

40. Acharya, "Regulatory Issues": 881; Kastor et al., "Clinical Electrophysiology";

DiMarco and Haines, "Sudden Cardiac Death": 209–12; Peter Chang-Sing and C. Thomas Peter, "Sudden Death: Evaluation and Prevention," *Cardiol Clin* 9 (November 1991): 653–64; Leonard N. Horowitz, "Clinical Cardiac Electrophysiology: History, Rationale, and Future," *Cardiol Clin* 4 (August 1986): 353–64, quotation at 355.

41. Horowitz, "Clinical Cardiac Electrophysiology"; Fye, *American Cardiology*, 305–6.

42. Gregory de Lissovoy and Thomas Guarneri, "Cost-effectiveness of the Implantable Cardioverter-Defibrillator," *Prog Cardiovasc Dis* 36 (November–December 1993): 209–10.

43. The first patent on the ICD was U.S. patent no. 3,614,954, "Electronic Standby Defibrillator," filed 9 February 1970, awarded 26 October 1971 to Mieczyslaw Mirowski et al.; a number of patents followed. In the United States, a patent is good for 17 years from the date of the award.

44. Reid, "Automatic Implantable Defibrillator," 746–50; Seymour Furman, "Patient Longevity with Implanted Cardioverter-Defibrillators," editorial in *Pacing Clin Electrophysiol* 13 (October 1990): 1219; Seymour Furman, "Evaluation of ICD Therapy," editorial in *Pacing Clin Electrophysiol* 15 (May 1992): 713–14.

45. John C. Schuder, "The ICD — Progress, Prospects, and Problems," *Pacing Clin Electrophysiol* 20 (October 1997, pt. 1): 2367–70; Roger A. Winkle, "State-of-the-Art of the AICD," *Pacing Clin Electrophysiol* 14 (May 1991, pt. 2): 964–65.

46. Winkle, "State-of-the-Art"; Hidefjäll, "Pace of Innovation," 268.

47. J. Thomas Bigger Jr., "Prophylactic Use of Implantable Cardioverter Defibrillators," *Pacing Clin Electrophysiol* 14 (February 1991, pt. 2): 376–80; "Ventritex, Inc." (San Francisco: Robertson, Stephens, industry report, 1992), 5–7; "Ventritex, Inc." (San Francisco: Hambrecht & Quist, industry report, 1992), 23.

48. "A Heartbeat Away from Approval" (San Francisco: Hambrecht & Quist, industry report, 1993), 3.

49. Dwight E. Harken (1969) as quoted in Mower, "Building the AICD": 928; Fisher (1990) quoted in Seah Nisam, Morton Mower, and Suzan Moser, "ICD Clinical Update: First Decade, Initial 10,000 Patients," *Pacing Clin Electrophysiol* 14 (February 1991, pt. 2): 260.

50. "Hospital Supply Industry: Implantable Defibrillators" (New York: PaineWebber, industry report 1989), 3.

51. Michael G. Mullen et al., "Tachyarrhythmia Management" (New York: Cowen, industry report, 1992), 15; "Ventritex, Inc." (Hambrecht & Quist), 16.

52. "Tachyarrhythmia Management," 12; "Medtronic" (New York: Shearson Lehman Brothers, industry report, 1992), 6.

53. Albert W. Preston Jr. and Thomas A. Preston, "The Patent System and Cardiac Pacing: Is the System Serving Its Users?" *Pacing Clin Electrophysiol* 8 (July–August 1985): 476–83.

54. "Ventritex, Inc." (Hambrecht & Quist), 25.

55. *Eli Lilly & Co. v. Medtronic, Inc.*, 496 U.S. 661 (1990).

56. Jonathan Kalstrom, "Medtronic, Lilly Settle Costly Legal Battle Behind Closed Doors," *Law & Politics* (Minneapolis), June 1991: 7; Hidefjäll, "Pace of Innovation," 265.

57. Preston and Preston, "Patent System"; Winkle, "State-of-the-Art": 963–64; Seymour Furman, "Pacemakers and Patents," editorial in *Pacing Clin Electrophysiol* 8 (July–August 1985): 475.

58. This case history appears in *Cardiovascular Network News* 4 (November 1997): 12–13.

59. David S. Cannom, "Implantable Cardioverter Defibrillator: The Promise and Perils of an Evolving Technology," editorial in *Pacing Clin Electrophysiol* 15 (January 1992): 1–3.

60. J. Thomas Bigger Jr., "Future Studies with the Implantable Cardioverter Defibrillator," *Pacing Clin Electrophysiol* 14 (May 1991, pt. 2): 883–89; Soo G. Kim et al., "Benefits of Implantable Defibrillators Are Overestimated by Sudden Death Rates and Better Represented by the Total Arrhythmic Death Rate," *J Am Coll Cardiol* 17 (June 1991): 1587–92; Richard W. Henthorn, Theodore J. Waller, and Loren F. Hiratzka, "Are the Benefits of the Automatic Implantable Cardioverter-Defibrillator (AICD) Overestimated by Sudden Death Rate?" ibid.: 1593–94.

61. NIH, National Heart, Lung, and Blood Institute, "NHLBI Stops Arrhythmia Study — Implantable Cardiac Defibrillators Reduce Deaths," news release, 14 April 1997 (www.nih.gov/news/pr/apr97/nhlbi-14.htm, visited 3 June 1997); "Technology, Trials Stirring Things Up in Rhythm Management," *Cardiovascular Network News* 4 (November 1997): 1–2. In spring 1996, NHLBI had stopped another clinical trial called MADIT (Multicenter Automatic Defibrillator Implantation Trial). MADIT focused on ICD patients who had had heart attacks: "MADIT Makes It a Bigger Market for Defibrillators," *Cardiovascular Network News* 3 (November 1996): 1–3. For an overview, see Cynthia M. Tracy, "Current Review of Arrhythmia Trials," *Cardiology Special Edition* (winter/spring 1997): 10–16. A recent and important report on ICD therapy is Alfred E. Buxton et al., "A Randomized Study of the Prevention of Sudden Death in Patients with Coronary Artery Disease," *N Engl J Med* 341 (16 December 1999): 1882–90.

62. Judith L. Wagner and Michael Zukhoff, "Medical Technology and Hospital Costs," in *Technology and the Future of Health Care*, ed. John B. McKinlay (Cambridge: MIT Press, 1982), 104–30; M. F. Drummond, "Economic Evaluation and the Rational Diffusion and Use of Health Technology," *Health Policy* 7 (June 1987): 309–24; Louise B. Russell et al., "The Role of Cost-Effectiveness Analysis in Health and Medicine," *J Am Med Assoc* 276 (9 October 1996): 1172–77; Alan Maynard, "Is High-Technology Medicine Cost-Effective?" in *Bicycling to Utopia: Essays on Science and Technology*, ed. P. Day and C. R. A. Catlow (Oxford: Oxford Univ. Press, 1995), 159–69; T. Joseph Reeves, "Advances in Cardiology and Escalating Costs," *Circulation* 71 (April 1985): 637–41.

63. Drummond, "Economic Evaluation"; Russell et al., "Role of Cost-Effectiveness Analysis"; Frank A. Sloan, *Valuing Health Care: Costs, Benefits, and Effectiveness of Pharmaceuticals and Other Medical Technologies* (New York: Cambridge Univ. Press, 1995).

64. Miriam Kupperman et al., "An Analysis of the Cost Effectiveness of the Implantable Defibrillator," *Circulation* 81 (January 1990): 94; De Lissovoy et al., "Cost-Effectiveness": 210.

65. Kupperman et al., "Analysis": 95–96.

66. De Lissovoy et al., "Cost-Effectiveness"; Kupperman et al., "Analysis"; Joel Kupersmith et al., "Evaluating and Improving the Cost-Effectiveness of the Implantable Cardioverter-Defibrillator," *Am Heart J* 130 (September 1995): 107–15; Medtronic, "ICD Therapy Cost-Effectiveness Evaluation" (brochure, 1997), citing numerous studies; R. N. Hauer, R. Dirkse, and E. F. Wever, "Can Implantable Cardioverter-Defibrillator Therapy Reduce Healthcare Costs?" *Am J Cardiol* 78 (12 September 1996): 134–39.

67. De Lissovoy et al., "Cost-Effectiveness": 212–13.

68. Nick Bosanquet, "Health Economics: Finance, Budgeting, and Insurance," in *Companion Encyclopedia of the History of Medicine*, vol. 2, ed. W. F. Bynum and Roy Porter (London: Routledge, 1994), 1386; Anne Elixhauser and Joan M. Wechsler, *The Cost Effectiveness of the Primary Prevention of Cardiovascular Disease: A Review of the Literature* (Washington, D.C.: Battelle Medical Technology Assessment and Policy Research Center, 1990); Theodore M. Porter, *Trust in Numbers: The Pursuit of Objectivity in Science and Public Life* (Princeton: Princeton Univ. Press, 1995), ch. 7.

69. Medtronic, "ICD Therapy"; cf. G. C. Larsen et al., "Cost-Effectiveness of the Implantable Cardioverter-Defibrillator: Effect of Improved Battery Life and Comparison with Amiodarone Therapy," *J Am Coll Cardiol* 19 (May 1992): 1323–34; and D. K. Owens et al., "Cost-Effectiveness of Implantable Cardioverter Defibrillators relative to Amiodarone for Prevention of Sudden Cardiac Death," *Ann Intern Med* 126 (1 January 1997): 1–12; M. F. Drummond, "Evaluation of Health Technology: Economic Issues for Health Policy and Policy Issues for Economic Appraisal," *Soc Sci Med* 38 (June 1994): 1593–600; de Lissovoy et al., "Cost-Effectiveness": 212.

70. Schuder, "The ICD": 2369.

71. "American Heart Association Meeting" (New York: Cowen, industry report, 1996), 59 (www.cowen.com, visited 18 August 1999).

72. Fye, *American Cardiology*, 301–6.

**CHAPTER ELEVEN. THE 1990S AND BEYOND**

1. Compare "Guidant (GDT): Re-Initiating Coverage" (New York: Morgan Stanley, industry report, 1995), 11–12; "American Heart Association Meeting" (New York: Cowen, industry report, 1996), 59, 68 (www.cowen.com, visited 18 August 1999); T. A. Hodgson and A. J. Cohen, "Medical Care Expenditures for Selected Circulatory Diseases," *Medical Care* 37 (October 1999): 994–1012, and the data presented in table 11.1.

2. U.S. Bureau of the Census, Current Population Reports, Special Studies, *65+ in the United States* (P23-190, Washington, D.C.: GPO, 1996); "Pacemakers and Implantable Defibrillators," *Cardiovascular Network News* 1 (November 1994): 1, 3–7; Alan D. Bernstein and Victor Parsonnet, "Preliminary Results: Survey of Cardiac Pacing and ICD Practice Patterns, 1997 Results" (unpublished).

3. Bernstein and Parsonnet, "Preliminary Results"; Joel Kupersmith et al., "Cost-Effectiveness Analysis in Heart Disease," part 3: "Ischemia, Congestive Heart Failure, and Arrhythmias," *Prog Cardiovasc Dis* 37 (March–April 1995): 334–35. The new indication for pacemaker implantation was heart block deliberately induced by catheter ablation of conduction cells, a procedure to be discussed below.

4. Bernstein and Parsonnet, "Preliminary Results"; T. Bruce Ferguson Jr., Bruce D. Lindsay, and John P. Boineau, "Should Surgeons Still Be Implanting Pacemakers?" *Ann Thorac Surg* 57 (March 1994): 588–97.

5. H. David Banta, Clyde J. Behney, and Jane Sisk Williams, *Toward Rational Technology in Medicine* (New York: Springer, 1981), 34–35; Victor R. Fuchs, "Economics, Values, and Health Care Reform," *American Economic Review* 86 (March 1996): 1–24.

6. U.S. Congress, OTA, *Diagnosis Related Groups (DRGs) and the Medicare Program: Implications for Medical Technology*, OTA-TM-H-17 (July 1983); John K. Iglehart, "Medicare Begins Prospective Payment of Hospitals," *N Engl J Med* 308 (9 June 1983): 1428–

32; John K. Iglehart, "The American Health Care System: Private Insurance," *N Engl J Med* 326 (18 June 1992): 1715–20.

7. John K. Iglehart, "The Struggle over Physician-Payment Reform," *N Engl J Med* 325 (12 September 1991): 823–28.

8. Jim Chandler, "The United States of America," in *Health Care Systems in Liberal Democracies*, ed. Ann Wall (London: Routledge, 1996), 169; Uwe E. Reinhardt, "The United States: Breakthroughs and Waste," *J Health Polit Policy Law* 17 (winter 1992): 637–66; David Wilsford, "States Facing Interests: Struggles over Health Care Policy in Advanced, Industrial Democracies," *J Health Polit Policy Law* 20 (fall 1995): 571–613; Robert Harrison, *State and Society in Twentieth Century America* (London: Longman, 1997).

9. Elizabeth W. Hoy, R. E. Curtis, and T. Rice, "Change and Growth in Managed Care," *Health Affairs* 10 (winter 1991): 18–36; John K. Iglehart, "The American Health Care System: Managed Care," *N Engl J Med* 327 (3 September 1992): 742–47; Allen Buchanan, "Managed Care: Rationing without Justice, But Not Unjustly," *J Health Polit Policy Law* 23 (August 1998): 617–34. The growth of HMOs may be dated to passage of the Health Maintenance Organization Act of 1973; for background, see Paul Starr, *The Social Transformation of American Medicine* (New York: Basic Books, 1982), 395–405.

10. Michael M. Weinstein, "In Health, Be Careful What You Wish For," *New York Times*, 31 May 1998; Alain Enthoven, quoted in Michael M. Weinstein, "Managed Care's Other Problem: It's Not What You Think," ibid., 28 February 1999.

11. Iglehart, "The American Health Care System: Managed Care"; Jonathan P. Weiner and Gregory de Lissovoy, "Razing a Tower of Babel: A Taxonomy for Managed Care and Health Insurance Plans," *J Health Polit Policy Law* 18 (spring 1993): 75–103; W. Bruce Fye, *American Cardiology: The History of a Specialty and Its College* (Baltimore: Johns Hopkins Univ. Press, 1996), 327–39.

12. Louise B. Russell and Jane E. Sisk, "Medical Technology in the United States: The Last Decade," *Int J Tech Assess in Health Care* 4 (1988): 269–86; John K. Iglehart, "The American Health Care System: Medicare," *N Engl J Med* 327 (12 November 1992): 1467–72; Greg Freiherr, "Growing Pains: The Medical Device Industry in the 1980s," *MD&DI* 14 (June 1992); Regina E. Herzlinger, "The Managerial Revolution in the U.S. Health Care Sector: Lessons from the U.S. Economy," *Health Care Management Review* 23 (summer 1998): 19–29.

13. Rosemary Stevens, *In Sickness and in Wealth: American Hospitals in the Twentieth Century* (New York: Basic Books, 1989), 302, 337; Thomas J. Duesterberg, David J. Weinschrott, and David C. Murray, *Health Care Reform, Regulation, and Innovation in the Medical Device Industry* (Indianapolis, Ind.: Hudson Institute, 1994), 39–41; Greg Freiherr, "Equipment-Management Deal Signals New Era in Vendor-Customer Relations," *MD&DI* 17 (July 1995); Herzlinger, "Managerial Revolution."

14. Richard Sanders (Sulzer Intermedics), interview by author, 16 March 1997, Anaheim, Calif.; "Device Costs as Percent of Payment," *Cardiovascular Network News* 2 (November 1995): 13; "Docs' Purchasing Clout Fizzling," *Hospitals & Health Networks* 69 (5 June 1995): 22; Jane Sarasohn-Kahn, "Adapting to a Managed-Care Environment," *MD&DI* 16 (May 1994); Greg Freiherr, "Merger Mania Strikes Device Industry," *MD&DI* 18 (January 1996); Candace Littell, "Redefining Product Value in Today's Market," ibid. (July 1996); Robert R. Dunford, "What Manufacturers Need to Know about Managed Care," *MD&DI* 19 (July 1997).

15. "Setting the Pace in Negotiations," *Cardiovascular Network News* 3 (November 1996): 12–14. The article does not name the manufacturers.

16. Suzanne Houck, "Revolution in Managed-Care Purchasing: Effects of Capitation on Health-Care Marketing," *MD&DI* 16 (December 1994); Susan Gilbert, "First Read the Directions, Then Do No Harm," *New York Times*, 22 November 1998; Karen Padley, "Industry Pulse Quickens with Unveiling of Pacemakers," *St. Paul Pioneer Press*, 14 January 1999; Karen Padley, "Gone in a Heartbeat," *St. Paul Pioneer Press*, 18 July 1999; Patrick K. Edeburn (Medtronic), telephone interview by author, 28 June 1999.

17. Littell, "Redefining Product Value."

18. Thomas E. Holloran, interview by author, 4 March 1997, Minneapolis; A. Jay Graf, "Challenges to Competitiveness in a Changing Medical Device Market," *MD&DI* 19 (January 1997). Holloran, briefly CEO of Medtronic in the 1970s, has been on the company board since 1961; Graf was president of CPI, the cardiac-rhythm management business of Guidant Corporation.

19. Medtronic annual report, 1999; Steve Alexander, "The Wait Is Over: St. Jude Announces Big Acquisition," *Minneapolis Star Tribune*, 29 June 1994; "Companies Review Past, Eye the Future at Two Conferences," *Cardiovascular Device Update* 1 (December 1995): 6–9; Lawrence M. Fisher, "$505 million St. Jude Deal for Maker of Defibrillators," *New York Times*, 24 October 1996; Winston R. Wallin, talk at the Bakken Library and Museum, 21 October 1997, transcript in author's possession.

20. Wallin talk.

21. "Medtronic Inc." (New York: Dain Bosworth, industry report, 1990); Steve Gross, "Medtronic Appears Ready for Growth," *Minneapolis Star Tribune*, 8 April 1991.

22. Winston R. Wallin, letter to employees, 21 July 1986, and "Medtronic's Long-Range Plan," both in *Medtronic Pulse* 4 (summer 1986): 2–5; Wallin talk; Medtronic annual reports, 1986–97; Alyssa A. Lappen, "Took a Licking but Keeps on Ticking," *Forbes*, 3 October 1988, 176–77; Wallin interview by David J. Rhees, 3 April 1998, Minneapolis. Wallin continued as nonexecutive board chairman until 1996.

23. Medtronic annual reports, 1985 and 1996; Karen Padley, "Work in Progress," *St. Paul Pioneer Press*, 20 December 1998.

24. Guidant Corp., 10-K filings dated 17 March 1998 and 24 March 1999; David J. Morrow, "Consolidation among Medical-Device Makers," *New York Times*, 22 September 1998.

25. "Guidant to Purchase Sulzer Medica's Electrophysiology Business for up to $850 Million," Guidant Corp. news release, 21 September 1998; Karen Padley, "Guidant to Acquire Sulzer Unit," *St. Paul Pioneer Press*, 22 September 1998; Morrow, "Consolidation."

26. "Heart Valve Procedures Increase," *Cardiovascular Network News* 3 (February 1997): 6–9; St. Jude Medical annual report, 1993.

27. St. Jude Medical annual reports, 1995 and 1996; Steve Alexander, "The Wait Is Over: St. Jude Announces Big Acquisition," *Minneapolis Star Tribune*, 29 June 1994; Fisher, "$505 million St. Jude Deal"; Judith Yates Borger, "From Rags, to Ruin to Riches," *St. Paul Pioneer Press*, 23 June 1996.

28. Terry Fiedler, "New CEO, Old Problems," *Minneapolis Star Tribune*, 3 May 1999; Edeburn interview by author; "American Heart Association Meeting."

29. Medtronic annual report, 1976; Bobby I. Griffin, interview by author and Patrik Hidefjäll, 30 November 1995, Fridley, Minn.

30. Victor Fuchs, "The New Technology Assessment," *N Engl J Med* 323 (6 Septem-

ber 1990): 673–77; Joel Kupersmith et al., "Cost-Effectiveness Analysis in Heart Disease," part 1: "General Principles," *Prog Cardiovasc Dis* 37 (November–December 1994): 161–84; Graf, "Challenges to Competitiveness"; John Bethune, "The Cost-Effectiveness Bugaboo," *MD&DI* 19 (April 1997); Holloran interview by author.

31. Confidential industry sources.

32. "American Heart Association Meeting," 66–71; Medtronic and Guidant annual reports, 1996–98; conversations with sales reps in the pacemaker industry.

33. Richard Sutton and Ivan Bourgeois, *Foundations of Cardiac Pacing, Part 1* (Mount Kisco, N.Y.: Futura, 1991), chs. 9, 11, 12; Philippe Ritter, Serge Cazeau, and Jacques Mugica, "Do We Really Need Automatic Pacemakers?" in *New Perspectives in Cardiac Pacing*, ed. S. Serge Barold and Jacques Mugica (Mount Kisco, N.Y.: Futura, 1993), 337–46; Medtronic news release, 24 June 1999.

34. Product information on the Guidant Model 2901 Programmer/Recorder/Monitor at www.guidant.com and on the Medtronic 9790 Programmer at www.medtronic.com; "Increasing Efficiency in the Pacemaker Clinic," panel discussion in *Guidant/CPI Pacing Dynamics* (first quarter 1998), 6–7, 12.

35. "Sweet Tip Rx Pacing Leads" (product brochure, Guidant Corp., 1998); "Locator Steerable Stylet" (product brochure, St. Jude Medical, 1997).

36. Freiherr, "Merger Mania"; John Bethune, "The Device Industry Looks Up — and to the East," *MD&DI* 18 (March 1996); Stephen Lipin, "Wave of Corporate Mergers Continues to Gather Force," *Wall Street Journal*, 26 February 1997.

37. M. R. Chassin et al., "Variations in the Use of Medical and Surgical Services by the Medicare Population," *N Engl J Med* 314 (30 January 1986): 285–90; William B. Stason, Charles A. Sanders and Hugh C. Smith, "Cardiovascular Care of the Elderly: Economic Considerations," *J Am Coll Cardiol* 10 (August 1987): 18A–21A; W. P. Welch et al., "Geographic Variation in Expenditures for Physicians' Services in the United States," *N Engl J Med* 328 (4 March 1993): 621–27; Sanjeev Saksena, "Practice Guidelines for Management of Cardiac Dysrhythmias: Is It Now Time?" *Pacing Clin Electrophysiol* 16 (June 1993): 1340–41; Fye, *American Cardiology*, 299–301.

38. Arnold M. Epstein and David Blumenthal, "Physician Payment Reform: Past and Future," *Milbank Q* 71 (1993): 193–215; Fye, *American Cardiology*, 322–27.

39. Robert Pear, "Medicare Lays Out Revised Schedule for Doctors' Fees," *New York Times*, 31 May 1991; John K. Iglehart, "Payment of Physicians under Medicare," *N Engl J Med* 318 (31 March 1988): 863–68; John K. Iglehart, "The New Law on Medicare's Payments to Physicians," *N Engl J Med* 322 (26 April 1990): 1247–52; John K. Iglehart, "The Struggle over Physician-Payment Reform," *N Engl J Med* 325 (12 September 1991): 823–28; Sanjeev Saksena, "Physician Payment Reform: An Arctic Wind or a Hint of Spring?" *Pacing Clin Electrophysiol* 16 (March 1993, pt. 1): 488–90; "Medicare Physician Payments Decrease," *Cardiovascular Network News* 3 (February 1997): 14–15; Lynn Etheredge, "The Medicare Reforms of 1997: Headlines You Didn't Read," *J Health Polit Policy Law* 23 (June 1998): 573–79.

40. Fye, *American Cardiology*, 325–26; Daniel E. Nickelson in a session on "The Medicare Physician Fee Schedule," NASPE 20th Annual Scientific Sessions, Toronto, 14 May 1999.

41. Daniel Callahan in "The Future of Health Care," roundtable discussion in *Harvard Magazine*, March–April 1999, 45–46 (italics added); Richard Gorlin, "Must Cardiology Lose Its Heart?" *J Am Coll Cardiol* 19 (June 1992): 1636–37 (italics added).

42. Penny S. Mills and Marie E. Michnich, "Managed Care and Cardiac Pacing and Electrophysiology," *Pacing Clin Electrophysiol* 16 (August 1993): 1746–50; Marie E. Michnich, "The Cardiovascular Specialist's Response to Managed Care," in a session on "Managed Care in Pacing and Electrophysiology," NASPE 17th Annual Scientific Sessions, Seattle, 18 May 1996; Caroline Poplin, "Mismanaged Care," *Wilson Quarterly* 20 (summer 1996): 12–24.

43. Heather Wood Ion, "Ethical Dilemmas in Managed Care," in *Managed Care and the Cardiac Patient,* ed. Richard Ott, Teresa Tanner, and Bryn Henderson (Philadelphia: Hanley & Belfus; St. Louis: Mosby, 1995), 115–29, esp. 122. I do not mean to imply that Ion is an enthusiastic advocate of managed care.

44. "The Hippocratic Oath," in *Medicine and Western Civilization,* ed. David J. Rothman, Steven Marcus, and Stephanie A. Kiceluk (New Brunswick, N.J.: Rutgers Univ. Press, 1995), 261–62.

45. David Mechanic, *From Advocacy to Allocation: The Evolving American Health Care System* (New York: Free Press, 1986), ch. 3; John K. Iglehart, "The American Health Care System — Expenditures," *N Engl J Med* 340 (7 January 1999): 70–76; Gorlin, "Must Cardiology Lose Its Heart?": 1637; W. Bruce Fye, "Managed Care and Patients with Cardiovascular Disease," editorial in *Circulation* 97 (19 May 1998): 1895–96.

46. W. Bruce Fye, "Cardiology Past, Present, and Future" (the Simon Dack Lecture, ACC, 46th Annual Scientific Session, Anaheim, Calif., 17 March 1997); Thomas J. Ryan et al., "Cardiovascular Disease as an Increasing Global Health Issue" (symposium, AAC, 46th Annual Scientific Session, Anaheim, Calif., 19 March 1997).

47. Walter J. Unger, "Implications of Healthcare Reform for Cardiologists," *Journal of Invasive Cardiology* 6 (March 1994): 36–41.

48. Sanger Clinic web site (www.sangerclinic.com, visited 26 August 1999); Stephen L. Wagner, interview by author, 8 August 1999, Maui, Hawaii. Wagner is CEO of the Sanger Clinic.

49. Regina Herzlinger, *Market Driven Health Care* (Reading, Mass.: Perseus Books, 1997); Herzlinger, "Managerial Revolution."

50. Robert W. Emery et al., "Experience in the Development of Continuous Quality Improvement and Managed Care at the Minneapolis Heart Institute," in *Managed Care and the Cardiac Patient,* ed. Ott, Tanner, and Henderson, 133–51; J. E. Moller, E. H. Simonsen, and M. Moller, "Impact of Continuous Quality Improvement on Selection of Pacing Mode and Rate of Complications in Permanent Pacing," *Heart* 77 (April 1997): 357–62.

51. Rhonda L. Rundle, "Can Managed Care Manage Costs?" *Wall Street Journal,* 9 August 1999; Eli Ginzberg, "The Uncertain Future of Managed Care," *N Engl J Med* 340 (14 January 1999): 144–46; Peter D. Jacobson, "Legal Challenges to Managed Care Cost Containment Programs: An Initial Assessment," *Health Affairs* 18 (July–August 1999): 69–85.

52. Richard Caso, "Cardiology Practice — Past, Present, and Future," in *Managed Care and the Cardiac Patient,* ed. Ott et al., 221; Fye, "Cardiology Past, Present, and Future."

53. Richard M. Levy, "A Vision for the U.S. Health-Care System: The Leveraging of Medical Device Technology," *MD&DI* 16 (November 1994).

54. "Electrophysiology Market Surging on Developments in Three Segments," *Cardiovascular Device Update* 2 (June 1996): 1–4.

55. "Cardiology Market Expanding Despite Limits of Managed Care," *Cardiovascular Device Update* 2 (May 1996): 1–7; "Electrophysiology Market Surging"; Ryan et al., "Cardiovascular Disease as an Increasing Global Health Issue."

56. Duesterberg, Weinschrotti, and Murray, *Health Care Reform, Regulation, and Innovation*, ch. 3; Jerry C. Griffin, interview by author, 8 May 1998, San Diego, Calif. Griffin, an electrophysiologist and former president of NASPE, was an executive with InControl.

57. Patrik Hidefjäll, "Can Pacemaker Innovation Go on Forever?" (unpublished, 1999; cited by permission).

58. Byron J. Allen and Michael A. Brodsky, "Cost-Effective Management of Congestive Heart Failure and Cardiac Arrhythmias in the Managed Care Setting," in *Managed Care and the Cardiac Patient*, ed. Ott, Tanner, and Henderson, 229–34; AHA web site (www.amhrt.org/statistics/, data for 1996).

59. Lawrence K. Altman, "New Therapy Combats 2 Kinds of Racing Heart," *New York Times*, 6 June 1991; Melvin M. Scheinman, "North American Society of Pacing and Electrophysiology (NASPE) Survey on Radiofrequency Catheter Ablation: Implications for Clinicians, Third Party Insurers, and Government Regulatory Agencies," *Pacing Clin Electrophysiol* 15 (December 1992): 2228–31; Lemaitre, "American Heart Association Meeting," 77–80; Donna M. Gallick, "Radiofrequency Catheter Ablation for the Treatment of Cardiac Arrhythmias," *Cardiology Special Edition* 3 (winter/spring 1997): 75–77.

60. Final Program, NASPE 20th Annual Scientific Sessions, 12–15 May 1999, Toronto; J. C. Daubert et al., "Permanent Left Ventricular Pacing with Transvenous Leads Inserted into the Coronary Veins," *Pacing Clin Electrophysiol* 21 (January 1998, pt. 2): 239–45; Leslie A. Saxon, "Pacing to Treat Congestive Heart Failure," *Guidant Pacing Dynamics* (first/second quarter 1999), 10–12.

61. "Guidant Announces First Implant of CONTAK CD/EASYTRAK System for Treatment of Congestive Heart Failure," news release, 25 February 1999 (www.guidant .com); "St. Jude Medical Announces First U.S. Implant of Aescula LV lead," news release, 23 June 1999 (www.sjm.com); "Medtronic Announces First Implant of New Device for Heart Failure Patients at Risk of Sudden Cardiac Death," news release, 28 June 1999 (www.medtronic.com); Karen Padley, "A Race for Life," *St. Paul Pioneer Press*, 16 January 2000. The device manufacturers' web sites highlight the progress of new technology from early trials to market release.

62. David L. Hayes, "Advances in Pacing Therapy for Bradycardia," *International Journal of Cardiology* 32 (August 1991): 183.

63. Seymour Furman et al., "Application of High Technology in the Diagnosis and Treatment of the Elderly," *J Am Coll Cardiol* 10 (August 1987): 22A–24A.

64. Eugene Braunwald, "The Golden Age of Cardiology," in *An Era in Cardiovascular Medicine*, ed. Suzanne B. Knoebel and Simon Dack (New York: Elsevier, 1991); Howell, "Changing Face of Twentieth-Century American Cardiology"; Fye, *American Cardiology*, ch. 8.

65. Furman et al., "Application of High Technology": 23A; Hans J. Biersack, Berndt Lüderitz, and Hans Schild, "New Frontiers of High Technology Medicine," *Pacing Clin Electrophysiol* 22 (April 1999, pt. 1): 658–63.

66. Medtronic annual report, 1999.

# BIBLIOGRAPHICAL NOTE

My study traces the careers of a pair of artifacts, the pacemaker and the implantable defibrillator, and the ideas and aspirations of the social groups that have created and managed them. Historians of medicine and technology told us relatively little about the invention and social shaping of medical devices such as these until interest in the subject began to burgeon in the last twenty years as an aspect of the broader societal debate over the benefits and costs of new medical technology.

For the reader seeking an introduction to the study of technological change, the most helpful textbook is Ron Westrum, *Technologies and Society: The Shaping of People and Things* (Belmont, Calif.: Wadsworth, 1991). Other works I have found useful include John M. Staudenmaier, *Technology's Storytellers: Reweaving the Human Fabric* (Cambridge: MIT Press, 1985) and *Does Technology Drive History?* ed. Merritt Roe Smith and Leo Marx (Cambridge: MIT Press, 1994). The essays in *The Social Construction of Technological Systems*, ed. Wiebe E. Bijker, Thomas Hughes, and Trevor Pinch (Cambridge: MIT Press, 1987), have strongly influenced recent scholarly thinking.

Thomas Hughes's concepts of technological systems and technological momentum have helped me organize this study of cardiac-rhythm management. See "Conservative and Radical Technologies," in *Managing Innovation: The Social Dimensions of Creativity, Invention and Technology*, ed. Sven B. Lundstedt and E. William Colglazier Jr. (New York: Pergamon, 1982), 31–44; "Machines and Medicine: A Projection of Analogies between Electric Power Systems and Health Care Systems," *Int J Tech Assess in Health Care* 1 (1985): 285–95; "The Evolution of Large Technological Systems," in *The Social Construction of Technological Systems*, ed. Bijker, Hughes, and Pinch, 51–82; and "Technological Momentum," in *Does Technology Drive History?* ed. Smith and Marx, 101–13.

On medical technology specifically, Stanley Joel Reiser presents a rich overview in *Medicine and the Reign of Technology* (Cambridge: Cambridge Univ. Press, 1978). Reiser shows how the use of artifacts such as the stethoscope has facilitated the scientific framing of formal, discrete disease entities, reinforced the authority of the physician, and added satisfying new rituals to the physician-patient encounter. See also Reiser's essay "The Machine at the Bedside: Technological Transformations of Practices and Values," in *The Machine at the Bedside: Strategies for Using Technology in Patient Care*, ed. Reiser and Michael Anbar (Cambridge: Cambridge Univ. Press, 1984), 3–19.

Historians of technology and of medicine, including Reiser, have told us more about the invention, dissemination, and societal impact of diagnostic technologies than about modern technologies of treatment such as the pacemaker. Important works on imaging technologies have appeared in recent years, particularly Ellen B. Koch, "The Process of

Innovation in Medical Technology: American Research on Ultrasound, 1947–1962," Ph.D. diss., University of Pennsylvania, 1990; Stuart S. Blume, *Insight and Industry: On the Dynamics of Technological Change in Medicine* (Cambridge: MIT Press, 1992); and Bettyann Holtzmann Kevles, *Naked to the Bone: Medical Imaging in the Twentieth Century* (New Brunswick, N.J.: Rutgers Univ. Press, 1996). Koch and Blume are interested in the way groups of specialists stake claims to the control of new technologies and in how the technologies in turn induce specialists to redefine their activities. Kevles also addresses these questions, but in addition explores how professionals and the broader public have ascribed cultural meanings to X rays and other imaging technologies. Nancy Knight, " 'The New Light': X-Rays and Medical Futurism," in *Imagining Tomorrow: History, Technology, and the American Future*, ed. Joseph J. Corn (Cambridge: MIT Press, 1986), 10–34, is a case study of the twentieth-century belief that new machines of diagnosis and treatment will refashion the conditions of life and health.

The historical literature on twentieth-century artifacts of treatment is not a large one. Lewis Thomas's thoughts on "halfway technologies" (which manage but do not cure a disease) are a good place to begin. Thomas's essays are collected in *The Lives of a Cell* (1974; reprint, New York: Penguin, 1995), *The Medusa and the Snail* (1980; reprint, New York: Penguin, 1995), and *Late Night Thoughts on Listening to Mahler's Ninth Symphony* (1983; reprint, New York: Penguin, 1995). He presents his ideas about the technology of medicine more formally in "Report of the Overview Cluster: The Place of Biomedical Science in Medicine," in *Report of the President's Biomedical Research Panel:* appendix A, *The Place of Biomedical Science in Medicine and the State of Science* (DHEW Publication No. [OS] 76-501, 30 April 1976), 1–22. James H. Maxwell takes issue with Thomas in "The Iron Lung: Halfway Technology or Necessary Step?" *Milbank Q* 64 (1986): 3–29, and Thomas replies in the same issue at 30–33. Joseph D. Bronzino et al., *Medical Technology and Society: An Interdisciplinary Perspective* (Cambridge: MIT Press, 1990), offers an overview that includes an insightful discussion of Thomas's idea.

Scholars today are inclined to historicize our very definitions of diseases by suggesting that our changing understanding of subtle conditions, including rhythm disorders such as the sick sinus syndrome or atrial fibrillation, is not merely the consequence of accumulating more and more laboratory and clinical knowledge about a fixed entity. Instead, the machines and devices that specialists employ and the kind of training they draw upon tend to yield discontinuous understandings of diseases over time. As Keith Wailoo shows in his recent study of hematology (*Drawing Blood: Technology and Disease Identity in Twentieth-Century America* [Baltimore: Johns Hopkins Univ. Press, 1997]), hematologists have repeatedly and radically redefined the conditions they attempt to treat, so that contemporary blood diseases bear little connection to those discussed and treated a century ago. Charles E. Rosenberg generalized the point in his introduction to an important collection of essays that he coedited with Janet Golden entitled *Framing Disease: Studies in Cultural History* (New Brunswick, N.J.: Rutgers Univ. Press, 1992). Another helpful discussion in this vein is Robert A. Aronowitz's *Making Sense of Illness: Science, Society, and Disease* (Cambridge: Cambridge Univ. Press, 1998).

Anyone interested in the history of specific diseases and their treatments will encounter numerous works that discuss the subject as one of heroic achievement in the face of scientific misunderstanding or professional indifference. Some of these are quite good — for example, Mickey S. Eisenberg's *Life in the Balance: Emergency Medicine and the Quest to Reverse Sudden Death* (New York: Oxford Univ. Press, 1997). But the assumptions about disease and scientific progress that undergird this kind of history no longer enlist our ready assent. Americans are less inclined than they once were to regard laboratory scientists and heart surgeons as gods walking upon the earth. I have tried to acknowledge the drama of open-heart surgery and the achievements of inventors and surgeons in the 1950s; but it also seems appropriate to treat postwar medical treatments and technologies as matters of contention, activities making for "full life" (in Earl Bakken's phrase) but also for expense and controversy.

Regarding heart medicine specifically, those wishing to delve into this complex and tumultuous field that has become central to our national discussions of health policy since World War II should begin with the books and articles of W. Bruce Fye, Joel D. Howell, and Christopher Lawrence. Fye has compiled two helpful bibliographies: *The History of Cardiology: A Bibliography of Secondary Sources* and *American Contributions to Cardiovascular Medicine and Surgery* (both Bethesda, Md.: National Library of Medicine, 1986).

Fye is both a practicing cardiologist and a trained historian of medicine. His book *American Cardiology: The History of a Specialty and Its College* (Baltimore: Johns Hopkins Univ. Press, 1996) shows how the various subspecialties of cardiology came to define themselves largely in terms of their use of specific technologies of diagnosis and treatment and traces the consequences of cardiologists' growing dependence on federal funding for research, training, and patients' fees in the postwar United States. For cardiologists' changing understanding of specific disorders of the heart (a secondary theme in his book), Fye's articles are full of insight; see particularly "Ventricular Fibrillation and Defibrillation: Historical Perspectives," *Circulation* 71 (May 1985): 858–65; "Acute Myocardial Infarction: A Historical Summary," in *Management of Acute Myocardial Infarction*, ed. Bernard J. Gersh and Shahbudin H. Rahimtoola (New York: Elsevier Science, 1991), 3–13; "A History of Cardiac Arrhythmias," in *Arrhythmias*, ed. John A. Kastor (Philadelphia: Saunders, 1994), 1–24; and "A Historical Perspective on Atherosclerosis and Coronary Artery Disease," in *Atherosclerosis and Coronary Artery Disease*, ed. Valentin Fuster, Richard Ross, and Eric J. Topol (Philadelphia: Lippincott-Raven, 1996), 1–12.

Joel D. Howell specializes in internal medicine and, like Fye, has trained in the history of medicine. Many of Howell's papers on cardiology show how physicians have repeatedly redefined diseases of the heart and their own professional roles and activities; see especially "Early Perceptions of the Electrocardiogram: From Arrhythmia to Infarction," *Bull Hist Med* 58 (spring 1984): 83–98; " 'Soldier's Heart': The Redefinition of Heart Disease and Speciality Formation in Early Twentieth-Century Great Britain," *Med Hist* (1985, supp. 5): 34–52; "The Changing Face of Twentieth-Century American Cardiology," *Ann Intern Med* 105 (November 1986): 772–82; "Diagnostic Technologies: X-Rays, Electrocardiograms, and CAT Scans," *Southern California Law Review* 65 (November 1991): 529–64; and "Concepts of Heart-Related Diseases," in *Cambridge World History of Human Disease*, ed. Kenneth F. Kiple (Cambridge: Cambridge Univ. Press, 1993), 91–102. Howell's *Technology in the Hospital: Transforming Patient Care in the Early Twentieth Century* (Baltimore: Johns Hopkins Univ. Press, 1995) does not treat cardiology and its technologies explicitly but shows how several other new technologies eventually altered the daily routines of care in hospitals.

Christopher Lawrence's seminal papers influenced Howell and Fye and are fundamental for anyone working on the history of heart medicine. See especially "Moderns and Ancients: The 'New Cardiology' in Britain, 1880–1930," *Med Hist* (1985, supp. 5):

1–33; "Incommunicable Knowledge: Science, Technology, and the Clinical Art in Britain, 1850–1914," *Journal of Contemporary History* 20 (October 1985): 503–20; and " 'Definite and Material': Coronary Thrombosis and Cardiologists in the 1920s," in *Framing Disease*, ed. Rosenberg and Golden, 50–82. Another paper of great importance is Robert G. Frank Jr., "The Telltale Heart: Physiological Instruments, Graphic Methods, and Clinical Hopes, 1854–1914," in *The Investigative Enterprise: Experimental Physiology in Nineteenth-Century Medicine*, ed. William Coleman and Frederick L. Holmes (Berkeley: Univ. of California Press, 1988), 211–90.

On the history of cardiology, P. R. Fleming, *A Short History of Cardiology* (Amsterdam: Editions Rodopi, 1997), a volume in the Wellcome Institute Series in the History of Medicine, is worth consulting. On electrotherapy for the heart, Margaret Rowbottom and Charles Susskind, *Electricity and Medicine: History of Their Interaction* (San Francisco: San Francisco Press, 1984), is the basic source. See also David C. Schechter, "Early Experience with Resuscitation by Means of Electricity," *Surgery* 69 (March 1971): 360–72, and Schechter's seven-part "Background of Clinical Cardiac Electrostimulation," which appeared in *N Y State J Med*, vols. 71 and 72 (November 1971 through May 1972).

The growth of interventional cardiology and clinical electrophysiology as subfields of cardiology are documented in the surveys of pacing practice that Victor Parsonnet inaugurated in 1969 and has continued with Alan D. Bernstein as his co-author. I have cited these surveys in my chapter references. Parsonnet has also published a series of thoughtful papers that view the changes and tensions in the field. Among them are Parsonnet and Marjorie Manhardt, "Permanent Pacing of the Heart: 1952 to 1976," *Am J Cardiol* 39 (February 1977): 250–56; Parsonnet, "The Proliferation of Cardiac Pacing: Medical, Technical, and Socioeconomic Dilemmas," *Circulation* 65 (May 1982): 841–45; Parsonnet and Alan D. Bernstein, "Cardiac Pacing in the 1980s: Treatment and Techniques in Transition," *J Am Coll Cardiol* 1 (1983): 339–54; and Parsonnet, "Cardiac Pacing as a Subspecialty," *Am J Cardiol* 59 (15 April 1987): 989–91. See also Leonard N. Horowitz, "Clinical Cardiac Electrophysiology: History, Rationale, and Future," *Cardiol Clin* 4 (August 1986): 353–64.

For the historian interested in cardiology and its technologies, the small library at Heart House, the headquarters of the American College of Cardiology in Bethesda, Maryland, has a good collection of texts, journals, and instructional videotapes. Like the libraries at many major schools of medicine in the United States, the Biomedical Library at the University of Minnesota–Twin Cities, where I did much of my research, has a collection covering every aspect of heart medicine, including old textbooks and full runs of numerous journals.

The Bakken Library and Museum of Electricity in Life in Minneapolis houses a rich collection of publications on electricity and medical therapeutics from the eighteenth century to the early twentieth, as well as historic artifacts and some materials on the twentieth-century medical-device industry. The Bakken also holds the original videotapes of Earl Bakken's "Pioneers in Pacing" interviews with physicians and engineers done between 1975 and the early 1980s.

The Oral History Collection at the offices of the North American Society of Pacing and Electrophysiology (NASPE), in Natick, Massachusetts, now comprises more than 200 interviews, most of them transcribed. The collection emphasizes developments in heart surgery and cardiology, especially electrophysiology, since World War II; it also includes copies of Earl Bakken's "Pioneers in Pacing" interviews. Many of the physicians, engineers, and entrepreneurs who have been interviewed recall the dates of events only vaguely, but the tapes contain numerous revealing details and anecdotes about research and early clinical cases.

A number of books and theoretical papers aided my understanding of the medical-device industry. R. D. Peterson and C. R. MacPhee, *Economic Organization in Medical Equipment and Supply* (Lexington, Mass.: Lexington Books, 1973), though a generation old, offers insights applicable to the device industry of that period. Louis Galambos, with Jane Eliot Sewell, *Networks of Innovation: Vaccine Development at Merck, Sharp & Dohme, and Mulford, 1895–1995* (Cambridge: Cambridge Univ. Press, 1995) applies the important concepts of networks and cycles of innovation. Also worthy of mention are James M. Utterback, "Innovations in Industry and the Diffusion of Technology," *Science* 183 (15 February 1974): 620–26; Utterback and Fernando Suárez, "Innovation, Competition, and Industry Structure," *Research Policy* 22 (1993): 1–21; W. J. Abernathy and Utterback, "Patterns of Industrial Innovation," *Technology Review* (July 1978): 40–47; and Utterback, *Mastering the Dynamics of Innovation* (New York: Free Press, 1994). Patrik Hidefjäll critiques this literature and tests the applicability of various theories to the heart-rhythm management industry in "The Pace of Innovation: Patterns of Innovation in the Cardiac Pacemaker Industry," Ph.D. diss., Linköping University, Sweden, 1997, a work written in English.

The medical-device industry is a highly competitive and litigious one, and none of the manufacturing firms allowed me to examine corporate records. The documents collected during patent lawsuits are customarily sealed by court order. I therefore had to rely on corporate filings with the Securities and Exchange Commission (SEC) (in the case of publicly held companies), internal newsletters for employees, and interviews with current and former managers and executives. Congressional investigations and hearings from the era of the pacemaker scandals (cited in chapter 8) provided a rich lode of inside information about Cordis, Intermedics, and Pacesetter during the late 1970s and early 1980s. For the same period, the weekly magazine *Medical World News* proved useful. Recent annual reports, press releases, product announcements, and SEC filings are available at company web sites or at www.FreeEdgar.com. Reports by industry analysts in the financial community offered useful discussions of sales, profit, and corporate strategy; but older reports are difficult to locate and most tend to overstate the wonders of each innovation in heart-rhythm technology. I found the trade journal *Medical Device & Diagnostic Industry* to be a revealing source for the broad strategies and worldviews of top corporate leaders in the 1990s. Recent and older issues can be read at www.devicelink.com/mddi/. The "Gray Sheet" (F-D-C-Reports, Inc., Chevy Chase, Md.), a weekly newsletter, covers companies' briefings to financial analysts, product launches, trade association conventions, and events on the regulatory and legal fronts involving the device industry. In the Twin Cities, reporters Terry Fiedler of the *Minneapolis Star Tribune* and Karen Padley of the *St. Paul Pioneer Press* have for years written accurate and probing analyses of St. Jude Medical, Guidant/CPI, and Medtronic. Patrik Hidefjäll's "The Pace of Innovation," cited above, covers the European side of the pacemaker/defibrillator industry.

Most books on the FDA are highly polemical, but C. F. Larry Heimann's *Acceptable Risks: Politics, Policy, and Risky Technologies* (Ann Arbor: Univ. of Michigan Press, 1998) is a refreshing exception. Heimann discusses the conflicting political pressures that come down on the agency: it is asked to conduct searching reviews of new products so that technologies reaching the market will be free of risk, yet at the same time is expected to work quickly and evaluate generously so as to encourage innovation and support American competitiveness. Also useful are OTA, *Federal Policies and the Medical Devices Industry*, OTA-H-230 (July 1983); and Susan Bartlett Foote, *Managing the Medical Arms Race* (Berkeley: Univ. of California Press, 1992). Harry M. Marks, *The Progress of Experiment: Science and Therapeutic Reform in the United States, 1900–1990* (Cambridge: Cambridge Univ. Press, 1997), discusses efforts to assess the safety and efficacy of new drugs; the latter part of the book focuses on the FDA. A good historical treatment of the agency's quandaries in trying to oversee the introduction of new medical devices under the terms of the Medical Device Amendments of 1975 and the Safe Medical Devices Act of 1991 is much needed.

For physicians' discussions of many of the nonclinical issues I have touched on in this book (e.g., subspecialty formation and credentialing; the appropriate uses of high-tech treatments; relationships with manufacturing firms) the leading medical journals are a good place to start—above all, the *New England Journal of Medicine*, the *Journal of the American Medical Association*, the *Journal of the American College of Cardiology*, *Circulation*, and *Pacing and Clinical Electrophysiology*.

I have argued that cardiac pacing was born during an era of enthusiastic public support for interventional medicine that lasted roughly from World War II until the beginning of the 1970s. Since then, physicians and business leaders in heart-rhythm management have been adapting to an era of greater public uneasiness about the cost and possible dangers of advanced medical technologies. The best introduction to the history of U.S. health policy in this period remains Paul Starr's *The Social Transformation of American Medicine* (New York: Basic Books, 1982). Daniel M. Fox, *Health Policies, Health Politics: The British and American Experience, 1911–1965* (Princeton: Princeton Univ. Press, 1986), offers a comparative analysis; Fox presents some of his main points in "The Consequences of Consensus: American Health Policy in the Twentieth Century," *Milbank Q* 64 (1986): 76–99. Rosemary Stevens's *American Medicine and the Public Interest: A History of Specialization* (1971; updated edition, Berkeley: Univ. of California Press, 1998), and *In Sickness and in Wealth: American Hospitals in the Twentieth Century* (New York: Basic Books, 1989) are important institutional studies. David Mechanic, *From Advocacy to Allocation: The Evolving American Health Care System* (New York: Free Press, 1986) argues that efforts to constrain costs often have the effect of forcing doctors to make allocational decisions, thereby undermining their role as their patients' advocates. See also Daniel M. Fox's review essay "Wealth and the Care of Sick Strangers: Rosenberg, Stevens, and the Uses of History for Health Policy," *J Health Polit Policy Law* 16 (spring 1991): 169–76.

Numerous studies, some of them already cited, have examined the cultural and political causes for the "technological imperative" in American health care. Audrey B. Davis, *Medicine and Its Technology: An Introduction to the History of Medical Instrumentation* (Westport, Conn.: Greenwood Press, 1981) compares Western medicine's eager acceptance of new instruments and other technologies with the indifference to them found in Asian medicine. H. David Banta, Clyde J. Behney, and Jane Sisk Willems, *Toward Rational Technology in Medicine: Considerations for Health Policy* (New York: Springer, 1981) offers a good review of the growing dependence of medical care on advanced technology since 1940; see also Banta's essay "Embracing or Rejecting Innovations: Clinical Diffusion of Health Care Technology," in *The Machine at the Bedside*, ed. Reiser and Anbar, 65–91. The essays in *New Medical Devices: Invention, Development, and Use*, ed. Karen B. Ekelman (Washington, D.C.: National Academy Press, 1988), are a good introduction to the policy-oriented literature on medical technology. David J. Rothman, *Beginnings Count: The Technological Imperative in American Health Care* (New York: Oxford Univ. Press, 1997) discusses in historical context the U.S. preference for policies that emphasize supply of ever more medical goods and services while finessing the problem of distribution among different social groups.

The tortured history of Medicare is a huge subject; unfortunately, we have few careful historical studies of this program thus far. Theodore R. Marmor, *The Politics of Medicare* (Chicago: Aldine, 1973), explains the political and legislative struggles that led to the Medicare law of 1965; Starr's book *The Social Transformation of American Medicine*, cited above, is also helpful. John K. Iglehart's articles in the *New England Journal of Medicine* offer overviews helpful to the general reader; see especially "The American Health Care System: Medicare," *N Engl J Med* 327 (12 November 1992): 1467–72, and "The Struggle to Reform Medicare," *N Engl J Med* 334 (18 April 1996): 1071–75. Successive efforts to control costs in Medicare through PROs, the DRG and RBRVS systems, and other measures are examined and debated in journals on health policy such as the *Milbank Quarterly*, *Journal of Health Politics, Policy and Law*, and *Health Affairs*. The *Social Security Bulletin* regularly publishes statistical information and analysis of trends in Medicare. On the DRG prospective-payment reform of the 1980s, see Louise B. Russell, *Medicare's New Hospital Payment System: Is It Working?* (Washington, D.C.: Brookings, 1989). Ronald J. Vogel, *Medicare: Issues in Political Economy* (Ann Arbor: Univ. of Michigan Press, 1999), is a valuable assessment of the problems facing Medicare, and Rosemary Stevens, *Medicare and the American Social Contract* (Washington, D.C.: National Academy of Social Insurance, 1999), is a sophisticated and responsible piece of advocacy.

As part of the new historiography that examines the domestic consequences of the cold war, scholars have written extensively about federal sponsorship of scientific and medical research during and after World War II. Important studies include Daniel J. Kevles, *The Physicists: The History of a Scientific Community in Modern America* (New York: Knopf, 1978); David Dickson, *The New Politics of Science* (1984; reprint, Chicago: Univ. of Chicago Press, 1988); G. Pascal Zachary, *Endless Frontier: Vannevar Bush, Engineer of the American Century* (New York: Free Press, 1997); Daniel M. Fox, "The Politics of the NIH Extramural Program, 1937–1950," *J Hist Med* 42 (October 1987): 447–66; and the books by Paul Starr and W. Bruce Fye cited earlier.

In the industrial democracies of western Europe and North America, the social authority of physicians and their institutions depends in good part on their claim to follow scientific methods in discovering, defining, and treating diseases. In *Science and the Practice of Medicine in the Nineteenth Century* (Cambridge: Cambridge Univ. Press, 1994), W. F. Bynum argues that the redefinition of medicine as a science took place between about 1800 and 1914. J. Rosser Matthews, *Quantification and the Quest for Medical Certainty* (Princeton: Princeton Univ. Press, 1995), traces the growing importance of clinical trials in modern medicine and follows physicians' debates over this trend: some champion the new statistical methods and others warn that statistical authority could supplant physicians' own authority rooted in their professional experience and judgment. Marks's *The Progress of Experiment*, cited above, can be read as a case study of these trends. Theodore M. Porter's *Trust in Numbers: The Pursuit of Objectivity in Science and Public Life* (Princeton: Princeton Univ. Press, 1995) also briefly discusses the use of statistics in clinical trials.

Page numbers for illustrations are in italics

Ablation, catheter, 8, 9, 286
Abrams, Leon D., 89
Activitrax pacemaker (Medtronic), 228–33
Adams, Robert, 15–17
Adlanco, subsidiary of Siemens, 21, 33
AEC. *See* Atomic Energy Commission
AICD. *See* ICD
American College of Cardiology (ACC), 33, 210, 214
American Heart Association (AHA), 59, 210, 240, 286
American Optical Co., 132, 134, 149; as American Pacemaker, 168; Medtronic and, 156–58, 221; subsidiary of Warner-Lambert, 157
American Optical Co., products of: "bifocal" pacemaker, 220; "demand" (noncompetitive) pacemaker, 155–59
Anderson, Ken, 229–31, 233
"April," ICD recipient, 257
Arco Medical, 115, 246
Asynchronous pacing. *See* Fixed-rate (VOO) pacing
Atomic Energy Commission (AEC)/ Nuclear Regulatory Commission (NRC), nuclear pacemaker and, 115, 116
Atricor pacemaker (Cordis Corp.), 145–46, 219
Atrioventricular node, 23, 24, 144
Ausubel, Kalman, 221
AV block. *See* Heart block
AVID clinical trial, 258–59
AV node, 23, 24, 144
AV synchronous (VAT) pacing: Atricor pacemaker and, 144–46, 219; Rockefeller Conference and, 70–73

AV universal (DDD) pacing: introduction of, 221–26; pacemaker-mediated tachycardia and, 224–25; the pacemaker syndrome and, 220–21; practitioners and, 227–28; rate-responsiveness and, 233–34, 276–77; transvenous lead technology and, 221–22

Bakken, Earl E., 83–84, 182; on electrode development, 110; inventor of transistorized external pulse generator, 65–70; Medtronic co-founder, 66, 137–38; Medtronic mission statement and, 142–43; Mirowski and, 239; on pacemaker batteries, 170–71; sale of company and, 138; on spiritual restoration of pacemaker patients, 81; on Xytron recalls, 190
Balanced Budget Act of 1997 (BBA), 280
Beck, Claude, 41–42, 49, 241
Becton, Dickinson Corp., 147, 168
Belgard, Alan, 55, 96, 113, 147
Bell, Susan, 10
Bellott, Peter H., 226
Berkovits, Barouh V., 242; anti-tachycardia pacing and, 233–34; inventor of "bifocal" pacemaker, 220, 338n 30; inventor of "demand" pacemaker, 132–34, 149, 155–56; physiological pacing and, 133, 326n 57
Bernstein, Alan D., 70, 124, 263–64
Beth Israel Hospital [Beth Israel Deaconess Medical Center], Boston, 49, 52, 93, 96
Beutel, Albert, 180, 199
"Bifocal" (DVI) pacing, 219–20
Bigelow, Wilfred G.: hypothermia and, 44–45; laboratory external pacemaker and, 51, 54; Zoll and, 51, 55–56
Bigger, J. Thomas, Jr., 258

Bilitch, Michael: NASPE founding and, 214; interviewing of practitioner pacemaker implanters, 167–68; on pacemaker survival rates, 188–89

Biotronik GmbH, 7, 275

Blalock, Alfred, 43–44

Blume, Stuart S., 84

Blumgart, Herrman L., 49

Booth, Edgar H., 28

Bosanquet, Nick, 261

Bradycardia, 8, 9, 163

Brady pacing. *See* Cardiac pacing

Braun, Elmer A., pacemaker recipient, 1–5, 12–13

Briller, Stanley, 7

Brockman, Stanley, 73

Brownlee, Robert R., 181

Brumwell, Dennis, 230–31

Butler, Edmund, victim of Stokes-Adams attacks, 14–15, 17, 23–24

Callaghan, John C.: attempts to pace dying patients, 55; hypothermia and, 44–45; laboratory external pacemaker and, 51, 54; Zoll and, 51

Cammilli, Leonardo, 88–89, 341n 56

Cannon, Walter B., 28

Capillary electrometer, 19

Caplan, Arthur, 4

Cardiac arrest: Beck and, 41; medical thinking about, in 1930s, 31–32; open-chest heart massage and, 41–42; ventricular fibrillation as common mechanism of, 40; Zoll "solves" in theory, 55. *See also* Sudden cardiac death

Cardiac catheterization, 42–43; invention of transvenous pacemaker lead and, 118

Cardiac conduction system, *20;* atrioventricular node, 23, 24, 144; bundle branches, 22, 24; bundle of His, 22, 23, 24; open-heart surgery and, 61; Purkinje fibers, 21, 302n 15; sinus node, 18, 24, 72, 163–64

Cardiac pacemaker: as artifact of indeterminate meaning, 10–11, 36, 287; as engineering contribution to society, 103; as merit good, 5; as object of desire, 195–96, 285, 288–89; as representative artifact of postwar era, 159; as smart machine, 4–5. *See also* Cardiac pacing; Cardiac rhythm management industry; *and specific pacemaker models*

Cardiac Pacemakers, Inc. (CPI): acquisition by Eli Lilly, 217; founding of, 173; ICD and, 247–48, 250–51; market shares of, 218, 276; pacemaker warranty and, 196; rhythm-management division of Guidant, 273; sued by Cordis over pacemaker programmer, 179. *See also* Guidant Corp.

Cardiac Pacemakers, Inc. (CPI), products of: dual-chamber pacemakers, 225; Microlith-P pacemaker, 162, 179; pacemaker with lithium battery, 173–74

Cardiac pacing: cost of, 205–6, 263; debate over rapid growth of, 202–5; Medicare prospective-payment system and, 206–8, 265–66, 280–81; nonlethal arrhythmias and, 152; open-heart surgery and, 105; rate-responsive pacemaker and, 232; research into heart rhythm disorders and, 81–82; set of doctrines and practices, 6; in western Europe, 165–66. *See also* Physiological pacing

Cardiac pacing, expansion of: after 1965, 151–55, 164; in 1970s, 165–66, 202–3, 294; in 1980s, 207, 295; in 1990s, 263–64, 294

Cardiac pacing, medical indications for: in 1960s, 134; in 1970s, 163–64, 326n 52; in 1980s, 232–34; in 1990s, 285–87

Cardiac pacing/EP community, 6–7; cost-containment pressures and, 266, 268, 279–84; dual-chamber pacing and, 209–10, 227–28; European component of, 7; expansion of, 154–55, 164–66, 203, 263–64; inception of, 108; manufacturers' sales representatives and, 142, 159, 184–85, 195–99, 268–70; NASPE founders and, 215–16; pacemaker follow-up and, 167–68; pacemaker programmability and, 182–85; pacemaker simplicity and, 166–68; Parsonnet and, 125; practitioner implanters *vs.* specialists in, 164–68, 202–4, 209, 211–13; pulse generator size and, 196. *See also* Clinical cardiac electrophysiology

Cardiac rhythm management industry, 8; advised by senior physicians, 214; catheter ablation and, 9, 286; consolidation during 1990s, 270–75; cost-containment pressures and, 12, 208, 268–71, 278; growth of, 151–55, 195; hospital purchasing departments and, 268–71; ICD and, 235,

253–56; inception of, 136–37, 149–51; market segments in, 218–19, 277; market shares for defibrillators in, 253–54, 275; market shares for pacemakers in, 168, 217, 275; Medicare prospective-payment system and, 207–8, 268–71; new companies in 1970s, 168–69; overview in 1960s, 137–49; overview in 1980s, 216–19; overview in 1990s, 284–87; pacemaker automaticity and, 277–78, 285; pacemaker commodification and, 160, 199, 228, 279, 285–86; patents in, 158–59, 179, 182, 254–56; physicians and, 9, 159, 184–85, 196–98, 200–201, 278–79; product cycles in, 276–77; sales representatives in, 141–42, 159, 197–98, 268–70, 273; Senate committee investigation, 200–202; technological momentum and, 11, 203, 287–89; Wallin on, 271. *See also* Innovation, technological; *and specific companies*
Cardioversion, 8, 9, 132; ICD and, 235
Caso, Richard, 284
Castellanos, Agustin, Jr., 133
Catalyst Research Corporation (CRC), 171
Catheterization. *See* Cardiac catheterization
Catheter pacemaker lead. *See* Transvenous pacing lead
Center, Sol, 144
Chardack, William: co-inventor of implantable pacemaker, 96–105; consultant to Medtronic, 138; early career, 97–98; emphasis on pacemaker reliability, 140; as innovator, 104–5; inventor of coiled-spring myocardial lead, 109–10; on pacing as a cause of VF, 131; transvenous pacing and, 123
Chardack-Greatbatch Pacemakers (Medtronic), *88, 112;* invention of, 96–105; licensed to Medtronic, 138
Citron, Paul, 221
Clinical cardiac electrophysiology (EP), 6, 305n 4; congestive heart failure and, 287; ICD and, 249–50, 252–53, 261–62; NASPE founding and, 214; transvenous pacemaker lead and, 124
Committee on Medical Research, 46
Competition, pacemaker, 86–87, 131–34
Complete heart block. *See* Heart block
Comroe, Julius H., Jr., 25, 45–46
Congestive heart failure (CHF), 286–87
Continuous Quality Improvement (CQI), in cardiology, 283

Cooley, Denton A., 111, 126
Cooper Committee, 192
Cordis Corporation: early history of, 143–47; FDA and, 188–89, 201–2; focus on innovative products, 145–46, 150–51; leading physicians and, 146; programmability technology of, 176–77, 179; production problems at, 146–47; recalls of defective pacemakers, 188–89, 202; withdrawal from pacemaker industry, 202, 225
Cordis Corporation, products of: Atricor pacemaker, 145–46, 219; lithium-copper sulfide battery, 174; Omnicor programmable pacemaker, 175–77; Sequicor and Gemini dual-chamber pacemakers, 222–24; "standby" and "triggered" pacemakers, 156
Coronary intensive-care units (CCUs), 49, 55, 243
Cost-containment in U.S. health care, 205–8, 264–66, 268. *See also* Managed care
Cost-effectiveness analysis of ICD, 259–61, 279–80
Cournand, André, 42–43, 49
Crafoord, Clarence, 90
Cranefield, Paul, 99
Cross-circulation, 60–61
Cuddihy, Charles, 141–42, 190, 199
CyberLith pacemaker (Intermedics), 180–81, 196, 219
Cycles of innovation. *See* Innovation, cycles of

Dann, Norman, 157
Davies, J. Geoffrey, 132
Day, Hughes W., 243
DeBakey, Michael E., 5, 111, 121; commission on heart disease and, 212, 336n 4
Defibrillation, 39–41, 132. *See also* ICD
Defibrillator, implantable. *See* ICD
Del Marco, Charles, 181
"Demand" pacing. *See* Noncompetitive (VVI, VVT) pacing
DeWall, Richard, 61
Diagnosis-related groups (DRGs). *See* Medicare
Dow Corning Corporation, 77, 101
Dual-chamber pacing. *See* AV synchronous (VAT) pacing; AV universal (DDD) pacing; "Bifocal" (DVI) pacing; Physiological pacing

Einthoven, Willem, 6, 19–20, 22, 27
Electrocardiograms: of complete heart block, 17; of normal heartbeat, 16; of ventricular fibrillation, 239
Electrocardiography: dissemination of, 26–27, 48; Einthoven and, 19–20, 22; Waller and, 19
Electrodyne Corp., 55, 104, 108, 242; declining market share of, 147; sold to Becton, Dickinson, 147; withdrawal from industry, 168
Electrophysiology. *See* Clinical cardiac electrophysiology
Elema-Schönander Company, 310n 21; acquisition by St. Jude Medical, 274; Elmqvist and, 90–92; introduction of implantable pacemaker, 104; introduction of transvenous lead, 122; Siemens subsidiary, 329n 25
Eli Lilly, Inc.: acquisition of CPI, 217; CPI sales force and, 218; ICD and, 247–48; patent dispute with Medtronic, 255–56; spin-off of medical device businesses, 273. *See also* Guidant Corp.
Elmqvist, Rune, 90–92, 104, 310n 21
Enthoven, Alain C., 266
Erlanger, Joseph, 22–23, 25, 45, 302n 15
Escher, Doris, 118, 169
External pacemaker: Hyman and, 30–34; Lidwill and, 27–28; Toronto group and, 51, 55–56; Zoll and, 50–57. *See also* Medtronic 5800

FDA. *See* Food and Drug Administration
Ferrer, Iréné, 164
Fiandra, Orestes, 92
Fibrillation, ventricular. *See* Ventricular fibrillation
Fischell, Robert, 177
Fisher, John D., 224, 252–53
Fixed-rate (VOO) pacing, 86–87, 106. *See also* Pacemaker competition
Food, Drug, and Cosmetic Act (1938), 191
Food and Drug Administration (FDA), 7, 11, 64, 186; ICD and, 236, 248–49; Medical Device Amendments and, 191–93; in 1990s, 284, 285; pre-1976, 189, 191; slow pace of regulatory work, 194
Foote, Susan Bartlett, 191
Ford, Gerald R., 193
Forssmann, Werner, 42
Foster, Michael, 18

Fox, Daniel M., 12
Frank, Howard A., 93, 95, 110
Frank, Robert G., Jr., 19
*Frankenstein* (novel and film), 33
Furman, Seymour, 82; advocate of cephalic vein for transvenous pacing, 123; champion of new technology, 130; co-inventor of devices to assess pacemaker function, 129; creator of NASPExAM, 216; early career, 118; editor-in-chief of journal *PACE*, 213; on FDA regulation of medical devices, 194; Hyman's pacemaker and, 35; on industrial control of pacing, 213; inventor of transvenous pacing lead, 118–20; on medical technology, 287–88; NASPE co-founder, 214; pacemaker-mediated tachycardia and, 224; Pacemaker Study Group and, 210–11; the pacemaker syndrome and, 220–21; on suicide of early patient, 54; transtelephone monitoring of pacemakers and, 129–30; Zoll and, 95, 120–21
Fye, W. Bruce, 152, 280, 282, 284

Gage, Andrew, 98
Galambos, Louis, 10
Gaskell, Walter, 18–19
General Electric Corporation, 104, 148, 168, 187–88
George, William W., 271
Gibson, G. A., 24
Glenn, William, 87–88
Gobeli, David, 196
Goldman, Bernard S., 215
Gorlin, Richard, 281, 282
Gott, Vincent, 63–64, 310n 10
Graf, A. Jay, 270
Grass Physiological Stimulator: Lillehei and, 63, 65–66; Zoll and, 51, 56
Greatbatch, Wilson: co-inventor of implantable pacemaker, 96–105; consultant to Medtronic, 138, 171; design and patent of noncompetitive ("demand") pacemaker, 156, 157; early career, 96–97; on early implantables, 107; as innovator, 104–5; lithium battery and, 171–73; Medtronic patent litigation with American Optical and, 158; on nuclear pacemaker, 116; on pacemaker reliability, 140
Griffin, Bobby, 229–32
Groedel, Franz, 33
Guidant Corp.: acquisition of Sulzer Inter-

medics, 273–74; sales and market share data (1998), 275; spin-off by Eli Lilly, 273. *See also* Cardiac Pacemakers, Inc.

Guidant Corp., products of: Model 2901 Programmer, *267;* pacemakers of 1990s, 277; Sweet Tip pacemaker leads, 278

Hagenson, Ronald T., 171–72

Harken, Dwight, 5, 49, 96; heart surgery on wounded soldiers and, 44; noncompetitive ("demand") pacemaker and, 134; on safety of medical devices, 252; on surgeons' reluctance to touch the heart, 43

Harthorne, Warren, 146, 213–16

Hartlaub, Jerry, 223–24, 229

Harvey, William, 5

Hayes, David L., 287

Health Care Financing Administration (HCFA). *See* Medicare

Health Industry Manufacturers Association (HIMA), 193

Health maintenance organizations (HMOs). *See* Managed care

Heart, beliefs about, 4–6

Heartbeat, electrophysiological understanding of, 18–20, 22–24

Heart block: clinicians and, 25–26, 83, 98, 315n 32; degrees of, 22–23; ECG of, 17; external pacemaker and, 49–55; implantable pacemaker and, 97, 98; open-heart surgery and, 60–64; physiologists and, 18–19, 22–25; Stokes-Adams attacks and, 14–18, 23–24, 137–38

Heart conduction system. *See* Cardiac conduction system

Heart disease: as global health problem, 284; hospitals and, 48–49; targeted for research after World War II, 46–47

Heart-lung machine, 45, 59, 61, 310n 10

Heilman, M. Stephen, 244, 247, 248

Heinz, John, 200

Hellerstein, Herman K., 71–72

Henefelt, Frank, pacemaker recipient, *88,* 103

Hermetic sealing of pacemakers, 162, 170, 173, 182; Xytron pacemaker and, 189–90

Hermundslie, Palmer, 137, 138

Hidefjäll, Patrik, 136, 223

Hill-Burton Act, 48–49, 59

His, bundle of, 22, 23, 24

His, Wilhelm, Jr., 22, 45

"H.N.," patient with transvenous lead, 121

Hoffman, Brian, 99

Holloran, Thomas E., 156–59, 172, 270

Hopps, Jack, 51, 54

Hospitals: as centers for acute heart care, 78; as diagnostic centers, 48–49; Medicare prospective-payment (DRG) system and, 206–8, 264–65, 268; new medical technology and, 48–49; purchasing alliances among, 268–70

Hughes, Thomas, 11

Hunter, Samuel W.: bipolar platform electrode and, 75–77; implantation surgery and, 78; Mauston case and, 77–80; Johnson case and, 126–27

Hyman, Albert S., 50, 57; cardiac arrest and, 28–30; focus on reviving healthy hearts, 29, 34, 37; inventor of pacemaker, 30–31; licensing of invention to Siemens, 21, 33; present-day pacemaker physicians and, 34–35; secondary figure in U.S. cardiology, 32–33; undocumented claims of clinical cardiac pacing, 34

Hyman, Charles, 30

Hymanotor pacemaker, *21,* 33–34

Hypothermia: heart surgery and, 44–45; Toronto external pacemaker and, 51; University of Minnesota and, 59–60

ICD, 5, 8; animal experiment with, *238,* 241; competing models, 253–54; definition of, 235; development of, 247–48, 250–53; electrophysiologists and, 249–50, 252–53; history of, *vs.* pacemaker, 236; invention of, 237–39, 244–47; patents and, 254–56; patient selection for, 240, 248–50; social benefits and costs of, 256–62; tiered therapy and, 251–52

ICHD. *See* Pacemaker Study Group

Implantable [cardioverter-]defibrillator. *See* ICD

InControl, Inc., 273, 285

Innovation, technological, in cardiac rhythm management industry: in 1970s, 158–59, 161, 169, 185, 194–96, 198–99; in 1980s, 216–19, 225–26; in 1990s, 276–78, 285; normatively sanctioned, 83–84. *See also* Innovation, cycles of

Innovation cycles, in pacemaker technology: discrete transistorized components and, 65–68, 85, 97–98, 102, 145; hybrid circuitry and, 174–77; microprocessor and,

Innovation cycles (*cont.*)
222–24, 277–78; vacuum-tube era and,
56–57

Intec Systems. *See* Medrad, Inc.

Intermedics, Inc.: acquisition and closing by
Guidant, 273; founding of, 180; rate-
responsive pacing and, 276; sales force at,
199–200

Intermedics, Inc., products of: Cosmos
pacemaker, 225–26; CyberLith pacemaker, 180–81; Dash/Dart/Stride/Relay
family of pacemakers, 277

Inter-Society Commission for Heart Disease Resources (ICHD). *See* Pacemaker
Study Group

Ion, Heather Wood, 281–82

Johns Hopkins University: Blalock-Taussig
procedure and, 43–44, 58–59; ICD
development and, 246; myocardial pacemaker lead and, 73; rechargeable pacemaker battery and, 177

Johnson, Wesley, pacemaker battery failure
and, 126–27

Kantrowitz, Adrian, 129, 148

Karolinska Hospital/Institute, Stockholm,
90–91, 122

Kastor, John A., 237, 241

Keith, Arthur, 18, 23

Keller, J. Walter, 144–45, 147, 150, 176

Koeppen, Siegfried, 33–34

Kolenik, Steve A., 246

Kouwenhoven, William, 242

Kramp, Rich, 173

Kreher, Louise, early pacemaker recipient,
81

Laennec, René, 16

Lagergren, Hans, 122

Langer, Alois, 244

Larsson, Arne, first recipient of implanted
pacemaker, 11, 91–92, 314n 14

Leads. *See* Pacemaker electrodes and leads

Leatham, Aubrey, 122

Lemberg, Louis, 133–34, 338n 30

LEM Biomedica, 172

Levine, Samuel, 242

Lewis, F. John, 60

Lewis, Kenneth B., 177

Lewis, Thomas, 33

Lidwill, Mark C., 27–28, 30, 57

Lillehei, C. Walton, 5, 67, 68, 70, 83–84,
85; battery-powered external pacemaker
and, 64–66, 69–70; cross-circulation and,
60–61; early career of, 60; heart-lung
machine and, 61; myocardial pacing lead
and, 63–64; postsurgical heart block and,
61–64; Roth's platform electrode and, 76

Lilly. *See* Eli Lilly, Inc.

Linenthal, Arthur, 55, 102

Lithium battery, 116, 173–74, 178, 182,
329n 27; Greatbatch and, 171–73; hermetic sealing of pacemakers and, 173;
invention of, 171; introduction by Cardiac
Pacemakers, Inc., 172–74; LEM Biomedica and, 172

Lown, Bernard, 240–41

Lowrance, William W., 3

MacGregor, David, 146, 215

Mackenzie, James, 29, 32

MacWilliam, J. A., 25, 26, 40–41

MADIT clinical trial, 346n 61

Mallory Co., 99, 111, 138

Managed care, 7; cardiologists and, 279–84;
cost containment in 1990s and, 266, 268

Mann, Alfred E., 177–78

Matricaria, Ronald A., 274

Mauro, Alexander, 87–88

Mauston, Warren, *63*, 85; Hunter-Roth
electrode and, 75, 77–78; pioneer patient,
79–81

McFarlin, Whitney, 229

Medical Device Amendments (1976), 186,
191–94

Medical devices: Cooper Committee and,
192; definition of, 3–4; modern medicine
and, 287–88; public anxiety about, 191–
92; regulation of, 192–94. *See also* Food
and Drug Administration

Medicare: administered by Health Care
Financing Administration (HCFA), 7,
264; growth of pacing and, 151–54, 325n
48; Physician Review Organizations
(PROs) under, 207; prospective-payment
(DRG) system for hospitals under, 205–8,
268–69; prospective-payment (RBRVS)
system for physicians under, 264–65,
280–81

Medrad, Inc.: creation of Intec Systems subsidiary, 246; development of ICD and,
244–47; sale of ICD project to Eli Lilly/
CPI, 247–48

Medtronic, Inc.: acquisitions during 1990s, 272; Bakken and, 65–67, 137–38, 142–43; Chardack, Greatbatch, and, 138, 140–41; corporate motto of, 288; corporate refocusing in 1990s, 271–72; early history of, 66–67, 76, 137–38; growth and cash shortages in 1960s, 138–39; ICD and, 239–40, 255–56; lithium battery and, 171–72, 174; market dominance in 1960s, 149–50; market segmentation and, 218–19; Micro-Rel subsidiary of, 333n 8; mission statement of, 142–43; patent litigation with American Optical, 156–59; patent litigation with Eli Lilly/CPI, 255–56; practitioner implanters and, 139–42, 218; product reliability in 1960s, 139–41; programmability concept in 1960s, 175, 291; prosthetic technology in 1960s, 139; R&D expenses at, 195, 275; Roth and, 75–76; sales and market share data for, 139, 151, 168, 195, 217, 232, 272, 275; sales/support force of, 141–42, 150, 197–98, 269–70; staging of introduction of new pacemakers, 228–29, 277; Wallin and, 271–72; Xytron recalls and, 174, 189–90. *See also* Bakken, Earl E.; Cardiac rhythm management industry; Medtronic 5800

Medtronic, Inc., products of: Activitrax pacemaker, 229–33; Byrel pacemaker, 221, 222, 228; Chardack-Greatbatch Pacemakers, *88*, 104, *112*, 138; coiled-spring myocardial lead, 110; Elite/Thera/Kappa DDDR pacemakers, 276–77; Greatbatch "demand" (noncompetitive) pacemaker, 156; Mirel pacemaker, 228; polyurethane lead, 221–22; Spectrax pacemakers, 217, 219, 228; Symbios pacemakers, 224–25, 228–29; Synergist pacemaker, 233; TeleTrace transtelephone monitoring device, 159–60; transvenous lead, 122–23; Versatrax pacemakers, 222–25, 228; Xyrel pacemakers, 174, 180, 218, 228; Xytron pacemaker, 174, 189–90

Medtronic 5800: invention of, 65–68; long-term pacing and, 75, 81; technology of reassurance, 68–70; transvenous lead and, 121

Mercury battery, 91, 95–96, 98, 99, 104; follow-up and, 129–30; Medtronic and, 170–72; problems with, 111–12, 169–70

Microlith-P pacemaker (Cardiac Pacemakers, Inc.), *162*, 179–80, 196, 218

Minneapolis Heart Institute, 283

Mirowski, Mieczyslaw (Michel): character, 241, 247; on clinical trials, 247; early career, 237; invention of ICD and, 237–41; Medrad and, 244

Montefiore Hospital/Medical Center, 95, 113; early pacemaker program at, 122; invention of transvenous pacing and, 118–20; pacemaker clinic at, 128, 129–30, 169

Morse, Dryden, 110–11, 146, 214

Mower, Morton M., 237, 239–41, 248

Murphy, William P., Jr.: AV synchronous pacemaker and, 145; Cordis marketing strategy and, 150–51; Cordis sales reps and, 188; FDA inspectors and, 201–2; founding of Cordis Corp., 144; reed switch for programmable pacemaker and, 175

Myers, George H., 111

Myocardial pacing lead, 66, *112*; Chardack and, 109–10; in Chardack-Greatbatch pacemaker, 101–2; invention of, 62–64; problems with, 67–68, 85–86, 109; Roth and, 75–77

Nader, Ralph, 204

NASPExAM, 216

Nathan, David A., 144–45

National Institutes of Health: AVID trial and, 259; Mirowski and, 239; National Heart Institute within, 47; Newark Beth Israel Hospital and, 111; postwar medical research and, 47, 59

Newark Beth Israel Hospital/Medical Center: as center of cardiac pacing innovation, 110–11; pacemaker clinic at, 128–29

New York Cardiological Society, 33

Nixon, Richard M., 191

Noncompetitive (VVI, VVT) pacing: invention of, 131–34, 145; magnet mode and, 167; patent struggle over, 155–59; practitioners and, 227; "triggered" (VVT) alternative to "demand" (VVI) concept, 156. *See also* Pacemaker competition

North American Society of Pacing and Electrophysiology (NASPE), 35; congestive heart failure and, 286; founding of, 213–14; growth of, 215; Medicare fee schedules and, 280–81; policy conferences of, 215; practice guidelines of, 279

Nuclear Materials and Equipment Corporation (Numec), 115

Nuclear pacemaker, 115–17
Nuclear Regulatory Commission (NRC), nuclear pacemaker and, 115, 116
Numec, 115

Omnicor pacemaker (Cordis Corp.), 175–77; defective units, 188
Open-heart surgery: hypothermia and, 44–45; Medtronic 5800 and, 68–70; University of Minnesota and, 58–64; in World War II, 44

*PACE* (journal), founding of, 213
Pacemaker. *See* Cardiac pacemaker
Pacemaker competition, 86–87, 131–34
Pacemaker electrodes and leads, 1–2, 12–13, 31, 51–52, 170, 221, 278. *See also* Myocardial pacing lead; Transvenous pacing lead
Pacemaker follow-up: commercial monitoring services for, 212; inception of, 125–29; instrumentation for, 129–30; programmability and, 167, 183–84; telemetry and, 181–82
Pacemaker identification code, 210–11, 296
Pacemaker implantation procedure: Chardack and, 103; Furman and, 123; Parsonnet and, 124–25
Pacemaker industry. *See* Cardiac rhythm management industry
Pacemaker power sources. *See* Lithium battery; Mercury battery; Nuclear pacemaker; Radiofrequency and induction pacing; Rechargeable pacemaker batteries
Pacemaker programmability, 174–77; multiprogrammable pacers, 180–81; patent struggle over, 176; physicians and, 177, 182–85. *See also* Pacemaker programmers
Pacemaker programmers, *162*, 175, 179, 267, 278
Pacemaker Study Group: created by Inter-Society Commission for Heart Disease Resources (ICHD), 210; practice guidelines by, 211, 213, 215; pacemaker identification code proposal, 210–11
Pacemaker syndrome, the, 220–21
Pacemaker telemetry, 180–81, 217, 267, 278
Pacesetter Systems, Inc.: acquisition by St. Jude Medical, 274; early years of, 177–79; FBI investigation of, 201; in 1980s, 178–79
Pacesetter Systems, Inc., products of: DDDR pacers, 276; Programmalith pacemaker, 217, 219; rechargeable pacemaker, 177–78
*Pacing and Clinical Electrophysiology (PACE)*, founding of, 213
Parker, Bryan, 129
Parkinson, John, 37–38, 45
Parsonnet, Aaron, 110
Parsonnet, Victor, 70; advocate of dual-chamber pacing, 222, 232; bioenergy pacemakers and, 114–15; champion of new technology, 130; on clinical effectiveness of pacing, 107; on competence of practitioner implanters, 183–84, 216; Cordis Corp. and, 146; on growth of pacing in 1970s, 203; instruction of Morse by telephone, 110–11; inventor of pacemaker electrode, 127–28; leadership of pacemaker research group, 110–11; learning of implantation technique, 110; NASPE founding and, 214; nuclear pacemaker and, 115; pacemaker failure modes and, 125–26; pacemaker follow-up and, 125, 128–29; pacemaker power sources and, 111–12; Pacemaker Study Group and, 210–11; surveys of cardiac pacing practice and, 164–65, 166, 183–84, 263–64; transvenous pacing and, 122, 124–25
Patents, 158–59, 179, 182, 254–56
Payer, Lynn, 6
Physiological pacing: Atricor pacemaker and, 219; dual-chamber pacing and, 209, 219–20, 227, 338n 30; ideal of emulating nature and, 132–33
Pless, Ben, 253
Pozzi, Renato, 88–89
President's Commission on the Health Needs of the Nation, 46
Professional Review Organizations (PROs). *See* Medicare
Programmalith pacemaker (Pacesetter Systems), 217, 219
Prospective-payment (DRG, RBRVS) systems. *See* Medicare
"P.S.," patient with transvenous lead, *113*, 119–20
Public Citizen Health Research Group, 204

"R.A.," treated with Zoll external pacemaker, 52–53
Radio Corporation of America (RCA), 70
Radiofrequency and induction pacing, 87–90

Rate hysteresis, 170
Rate-responsive (VVIR, DDDR) pacing: invention of, 228–33, 341n 56; in 1990s, 276–77
Rechargeable pacemaker batteries, 90–92, 99, 177–78
Reiser, Stanley, 27
Resource-Based Relative Value Scale (RBRVS). See Medicare
Robinovitch, L. G., 303n 22
Rockefeller Conference, 70–75, 83, 86, 120, 145
Roth, Norman, 75–77, 79, 102
Ruben, Samuel, 99

Safe Medical Devices Act (1990), 334n 20
Samet, Philip, 144
Sanger Clinic, 282–83
Sarnoff, David, 83
Schechter, David C., 25–26, 31–32
Schivelbusch, Wolfgang, 31
Schlesinger, Monroe J., 49
Schuder, John C., 261
Schwedel, John B., 119, 120
Senate Select Committee on Aging: on growth of pacing, 203, 204; pacemaker industry investigation, 200–201
Senning, Åke, 90–92, 310n 21
Sequicor pacemaker (Cordis Corp.), 222–23
Shelley, Mary, 33
Shumway, Norman, 5, 59
Sick sinus syndrome, the, 154, 161, 163–64, 326n 52; dual-chamber pacing and, 220
Siddons, Harold, 89, 92, 122
Siemens Corp., 21, 33, 179, 274
Siemens-Elema. See Elema-Schönander Company
Siemens Pacesetter. See Pacesetter Systems, Inc.
Sinoatrial node. See Heart conduction system: sinus node
Sinus node, 18, 24, 72, 163–64
Sinus node disease. See Sick sinus syndrome
Smyth, Nicholas P. D., 210–11, 221
Social Security, 206–7
Sowton, Edgar, 131
Spectrax pacemaker (Medtronic), 217, 219
St. George's Hospital, London, 122, 131, 132
St. Jude Medical, Inc.: acquisition of rhythm-management companies, 274; heart-valve manufacturer, 274; sales and

market share data (1998), 275; steerable stylet of, 278. See also Pacesetter Systems, Inc.; Ventritex, Inc.
Staewen, William, 237
Stark, Fortney, 280
Stevens, Rosemary, 153
Stokes, Ken, 221–22
Stokes, William, 14–17, 23–24, 45
Stokes-Adams attacks. See Heart block
Stotts, Larry, 225–26
String galvanometer, 16; invention of, 22
Sudden cardiac death (SCD): causes of, 242; definition of, 341n 2; epidemiology of, 243; medical responses to, 242–44. See also Cardiac arrest; ICD; Ventricular fibrillation
Sulzer Intermedics, Inc. See Intermedics, Inc.
Sweeney, Mike, 253
Symbios pacemakers (Medtronic), 224–25
Szarka, George, 35

T. R. Mallory Co., 99, 111, 138
Tachycardia, ventricular. See Ventricular tachycardia
Tarawa, Sunao, 23, 302n 15
Tarjan, Peter, 144, 147, 222–23
Taussig, Helen, 43–44, 237
Technological momentum, 11, 203, 287–89
Technology: experience of sickness and, 4; "halfway," 3, 212; medical, 3–4; postwar hospital and, 48–49. See also Innovation, technological; Medical devices; Technological momentum
Telectronics Corp.: acquisition of General Electric pacemaker business, 168, 216; closing, 271; FBI investigation of, 201; in U.S. market, 216–17
Tepper, Fred, 171
Tetralogy of Fallot: Blalock-Taussig procedure and, 43–44; Lillehei and, 60–61
Thomas, Lewis, 3
Threshold of stimulation: Chardack, Greatbatch, and, 98–99, 101–2; definition of, 93, 314n 20; Senning and, 92; Zoll and, 93–95
Toffoli, Michael, 173, 224–25, 329n 25
Transtelephone monitoring of pacemakers, 2; commercial services for, 212; invention of, 129–30
Transvenous pacing lead: cephalic vein and, 123; consequences of, 124; diffusion of,

Transvenous pacing lead (*cont.*)
121–24; introducer kit and, 165; invention of, 117–20; Parsonnet and, 124–25; "P.S." and, 113, 119–20; as radical innovation, 117, 123–24; reduced mortality risk with, 123; Zoll and, 95, 120–21, 315n 27
Triggered pacing. *See* Noncompetitive (VVI, VVT) pacing
Tyers, Frank, 181

Unger, Walter J., 282
University of Minnesota Medical School, Department of Surgery: Bakken and, 66; heart-lung machine at, 61; heart surgery with hypothermia and cross-circulation at, 45, 59–61; myocardial pacing at, 62–66, 68–70, 75; power outage in 1957, 65; surgical research at, 58–64; Variety Club Heart Hospital at, 58–59

Variety Club Heart Hospital, 58–59
Ventricular fibrillation (VF), 8, 9; defibrillatory countershock to terminate, 40; fixed-rate pacing and, 131, 134; study by Mac-William, 26, 40–41; sudden cardiac death and, 241–43; vulnerable period and, 41. *See also* Cardiac arrest; Sudden cardiac death
Ventricular tachycardia (VT): Berkovits and, 233–34; ICD and, 249–51, 258; pacemaker-mediated, 224–25; Zoll and, 55
Ventritex, Inc., 285; acquisition by St. Jude Medical, 274; ICD and, 253–54
Versatrax pacemaker (Medtronic), 222–23; pacemaker-mediated tachycardia and, 224–25
VF. *See* Ventricular fibrillation

Villafaña, Manuel A., 172–73, 274
VT. *See* Ventricular tachycardia

Waller, Augustus, 19
Wallin, Winston R., 271–73
Wangensteen, Owen, 59, 60
Watkins, Levi, Jr., 246
Weirich, William, 63–64, 70–71
Weldon, Norman R., 202
White, Paul Dudley, 32
Wiggers, Carl J., 41, 45, 131
Wingrove, Robert, 175

Xyrel pacemaker (Medtronic), 174, 180, 196, 218
Xytron pacemaker (Medtronic), 174, 189; recalls of, 189–90

Zacouto, Fred, 132
Zoll, Paul M., 10, 11, 14, 58, 63, 66, 70, 289; Bigelow, Callaghan, and, 51, 55–56; cardiac monitor and, 39, 55; conservative innovator, 83–84, 93, 96, 104–5, 147; cycles of innovation in cardiac pacing and, 56–57; demonstration of external pacemaker, 38; early career of, 49–50; external defibrillator and, 39; Grass Stimulator and, 51, 56; Harken and, during World War II, 50; Hyman and, 50–51; implantable pacemaker and, 92–96; inventor of external pacemaker, 36–37, 50–54; on pacing as a cause of VF, 131; poor background in electronics of, 51; "R.A." and, 52–54; rejection of Hunter-Roth pacemaker lead, 102; Rockefeller Conference and, 73–75; Stokes-Adams disease and, 39; transvenous pacing and, 95, 120–21, 315n 27; VF research and, 41